PETER GEACH: PHILOSOPHICAL ENCOUNTERS

SYNTHESE LIBRARY

STUDIES IN EPISTEMOLOGY,

LOGIC, METHODOLOGY, AND PHILOSOPHY OF SCIENCE

VOLUME 213

PETER GEACH: PHILOSOPHICAL ENCOUNTERS

Edited by

HARRY A. LEWIS

Department of Philosophy,
The University of Leeds

KLUWER ACADEMIC PUBLISHERS

DORDRECHT / BOSTON / LONDON

Library of Congress Cataloging-in-Publication Data

Peter Geach, philosophical encounters / [essays by Peter Geach ... et al.] ; edited by Harry A. Lewis.
p. cm. -- (Synthese library ; v. 213)
Includes bibliographical references and index.
ISBN 0-7923-0823-9 (alk. paper)
1. Philosophy. 2. Geach, P. T. (Peter Thomas), 1916- .
I. Geach, P. T. (Peter Thomas), 1916- . II. Lewis, Harry A.
III. Series.
B29.P416 1990
192--dc20 90-38724

ISBN 0-7923-0823-9

Published by Kluwer Academic Publishers
P.O. Box 17, 3300 AA Dordrecht, The Netherlands.

Kluwer Academic Publishers incorporates the publishing programmes of D. Reidel, Martinus Nijhoff, Dr W. Junk and MTP Press.

Sold and distributed in the U.S.A. and Canada by Kluwer Academic Publishers, 101 Philip Drive, Norwell, MA 02061, U.S.A.

In all other countries, sold and distributed by Kluwer Academic Publishers Group, P.O. Box 322, 3300 AH Dordrecht, The Netherlands.

Printed on acid-free paper

TABLE OF CONTENTS

PREFACE

The present volume owes its existence to a proposal of Dr Esa Saarinen. Our aim was to celebrate the work of a living philosopher by presenting it both from his own point of view, through the medium of a philosophical autobiography, and from that of his closest philosophical colleagues and adversaries. We felt that a philosophical career lived through vigorous controversy was best reflected not by adulation but in the spirit of that career – by open debate.

Contributors were not constrained in their choice of topic, but their contributions fell naturally into groups linked with some of Peter Geach's principal areas of interest, and we have so grouped them in the book. There is an interweaving of biographical and philosophical themes, not only in Peter Geach's philosophical autobiography, but also in the introductions he has contributed to each section. Professor W. V. O. Quine's contribution, which consists of extracts from his correspondence with Peter Geach, has been set apart as it forms a natural bridge between Peter Geach's autobiography and the contributions that follow. Their correspondence reproduced here throws new light on many familiar themes from the writings of both philosophers: among them, the objects of belief and other attitudes, issues in set theory, the nature of causality, and evolution in epistemology.

A feature of Peter Geach's work for which British philosophers of his generation have not all been noted is a scholarly, but also philosophically informed, concern for the writings of philosophers of the past. (Indeed, his first two books were translations of writings of such philosophers – from Frege, and from Descartes.) Our contributors here write in the same spirit, whether of Aristotle and Diodorus on time and necessity, of Aquinas, of Abelard, or of Wittgenstein.

Peter Geach, who crowned his career of university employment (his philosophical career is still in progress) with his appointment to a Chair of Logic at the University of Leeds, has been a vigorous proponent of logic both within philosophy and as a subject of educational value. He is also more aware than most that not all that goes by the honorific title of 'logic' is worthy of the name. (He has built up a notable

Harry A. Lewis (ed.), Peter Geach: Philosophical Encounters, ix–xi.

collection of Bad Logic Books.) His views are as pertinent as ever when the line between good and bad logic is blurred, if not in the name of 'rhetoric', then in the name of 'informal logic'. Philosophical logicians will find much to please and to intrigue them here. J. E. J. Altham moves forward the boundaries of formal study with his contribution on plural and pleonetetic quantification; G. E. M. Anscombe questions the applicability of already familiar logical principles using examples of actual argumentation. The relation of natural language to formal argument is further explored in the contributions from Jaakko Hintikka and James D. McCawley.

Readers are encouraged to read Peter Geach's replies (printed together at the end of the book: pp. 247–306) with the articles to which they refer. The sub-section of the group of papers on logical topics that deals with identity receives close attention in the replies (pp. 276–306). No discussion of Peter Geach's work would be complete without mention of his views on identity, which have caused much controversy; readers will not be disappointed. Anselm Müller and Michael Dummett address these views directly, while Christopher Williams considers the closely related topic of personal identity.

In Peter Geach's writings, logical themes are interwoven with almost all the topics he has considered. But his writings on the philosophy of religion, including now the two books of his Stanton lectures at Cambridge (1971–72 and 1973–74) form a significant corpus that nonetheless stands apart from his other work. He gives a hint of the reason for this asymmetry, if we may so label it, in his introduction (pp. 213–214) to the final group of essays. In those essays, Norman Malcolm takes his lead from a remark in one of the Stanton lectures, in which an analogy is offered between philosophical confusion and sin. Hugo Meynell presents another possible analogy, between revisionary metaphysics and revisionary theology, in the final contribution.

Practising professional philosophers will find their favourite topics quickly in the book. But Peter Geach's writings, in their own clarity and directness, to which the contributors respond in the same style, are accessible to a very wide audience; undergraduates and general readers will find much to interest and stimulate them here. Not least, the historical and personal framework of a career of one of our most distinguished senior philosophers is presented together with his ideas, in a way that reveals how closely they are connected.

We expected when the book was first planned that it would be at press very quickly; and our contributors were punctilious in producing their work in good time. It is a pleasure to make a public acknowledgment of their expedition. However other difficulties supervened, and it has been some ten years until the volume could be sent to press. In defence of our long-suffering contributors, I must ensure that the reader is aware of this delay. Readers will find hints of the passage of intervening time in certain parts of the book, but the contributions of authors other than Peter Geach himself have remained unaltered for ten years. In graciously allowing us to proceed albeit after such a long delay, our contributors have nonetheless wanted us to draw the reader's attention to this background; they are not to be held responsible for any failure to take into account here either more recent work on the subjects of their contributions, or any latter-day modifications of their own views.

Special acknowledgment is due to Peter Geach himself, whose work inspired this book; to Esa Saarinen for thinking of it first; and to Antoni Diller for assistance with the bibliography of Peter Geach's works.

Leeds, July 1990

PETER GEACH

A PHILOSOPHICAL AUTOBIOGRAPHY

In this account of my life I leave much unsaid; I am concerned with those facts and events that I see as having had a manifest influence on my career as a philosopher and with the way I came to know, in person or in their works, those philosophers who have most guided my thought.

I was born in Lower Chelsea, London, on March 29, 1916. My father, George Hender Geach, was at that time working in the Indian Educational Service; he became Professor of Philosophy at Lahore, and afterwards Principal of a training college for teachers at Peshawar. On furlough he had met and fallen in love with my mother, Eleonora Frederyka Adolfina Sgonina, the daughter of Polish emigrants: her father, a civil engineer, had rightly judged that he would prosper in England better than in his own country under the Prussian heel. My mother came back to England for my birth after a short time in India; the marriage had not been happy, and she never returned to my father. My earliest years were spent in Cardiff in my Polish grandparents' house; the novelist Doreen Wallace, an old friend of my mother's, told me that my grandmother never learned English well, so I must often have heard Polish spoken, though I lost all memory of the language. When I was four years old my father secured a court order, making me the ward of a Miss Tarr during his absence in India, and for me all contact with my mother and her parents ceased; Miss Tarr, a rather formidable elderly lady, had been my paternal grandfather's betrothed and the guardian of his children after his death. I remained in Miss Tarr's care until my father was once again in England, invalided out of the I.E.S.

The first philosophical reading I did was in this period. I could not resist the impulse of curiosity that led me to take down some of my father's books, stored in Miss Tarr's house, from the shelves. I made little of them, but I remember in particular Bradley's *Appearance and Reality*. The work bewildered me; there were hardly any words I did not know, and the sentences were simply constructed, but I could not tell what it all meant. However, I formed the general impression that

Harry A. Lewis (ed.), Peter Geach: Philosophical Encounters, 1–25.

the author was a wicked man who worshipped a false God called the Absolute; I was imbued at the time with Miss Tarr's narrow Protestant pieties. My adult views of Bradley are not so very different. Reading a few pages of *Appearance and Reality* now makes me feel as if I had drunk several pints of beer; and having read (much later) Bradley's essays 'Some Reflections on Punishment' and 'The Limits of Individual and National Self-Sacrifice' I must say that to my mind they fully justify my childish impression of great wickedness; some of Bradley's sentiments are strongly reminiscent of Nazi state-worship.

When I was eight years old, at my father's behest, I was sent away to be a boarder at the Llandaff Cathedral School. Shortly afterwards my father returned from India for good, and my philosophical education began.

My father was a very strange man. A trait of his that I have never encountered again in life, and indeed only once in fiction (in Rose Macaulay's novel *Told by an Idiot*), was his propensity to change his religious belief. While living with my mother he had been a Buddhist and a vegetarian ('Queer fellow, Geach, used to eat grass' was the way an Indian Army officer who had known him described this). He changed about three times a year, with no apparent agonies of mind attending his conversions—his library bore evidence of his successive beliefs. He always had persuasive arguments in favour of his latest belief, which he would bring out for his son's benefit. Our first relations were stormy: he could not endure the Nonconformist conscience that Miss Tarr had inculcated in me, and his response to my rebukes for drinking, smoking, Sabbath-breaking, etc. was simply to beat me. But when I was a little older, he decided that reason was now the right remedy and he got me to read McTaggart's work *Some Dogmas of Religion*. This had a speedy effect upon me; by the time I was thirteen I was emancipated from Miss Tarr's creed, and for some time followed my father through his various phases of faith. I was foolish enough to talk about the matter to my schoolfellows, having now left Llandaff Cathedral School for Clifton College; I well remember how a school friend greeted me once upon my return: 'Hullo, Geach! Good hols? Does God exist *this* term?'

My father had now secured for himself an apprentice to whom he could teach the disciplines of which he was a master. He had had a brilliant career in Moral Sciences at Trinity College, Cambridge, at a time of unique distinction for Cambridge in philosophy: Russell,

Moore, McTaggart, W.E. Johnson, and Neville Keynes were his teachers, Wittgenstein and C.D. Broad were his fellow-students. His teaching career in India had been prematurely cut short by a complication of illnesses that left him with a permanently shattered constitution: he could not secure a teaching position in England, for he applied only for Professorships and never with success, though he was several times on a short list. His only way of satisfying his desire to teach philosophy was to teach his son. He had high vicarious ambitions on my behalf; he had the good luck that I proved as eager to learn as he was to teach. From when my father's teaching of philosophy seriously began, I never felt any doubt that philosophy was what I wanted to do. But he died before I had any permanent teaching position.

He began by teaching me logic. Neville Keynes's *Formal Logic* was the first textbook I used, and a work for which I retain a great esteem. The skill in syllogistic manipulations that I gained from this work has not been lost, and has occasionally been directly useful: to any who may know Keynes, my proof that if '*p,q,r*' are uniformly read as categoricals, there are not two valid syllogistic schemata corresponding to '*p,q, ergo r*' and 'not *p*, not *q*, *ergo r*'[1] is recognizably Keynesian in style and spirit. When I had learned as much from Keynes as my father thought advisable, he said to me, 'Now, Peter, tomorrow we begin *Principia Mathematica*': and so we did. We worked through about twenty sections of the work in detail, and by then I had acquired a sufficient knowledge of the language of logic to read other works without discomfort.

Two seeds of thought that were then sown in my mind deserve mention. I found the Theory of Descriptions utterly clear and convincing; apart from a brief aberration, this has been my constant view; but I could not get up any enthusiasm for the Theory of Types or the no-class theory of classes, nor could I understand why e.g. 'the class of all non-self-membered classes' should not be regarded as simply a definite description, 'that class which . . .', and a demonstrably empty one. This made me ready to feed my mind on Quine's logical work when I came across it. The other seed was the puzzle I felt about identity. The *Principia* account of identity will not work unless it is possible to quantify over properties unrestrictedly, including properties themselves expressed by predicates containing just such quantification: and this brings in a threat of vicious circularity over truth-conditions.

Whitehead and Russell here execute a very odd manoeuvre; they introduce identity using only a restricted quantification over properties, but then they argue that this has the desirable consequence of using unrestricted quantification because of the Axiom of Reducibility. This could not but strike an innocent child as fishy. The seed of doubt took decades to grow but eventually flowered in the theorizing about relative identity that outrages so many of my contemporaries.

My father was specially concerned that I should learn this sort of logic before I went to Oxford; he rightly thought that I'ld have no chance to learn it there, and 'what they call logic at Oxford, Peter, is just a bad joke'. (It would be tedious to explain the financial considerations that dictated my seeking a scholarship to Oxford, rather than to Cambridge as my father would have preferred.) I think that as regards the 'logicians' who were taken seriously when I was at Oxford—Bosanquet, Joseph, Cook Wilson, etc.—my father was clearly right: by reading Keynes, mild as his style of criticism was, I had learned how to judge the reasonings they went in for. When my tutor told me I ought to read Cook Wilson, Providence directed my eye to a passage in which Cook Wilson wrote contemptuously about Russell's *Principles of Mathematics* and showed his own haziness about the difference between a member and a sub-class of a class: I shut the book, and for many years found no occasion to open it again.

Much earlier I had upset my form master at school by bringing logical terms of art into my essays, and showing that I cared about cogent argument and consistent thinking. He was a disciple of F.C.S. Schiller, the Oxford Pragmatist, and induced me to read Schiller's *Formal Logic*: I thought this a terribly silly book, and said so. My schoolmaster's daughter was later a pupil of Elizabeth Anscombe's, and told her that he claimed to have started my interest in logic: I was astounded at this claim, from a man who in fact tried to wean me from logic! But even Schiller can claim some part of my philosophical development. He regards the influence of logic on religion as specially noxious: people who take logic seriously will be attracted to a religion with fixed dogmas, claimed to be the very truth of God, and excluding the dogmas of other religions as false. What I on the contrary got out of my reading of Schiller was a lasting conviction that whether or not any religion was true, modernist varieties of Christianity are wholly incredible; this reinforced the effect of McTaggart's book, which I shall come on to later.

My father's teaching was not confined to logic. He got me to read a lot of philosophy besides: Berkeley's dialogues between Hylas and Philonous and Mill's *Utilitarianism* (both as awful examples of logical fallacies!) came in at an early age. At a slightly later date I read *Principia Ethica*. Those philosophers who know this work, rather than use it as a grimoire to banish the demon Naturalism, will recall the high claims it makes for its own clarity and rigour, and the scornful exposure of other men's confusions. Men of the highest distinction who lived in Moore's Cambridge milieu—McTaggart, Russell, Maynard Keynes—showed a complete acceptance of these claims, at any rate for a time; my father too had come under the spell, and succeeded in passing it on to me. For some years I was a convinced Moorean in ethics; while still at school I tried to rewrite parts of the work as a deductive system (in words rather than symbols, like Euclid), and flabbergasted my schoolmasters in essays by brisk demolitions of what I took to be Christian morality, using Moorean tools. The spell was unbroken until I was obliged to read the *Nicomachean Ethics* for Lit.Hum. at Oxford; then it was broken completely and for ever. I can no longer even imaginatively enter into the frame of mind in which Moore's concepts seemed to me sharp, his premises self-evident, his proofs irrefragable. In a portrait gallery of my philosophical heroes, Moore's portrait would be turned to face the wall.

McTAGGART

As I have said, apart from my childish plunge into *Appearance and Reality* my first philosophical reading was McTaggart's *Some Dogmas of Religion*. From then on I had towards him such feelings as Lucretius expresses towards Epicurus: he was my liberator from a miserable superstition, and my protector against the sort of Christianity with which my public school sought to indoctrinate me. Of his own positive contributions to philosophy I learned later, when I was sufficiently senior to be able to spend many hours in the school library. There were many philosophy books there, largely donated by a member of the Muirhead family, but also including the complete works of McTaggart, given by himself (or his widow) to his old school. I could not make much of his work on Hegel, but I read the first volume of *The Nature of Existence* with inexpressible delight: here was high metaphysics in clear language and a really rigorous style. I was simultaneously reading Spinoza's *Ethics*, and for a time fancied myself a Spinozist. Naturally I was wholly ignorant of the Jewish-Christian-Muslim theological debates about God and the soul that are indispensable background for one who would understand Spinoza; what I then made of Spinoza through my McTaggartian spectacles is not worth further discussion, but my interest in Spinoza has remained.

When I came on to the second volume, the bold paradoxes initially met with my outraged incredulity. I was expected to believe that time and material objects were self-contradictory chimeras; that sense-perception was almost wholly misperception; that introspection too was largely delusive; but that in all this delusion and confused perception two things were not delusions at all—persons, oneself and others, and the love of persons for one another; these things were not only real but eternal, and this eternal life of love bore to our actual miseries such a relation as is best conveyed in temporal terms by calling it *future*. My initial incredulity yielded, upon further study of the work, to what seemed the irresistible force of reasoning. I ceased to follow my father in his varieties of religious experience; for five or six years I was a convinced Mactaggartian—I wonder if anybody else ever has been.

Since my belief was based on reason, I did not shrink from arguing about it or from subjecting it to test. The appearance of Broad's *Examination of McTaggart's Philosophy* meant that I could study the strongest attack that a very ingenious and persistent hostile critic could mount. (Broad's hostile intention from the very first ought not to be a matter of serious doubt; a man of his learning would not otherwise have modelled his title on the one Mill chose for his demolition, in intention total, of Sir William Hamilton.) I was unimpressed; much of Broad's work struck me as elaborate misunderstanding; the few points he really did score against McTaggart were easily met by minor alterations in McTaggart's system. On this matter I have hardly altered my view in forty years, during which I have changed my mind on many things.

Apart from the healthy immersion in Plato and Aristotle that I owe to my tutor, Donald Allan, I owe far more to Balliol for the freedom of endless discussion with my peers than for any formal philosophical teaching. In retrospect I seem to have spent four years almost entirely in Balliol; I never went to philosophy lectures outside the College and knew hardly anybody in other Colleges. My Mactaggartian beliefs were honed to a sharp edge by controversy. Increasingly, as time went on, I found myself arguing with Catholics. I was certainly cleverer than they, but they had the immeasurable advantage that they were right—an advantage that they did not throw away by resorting to the bad philosophy and apologetics then sometimes taught in Catholic schools. One day my defences quite suddenly collapsed: I knew that if I were to remain an honest man I must seek instruction in the Catholic Religion. I was received into the Catholic Church on May 31, 1938.

THOMAS HOBBES: POLITICAL PHILOSOPHY

In the summer of 1937 Balliol appointed a man to give tuition in political philosophy—Hobbes, Locke, and Rousseau. As he presented these three thinkers, Hobbes was a crude philistine defender of the doctrine that might is right, Locke was a worthy upholder of sound British common sense, and Rousseau had inspired and inspiring ideas about the General Will. I reacted against this gentleman's ideas very strongly, partly through a clash of temperaments. Locke seemed to me to be an unconvincing rationalizer of his Whig patrons' successful usurpation of power, and Rousseau to have been mad, bad, and dangerous to know. I see no reason now to change my mind.

I soon noticed that my tutor's malign picture of Hobbes was drawn from just a few short passages of *Leviathan*; allured by Hobbes's magnificent English style, I soon devoured the whole book. I became convinced that Hobbes was a great and good man, who loved justice and mercy and the rule of law, and hated pride, arrogance, cruelty and other such vices of public life. I did not at this time feel able to enter sympathetically into his religion, but I never doubted that he sincerely believed the very unorthodox form of Christianity which, at no little risk, he openly professed. Hobbes should in fact not be seen as an isolated figure; he belongs in the history of the Socinian heresy, imported during his lifetime from Poland (where it was tolerated) to England (where several Socinians were burned); and the Christadelphian body today not only share many of his professed religious tenets, but even support them by the same sort of appeal to Scripture.

The political system in England in my youth seemed to many observers to be on the point of disintegration, though it has proved more durable than we young men ever dreamed. Asking myself what successor to the corrupt Parliament one ought to work for, I decided on Hobbist grounds that the best option would be the restoration of a strong Monarchy. No such thing could be hoped for, I thought, so long as the House of Windsor reigned; I judged them to be quite unfitted, by character, tradition, and training, to assume the role of Sovereign, instead of rubber-stamping Acts of Parliament. Only romantic folly made me ignore the question whether the family who would be *de jure* claimants if the Hanoverian usurpation were undone were at all likely to be better as Sovereigns. But the negative side of the Jacobitism I professed as an undergraduate has stayed by me. The recent antics of some living members of the House of Windsor give me no reason to repent this. As for our real rulers, I sadly watch how there are steady trends to the bad throughout changes of administration; and I continually marvel at the naivety of people who believe the automatic promises of whichever party may be in opposition to reverse these trends after coming to power. But since I know how prejudice and passion (in my case *saeva indignatio*) warp a man's judgment, political allusions in my writings are few.

1938–1939

My conversion to the Catholic Religion was not effected by finding a philosophy superior to McTaggart's, but of course I knew that I must now completely rethink my philosophical position. So I began reading Aquinas: since I was a classical scholar, it was no great toil to read his lucid prose in the original; I helped myself out as regards the subject-matter with the faithful Dominican translation. What then counted as Thomist philosophy was in a bad way, at least so it seems to my present mind; providentially I did not approach Aquinas in company of such blind guides. I have kept on reading Aquinas ever since; I hope I have continued to learn from him. Two pieces of my work—'Form and Existence' in *God and The Soul*[2] and 'Aquinas' in *Three Philosophers*[3]—have been directly concerned with Aquinas: his influence has been constant.

Throughout the year my philosophical position was in flux: fortunately not too much so for me to win a First in my finals. I was then awarded a Gladstone Research Studentship, tenable at St Deiniol's Library, Hawarden; I chose as the topic of my research the philosophy of McTaggart, to whom I felt and still feel gratitude for having set me standards of rigour and clarity and honesty by his example and for giving me a firm grasp of some truths and at worst saving me from fashionable errors far more pernicious than his own. I resolved to write a critically sympathetic account of McTaggart, to demolish Broad's monstrous mausoleum. In my year at St Deiniol's I actually finished the first draft of a commentary on McTaggart: if the MS had survived, I think this mixture of the new wine of my nascent Thomism with the old wine of Mactaggartianism would now be merely nauseous to me. This MS was scrapped during the war, but at intervals I returned to the work: eventually, forty years from its beginning, the task was finished, and the commentary was published under the title (borrowed from a dictum of McTaggart) *Truth, Love, and Immortality*.

While studying Aquinas at St Deiniol's, I could not help noticing that he is linguistically very self-conscious, in a way that McTaggart is not: again and again there is a careful discussion of logico-grammatical points, like the roles of nouns, adjectives, verbs, and particles; he uses the best contemporary work of logicians (*sophistae*), but when their work will not serve his ends he devises tools of his own to analyse the language of his theological arguments. The light he sheds for me upon the philosophy of logic is the very least of the debts I owe to Aquinas; but I shall dwell on the matter here, because of its bearing on later developments of my thought.

Brought up by my father on Cambridge logic, I had hitherto taken it for granted that any name is a singular term. My reading of Plato and Aristotle at Oxford did not reveal to me how extraordinary this thesis is, historically speaking; for I read little of Plato outside the *Republic* or of Aristotle outside the *Nicomachean Ethics*. But from my reading of Aquinas I came to a view of common nouns which I have ever since held and defended. In *many* uses a common noun is a name, a shared name: 'elephant' names each elephant impartially. But this is not true for *all* uses of common nouns: in predicative use, e.g. as complement of a verb like 'be', 'look', 'become', a common noun has not the role of a name; what it here answers to *in rebus* is a form, which is not *ens*, an object, but *entis*, *of* an object. I came to a clearer view of this matter when I started reading Frege and encountered his doctrine of object, concept, and function, but I did not abandon my original insight; I regarded and still regard as an unfortunate prejudice Frege's insistence that names must be singular. I was delighted, when I came to know the Polish language and the work of Polish logicians, to discover that the distinction between the two ways of using common nouns is often grammatically marked in Polish by a case-inflection, and that some Polish logicians accept proper and common names alike as names without blurring out the distinction between a name and a sentence-forming functor with name as argument (Frege's *Begriffsausdruck*).

Here I may mention that I think it useful for a philosopher to have mastered to a reasonable degree some language other than English. Though the Latin and Greek exercises I had to do at school and at Balliol no doubt produced pastiches that would have seemed ridiculous to native speakers, they had at least the benefit of forcing me to hack my way through the abstract verbiage to which one so readily resorts in English and find a plainer expression of the thoughts involved. Modern European languages like French and German have not this advantage, but even so they liberate the mind from a mere servile following of English idiom. It has made an enormous difference to my life that in my early twenties the sense of my Polishness, long repressed by my early education, strongly revived; I set out to learn Polish, and felt myself (I am sure not delusively) to be recovering a skill long out of practice rather than acquiring a new one.

The gloomy prospects of the year 1939 were for me considerably brightened by my being now engaged to be married.

ELIZABETH ANSCOMBE

As my time at Oxford approached its end, I was in Augustine's words 'in love with love': I desperately needed a girl to love and woo and marry. This is a dangerous state of mind, which often leads to humiliation or heartbreak or worse: by God's mercy I met Elizabeth Anscombe, whom I married in 1941. I find it quite impossible to say how much she and our children have meant to me; I have never got over being suddenly struck with amazement from time to time at my good fortune.

Although we have both followed a philosophical career, and have sometimes formally collaborated and often critically read each other's works, we think about different though overlapping topics, and in a noticeably different style; and either of us, when questioned about the thought of the other, will often not know the right answer. I am surprised that people find this surprising. The bond between us two in the first instance had comparatively little to do with shared philosophical thought: it was of course immensely important that both of us were Catholics and recent converts.

Elizabeth did not take her degree till three years after we met, and we did not get married till then. For those years, and for some time after our marriage, Elizabeth had a lot of philosophical teaching from me; I could see she was good at the subject, but her real development was to come only under the powerful stimulus of Wittgenstein's lectures and her personal conversations with him. Naturally she then moved away from my tutelage; I am afraid I resented that, but I could recognize this feeling as base and irrational, and soon overcame it.

As a mature philosopher, Elizabeth strikes me as a more adventurous thinker than I am: it is she who gets bold and at first sight merely zany ideas, to which I sometimes reacted with initial outrage. (*Cfr.* her papers 'The Intentionality of Sensation'[4] and 'The First Person'.[5]) Usually I come to think these bold ideas are more defensible than I had originally supposed. My mind works differently; the shocking theses I have defended in the philosophy of logic were reached not in bold leaps but by slow steps, with each step mentally tested against a multitude of examples and objections before the next step was taken. Both of us, I hope, have avoided two vices: frivolous change of mind, and adherence to past sayings in the desire to *have been* right rather than *be* right.

1939–1951

I spent the war years in the work of timber production. This open-air work undoubtedly saved my life; symptoms that showed themselves intermittently later turned out to have been tubercular, but the infected region of my lung has long since healed and my health since then has been uniformly good. Since on the Geach side I had an uncle and a great-uncle who died of TB in their twenties, I count myself very lucky.

The circumstances of the war gave me opportunity for many conversations in Polish, and in Italian too, with prisoners of war working in the forests. I acquired a fluent though ungrammatical Italian, and took pleasure in now being able to read Dante in the original. Of the Poles I came to know I must specially mention Casimir Komierowski, Zbigniew Jordan, and Casimir Lewy. Casimir Komierowski, at that time Polish Consul in Southampton, showed much kindness and hospitality to Elizabeth and me, and lent and gave me Polish books. Zbigniew Jordan, whom I met in the Polish Ministry of War in London, published a book in English about Polish logic, to show the world one aspect of the culture that the Nazis were trying to destroy—a culture then being maintained in Poland's underground Universities at the cost of torture and death. Komierowski and Jordan helped me with my efforts to join the Polish Army, as my Balliol contemporary Auberon Herbert had done: efforts that were vain.

Heu, nec defensor valui tuus esse, nec ultor.

Casimir Lewy on the other hand, forwarded my efforts to equip myself with philosophical Polish; he lent me Polish works and Polish translations of English philosophy, and he kindly read my first brash efforts at Polish philosophical writing. This work was not wasted; it has borne fruit in many lectures that I have given in Polish and in seven Polish articles published in Poland.

I was however to be frustrated in my hope that after the War I might go to Poland to serve in the work of reconstruction. Since I must make my livelihood in England and not in Poland, I was determined to do so as a philosopher; but there was no easy way for me into regular academic life. Elizabeth managed to secure the lease of a house in Cambridge, and from this base I could make many academic contacts: I regularly attended meetings of the Moral Sciences Club. (My work with Wittgenstein and Georg Henrik von Wright will be discussed later.) My long-term strategy was to get myself known in the philosophical world, both personally and by published writings, and then eventually apply for a teaching post. It was a good strategy; even while in Cambridge I did a little paid lecturing and supervision; but it took much longer to establish myself then I had expected. Some kind friends sometimes tried to find me employment of a non-philosophical sort; I hope I did not offend them, but I felt bound to ensure that I evaded what I saw as traps: such employment would ruin my strategy of doing philosophical work and getting it known, for it would leave me with no energy to do the work. But my resolution entailed a long period of comparative poverty for Elizabeth and me, which was to end only after our third child was born.

WITTGENSTEIN

Elizabeth Anscombe had close contact with Wittgenstein earlier than I did, though at that time I sometimes saw him, when he came to our house or at the Moral Science Club. Even from these slight contacts it was easy to recognise the unique power of the mind that had produced the *Tractatus*. I cannot now remember at what time Wittgenstein began to ask me to go for walks with him. Those walks were rewarding but very tiring; on a walk Wittgenstein never relaxed mentally for a moment, and required the same degree of concentration from me; attempts at light conversation were immediately quashed, and careless talk about philosophy was ruthlessly and devastatingly exposed.

Both Elizabeth and I had made serious efforts to understand the thoughts of other very great philosophers before we met Wittgenstein. I think some people were harmed by Wittgenstein because this was not true of them; it was as though they had never had a good view before of any really high mountain, and now stood so close to one mountain that other mountains were invisible.

No doubt some of Wittgenstein's work failed to impress me sufficiently because for long I had not felt the temptations for which he provided a remedy. The idea that the mental is in some sense essentially private, and inaccessible except to *one* observer, had long since been rooted out of my mind by my long meditations on McTaggart. (Enough to say here that in McTaggart's metaphysics each person not only directly perceives the mental contents of at least some other persons, but 'in absolute reality' perceives them with complete clarity and accuracy; his problem, one which he laboured mightily, was to explain how this is reconcilable with our very imperfect and confused views of other people's thoughts and feelings 'in present experience'.) No doubt there were other things in Wittgenstein's teaching that I similarly took too easily. The great philosophical confusions are ones that not only play a great part in the history of philosophy, but are ever liable to arise from our use of language. A real victory perhaps comes only to one who has deeply felt the temptations. People sometimes ask who the opponent is in the passages of Wittgenstein's writing that approach the dialogue style; obviously, himself; he is answering his own objections, fighting his own temptations.

One thing I learned from Wittgenstein, in part from the *Tractatus* but still more from personal contact, is that philosophical mistakes are often not refutable falsehoods but confusions; similarly the contrary insights cannot be conveyed in proper propositions with a truth-value. I offer as instances of such insights Frege's distinction between concept and object ('No concept is an object' has no translation into a well-constructed symbolism); Frege's view that numbers cannot attach to objects; *the* Frege point, that truth-value does not depend on there being an *assertion*. Such insights cannot be demonstrated as theses, but only conveyed dialectically; the dialectic process largely consists in the art, whose practice I have perhaps learned in some measure from Wittgenstein, of reducing to patent nonsense the buried nonsense that is found in attempts to reject these insights. We cannot refute nonsense by a straight-forward logical process; as Frege said, logic cannot deal with nonsense, but only characterise it as being nonsense.

It is not out of place that in writing of Wittgenstein I should write so much about Frege. It is a foolish legend that Wittgenstein came to think he had 'seen through' Frege. His own remark published at the end of *Zettel* ought to have been a warning against this legend. Shortly before Wittgenstein's death I often talked to him about Frege; he was pleased at my taking Frege seriously, and gave me much help and advice. I am gratified to learn, from a recently published letter of his to von Wright, that he welcomed my appointment to give a course of lectures on Frege in the Moral Science Faculty at Cambridge. The very last time I saw Wittgenstein we were talking about Frege; taking the book in his hands, he said slowly 'How I envy Frege. I wish I could have written like that.' Of each of them I would use the words Aristotle used of Plato: a man whom the base have no right even to praise. Hearing Wittgenstein on Frege was like hearing Aristotle on Plato.

Personally Wittgenstein was a trusty and generous friend. His practical advice was sound and often helpful; he wrote references both for me and for Elizabeth, when we were looking for academic work. (He remarked, which I am sure was true, that with some people *his* commendation might be the reverse of helpful.) He helped us financially when our second child was born, and devoted a lot of trouble to removing our young daughter's helpless perplexity over elementary arithmetic; both acts were in character.

FREGE AND WITTGENSTEIN'S *TRACTATUS*

I cannot remember when I first read the *Tractatus*. Elizabeth read it before I did, browsing in Blackwells, and bought a copy; I am sure I had read it before our marriage.

In my youth what has since been called the linguistic turn was all the fashion; forty years on, I think it has not lost all of its appeal; the label is meant for attempts to solve or dissolve philosophical problems by shifting from talk about non-linguistic entities to talk about language. The *Tractatus* made me aware that often this technique will not work, and why it will not; very often the problem supposedly dissolved by this shift only reappears at the level of language—not surprisingly, since language is just one part of the world. I suppose the oldest use of the linguistic turn is the attempt to combat Platonic realism about universals by setting up the distinction between general and singular terms as an *Ersatz* for the distinction between universals and individuals. But if the nominalist is willing to talk of *the* word 'pig', he is blind indeed if he fails to see that just the same problem arises about the relation of this word to its individual occurrences as about the relation of *the* pig to individual pigs. Of course there are sophisticated nominalists, like Nelson Goodman, who avoid this obvious trap; but many nominalists have fallen into it, from Ockham on.

Similarly, Frege is often accused of misconstruing the difference in mode of signification between functional *signs* like 'log' or 'sin' and numerical *signs* like '3': as if both sorts of sign signified the same way, but what is signified were now an object and now a 'mysterious' non-object, a function. But the distinction between a function on the one hand, and its argument and values on the other hand, needs to be made just as much on the level of language. Not only are the *numbers* 2^2 or 4, 3^3 or 27, 5^5 or 3125, values of a certain function for the *numbers* 2, 3, and 5 respectively as arguments: equally, the *numerical signs* '2^2', '3^3', and '5^5' are values of a certain function for the *numerals* '2', '3', and '5' as respective arguments. And here there is no piece of type whose role Frege could misconstrue as that of standing for a 'mysterious entity'. Of course similar examples occur in Frege's own writing.

What is here in question is the *Tractatus* notion of what *comes out, sich zeigt*, in language, but cannot informatively be explained in language, because the would-be explanation is either a metaphor or just another case of what we are trying to explain. I learned this lesson early; I tried to put across what I had learned in my very first published article 'Designation and Truth'[6]. I am still quite pleased with the article; its purport was that we must fail whom we attempt to ascribe to sentences something analogous to designation. If we try to introduce a sign meaning 'what sentence—in language *L* designates', the meaning of the sign simply collapses into 'sentence—in language

L is true' and the sign-designatum relation disappears. To my surprise, comments on the article ascribed to me a positive view on what sentences *do* designate; it illustrates Quine's remark that the plainest writing is not proof against stalwart reading.

Had I not been reading the *Tractatus* with some measure of understanding, I think I could never have profited by reading 'the great works of Frege'. Frege's *Grundlagen* came to me when my father died, but I came to read it only shortly before John Austin's translation appeared; I had only slight knowledge of German, but I persevered in my reading, writing out the best translation I could manage, by way of exercises, day by day. All the time my enthusiasm for Frege grew. I became eager to read his other works. Some of them I could find for myself in Cambridge University Library, others were lent me by Bertrand Russell and Gilbert Ryle; yet others I was told how to find by Wittgenstein. Before I began teaching at Birmingham I had collaborated with Max Black in producing a volume of translated selections from Frege, and had given a course of lectures on Frege in Cambridge.

Having come to Frege by way of the *Tractatus*, I could see that his difficulties in expressing himself about function, concept, and object were not from a muddled self-bemusement but from the nature of the case. The same influence led me to think that Frege's assimilation of sentences to names, even to complex names, was a mistake; though I thought many citics of Frege on this point started out from positions more radically mistaken. One of Frege's great feats was to say loud and clear that the same thought may occur now asserted, now unasserted, and has the same truth-value whether asserted or not; I have sometimes called this *the* Frege point. When critics of Frege's idea that sentences are names are muddled about *this* point, or attempt to reject it (an attitude which naturally they cannot long sustain), their case against him is not to be heard.

So far as the formal structure of *Grundgesetze* is concerned, it would be fairly easy to rewrite it so as to recognise the category-difference between names and sentences, as in the earlier *Begriffsschrift*. In fact, people who discuss Frege's theory of classes quite often do tacitly rewrite fragments of *Grundgesetze* in just this style. But in the philosophy of logic it is otherwise; Frege's view of sentences has been treated not as an error, but as a first step towards the construction of a semantic theory.

It was through a shared interest in Frege that I came to know

Michael Dummett. For a long time we had our family homes in Oxford, and discussed Frege and other logical topics continually; on very many matters we agreed. Dummett's first book on Frege is a masterpiece, but a flawed one; it is too long, there are manifest inconsistencies, and Dummett's careful scholarship and sure interpretation where Frege is concerned are offset by very reckless description of other writers. This last fault is most serious in regard to the *Tractatus*; sometimes, as when dealing with truth-functions, Dummett's account of this work is a mere travesty. Dummett and I agree that Frege went wrong in assimilating sentences to names; but Dummett's final view is that sentences *are* names, only not names of *objects* but of certain non-objects, the truth-values. For anyone with the *Tractatus* in his bones, that is a non-starter; it makes the wrong assimilation under the guise of rejecting it.

One essential difference between names and sentences is that sentences come in contradictory pairs, mutually replaceable *salva congruitate*, and names do not; names, said Wittgenstein, are like points, sentences are like arrows (with a reversible sense). One way to bring out this duality of sentences is this: we could construct a dictionary such that, translating by it, we shall get for every sentence *S* a sentence in an alternative language that is equiform to a contradictory of *S* in the usual language; under such translation names would be left unchanged. (The formal logic of duality shows the possibility of such a dictionary.) Dummett quite fails to grasp this fundamental thought of the *Tractatus*; it is to this that I must put down what I can only regard as his obsession with attempts to reject the law of double negation. Two propositions are each other's negations as two relations are each other's converses; to try to distinguish the senses of a proposition and its double negation (an attempt Frege will certainly not favour) is as nonsensical as to try to distinguish a relation from the converse of its converse.

It would be an endless task to spell out how much the *Tractatus* and Frege together have influenced my thought. My most serious departures from Frege are largely spelled out in my reply to Dummett on identity; apart from two theses, which I take for granted readers will use in order to understand that reply. As regards names, I differ from Frege (and from Dummett) in two ways: I reject complex names, and I accept shared names. The rejection of shared names by Frege and Russell has been enormously influential; somebody as little Fregean or

Russellian as Strawson regards the idea of shared names as a 'proposal' by—Peter Geach! From when Plato and Aristotle distinguished between *onomata*, names, and *rhemata*, verbs or predicables, words like 'horse' and 'gold' have regularly been counted as capable of use in a naming role; historically, it is the Frege-Russell restriction that is the 'proposal', and its motivation does not seem clear. (Frege did not yet observe this restriction in *Begriffsschrift.*) As regards the simplicity of names, I came round in the end to the view of the *Tractatus*: since a syntactically simple name *can* name, any complexity a name happens to have is irrelevant to its naming role. (A proposition about a pub called 'The Duke of Cambridge' says nothing about Cambridge.) Complex phrases' *seeming* to play a name-like role is a grammatical illusion; in *Reference and Generality*, and in a number of articles, I have tried to dispel this illusion. Of course the simplicity of names does *not* mean that the *things named* have to be simple; nor does it mean that names have to be *singular*, i.e. unshared. (Incredible as it seems, the latter confusion is perpetrated by some people who cannot be suspected of trying to throw dust in their readers' eyes.)

It may be well to mention one likely motive for the rejection of shared names: desire for a simpler theory. It is pretty clear that a one-word general term used predicatively is not a name; this is specially clear for predicative complements after verbs other than 'is' or 'is not'. If a man became a thief, or Mary looked an angel, it is nonsense to ask which thief or which angel. If we then have to regard the same vocable as having the role of a name, a logical subject, in other contexts, this may seem an undesirable complication. But systematically multiple use of identical vocables just is a feature of natural languages. Frege, as we all know, ascribed a special use to words coming in indirect-speech constructions; whether he was right or wrong about this, words in direct quotation unquestionably have a special use—so special that the bad vogue has come in of saying that this is *not* use of the words at all, but *mention*! Again, in very many natural languages a proper or common noun may be applied, not to the thing primarily so called, but to an image or picture of it. Since such shifts of use have to be recognised, there is not an overwhelming antecedent presumption against the view that a vocable may shift between a naming and a predicative role. But we must insist that in all such cases there *is* a change of sense. The idea of a 'term' which can, *without* change of sense, be now subject, now predicate, seems to me a disastrous muddle; and Aristotle's introduction of it is comparable to the Fall of Adam.

BIRMINGHAM, 1951–1966

While we lived in Cambridge I was submitting articles to various journals; mostly they were accepted, and my name was getting known. Gilbert Ryle, then Editor of *Mind*, and Austin Duncan-Jones, then Editor of *Analysis*, were particularly helpful. At the latter's suggestion I applied for a post at his University, Birmingham, and was appointed. I had already done some teaching, both lecturing and supervising undergraduates, for the Moral Science Faculty at Cambridge; some of my colleagues were clearly at first worried about whether I could settle down to a regular career as a University teacher, but I think their worries were soon dispelled. Upon my appointment the family moved to Oxford, to a house near Somerville College, where Elizabeth was already a Research Fellow. We stayed there till she was appointed Professor of Philosophy in Cambridge in 1970: three daughters, born in 1951 or later, became Somerville undergraduates. I travelled home at weekends.

I was quite quickly promoted at Birmingham, being successively Assistant Lecturer, Lecturer, Senior Lecturer, and Reader in Logic. I got on well with my new colleagues in the Department, apart from one unhappy man whose troubled mind eventually broke down. I am sure the regular practice of teaching, discussion, and examining has been very profitable to me as a philosopher. During this period I was quite productive as an author, and I began my concern with the production of *Analysis*, of whose Editorial Committee I have been successively Secretary and Chairman.

Unhappily I soon became aware of a pronounced hostility towards the Department on the Arts Faculty Board; at Birmingham this was a powerful body whose recommendations nearly always got ratified. Again and again motions that would benefit the Department were voted down and motions hostile to us were brought forward. Our situation was improved by the fact that so many of us were promoted to Readerships. This promotion did not lie within the Board's purview, but carried with it *ex officio* membership of the Board; when I left, three members of the Department were Readers. The prompt reaction to my own promotion was a discussion whether this 'undemocratic' *ex officio* membership should now be abolished; this proposal was narrowly defeated. If it had been carried, Duncan-Jones alone could have been at meetings to watch over our interests; his health was never very strong, and even as things were attendance was a great strain on him, given the attitudes he faced.

I watched the way things went with a growing resentment. The breaking-point came when the dominant clique rapidly pushed through a proposal to establish an Institute of Contemporary Culture; this meant research into Pop Art, and at this distance it is amusing to recall the argument that 'the evanescent nature of the research material' made it imperative to act without delay. I believe a Birmingham magnate had guaranteed the *first* year's costs; anyhow, the University was going to be saddled with a heavy permanent expense. I brooded on the matter, recalling the many times when the Department had been refused much smaller sums of money with mutterings about 'quinquennial plans' and the like; then I wrote a letter to the Dean saying that I had no wish to stay at a University that preferred Pop Art to Logic as a subject to endow research in.

I had burned my boats; I acted without any definite plan as to where I should next go. I knew that more than one University in the United States would welcome me; but I did not wish to spend the rest of my life, and bring up my children, in that country. I wrote to my friend James Cameron, then Professor of Philosophy at Leeds, asking for his advice; the proposed Chair of Logic at Leeds was still vacant, it appeared, and upon application I was appointed.

This sort of struggle against petty hostility can be bad for heart and soul. It was an added burden that, kind and helpful as my Department colleagues were, there were some things about which I felt I was shouting into the wind as I talked with them. For example, I found that the concept of mercy as a human virtue was quite alien to them; 'either it is wrong to punish or it is wrong to pardon' was their attitude; the idea that letting a man off whom one had a perfect right to punish could show an admirable trait appeared to them unintelligible. I could persuade Bernard Mayo, then Editor of *Analysis*, to publish an article in which this concept of mercy figured, only by rewriting a paragraph so that this concept was presented as part of an old feudal code![7] This deep lack of sympathy made me often feel isolated.

I found relief in the friendship of a group of Christadelphians. I have mentioned the Christadelphian body before in connexion with Thomas Hobbes. The Socinian doctrines that Hobbes expounds in his writings were and are widely believed in the Midlands; since in his time Socinians were liable to be burned, I think his profession of Socinian doctrines is rationally explicable only by his actually believing them; an atheist's tactics would rather be ironical deference to the local orthodoxy, like Voltaire's or Hume's. In Lichfield, not far from Birmingham, several Socinians were burned; Unitarianism based on Hobbist interpretations of the Scripture were for long prevalent in Birmingham; Unitarianism is now a different sort of thing, but Socinianism lives on under the new label of Christadelphianism, which has its main *ecclesia* in Birmingham.

The Christadelphian tradition is one not of fanatical enthusiasm but of quiet persuasion; they often cite the text 'Be ready always to give an answer to every man that asketh you a reason of the hope that is in you' (I Peter iii.15). My dialogue with my friends was, I hope, a matter of 'speaking the truth in love' on both sides; honesty about our differences served to bring out some measure of deep agreement. My publications in the field of philosophy of religion all show the influence of these talks. I dedicated one volume to the memory of L. G. Sargent, with whom friendship endured till his death. In Birmingham I often felt spiritually like a man shut in a hot fetid room; my Christadelphian friendships made it possible sometimes to get 'windows open towards Jerusalem' though I am afraid I did not constantly keep them open.

LEEDS, 1966–1981

My time at Leeds was far happier than my time at Birmingham had been. James Cameron and his successor Roy Holland, as heads of the Department, built up a team of talented philosophers. Such talent often goes with an excitable temperament that leads to clashes; I was no exception, and was sometimes involved in such clashes. But the mental stimulation of living in this Department was great. There was a great measure of agreement on certain fundamental matters: on the teaching of logic and the philosophy of logic, on the importance of Frege and Wittgenstein. I valued the chance to share in teaching each year a course on Ancient and Medieval Philosophy for undergraduates. (My colleague Timothy Potts made a valuable collection of medieval philosophical texts in photostat, and of translations of excerpts for the use of students.) In a valedictory speech I quoted to my colleagues the words of an old lyric: 'You may have been a headache, but you never were a bore'; I was overall very happy in my work. As a background to my work, there was my collegiate life as a staff member of a hall of residence, Lyddon Hall; the young men did me the honour of making me their Librarian. The shields of Leeds University and of Lyddon Hall hang over the door of my study: *estote perpetuae*!

POLAND, 1963–1985

As I write, I am in Poland as a Visiting Professor at the University of Warsaw. I wanted to come to Poland in 1945: possibly to teach, in any event to do something for Polish reconstruction. I had to wait till 1963 for my first visit. Before then I had met Ajdukiewicz at a conference in Helsinki; he sent me a general invitation to come to Warsaw, which I provisionally accepted. My chance came in 1963; it had emerged that my teaching commitments at Ann Arbor would be over a little earlier than the date the Arts Faculty Board in Birmingham had set for my return to that University. I applied for a Polish visaa, and flew to Warsaw at the end of the Ann Arbor term. Unhappily Ajdukiewicz had suddenly died before my arrival; but his colleagues in Warsaw made me feel much welcomed.

Since then I have kept up my contacts with Poland, and have made many Polish friends. It is an invidious task to decide whom to mention by name: if I tried to mention everybody to whom I owe thanks for great kindness and hospitality, and for intellectual stimulation as well, I should be afraid of committing the offence of the king in fairy-tales who fatally forgets to invite one of the fairies. So I shall here mention only Tadeusz and Janina Kotarbińscy, and Boguś and Nina Wolniewicz. It was a privilege to know Kotarbiński; I honoured him very much as a man and as a philosopher, who set standards for successive generations of Polish philosophers. I well remember how he conducted a philosophical discussion like a master of chess; a series of innocent-sounding questions, then a question that already put the speaker in check, then a final question that threatened instant checkmate. There was nothing eristic about this; he wanted truth, not victory. Of Wolniewicz I have already said something in my reply to his paper; I think his service to his country in Wittgenstein scholarship has been inestimably great; it is strange indeed that in a country where logic matters so much Wittgenstein had to wait so long for a worthy exponent.

It satisfies a deep need in me to be in Poland and hear Polish spoken all around me: the language my Sgonina grandparents will have spoken when I was with them in my earliest childhood—my mother was often away, as an undergraduate at Oxford. I come to Poland as often as I can. I feel now that it was indeed necessary for me to relearn Polish and eventually come to Poland; I am proud of my links with a people that has maintained culture and civilization with such courage in the face of adversities that have often appeared overwhelming.

I have written a number of papers in Polish, and I am in the habit of reflecting on whether an argument that I construct could be formulated effectively in Polish. Like any natural language, Polish has idiotisms of idiom, as Prior used to call them; it is awkward for learners that the syntax of a sentence has to be different if the numeral word for 4 is replaced by the numeral word for 5. But then no Pole has ever thought that this matters for logic. English philosophers on the other hand are, in my experience, often misled by English idiotisms of idiom. I have just read an essay in which the author argues that if 'ought' implies 'can', then 'I ought to have mown the lawn' implies 'I can have mown the lawn'! He simply did not notice that it is a matter of a former obligation to mow, not of a present obligation in respect of former mowing. His argument rests on the peculiar grammar of English auxiliaries, and is not to be reproduced in other European languages. English articles, which Poles tend to omit, are often noise and not message; none of the message is lost in the sentence with which a Polish professor of linguistics is said to have begun a course in the United States: 'In English language common noun in singular number invariably has article'. I remember that criticism of a paper of mine once turned on the question whether 'a book' or 'book' in a logical example, should be called the antecedent of a pronoun; but if a corresponding example and the discussion of it had come in a Polish paper of mine, the objection could not even have been stated. Some English theologians worry about whether it should be said of the Son of God that he became *a man*, or, became *man*!

There is a sort of linguistic chauvinism on the part of English-speaking philosophers which I find all the more offensive because it is quite unconscious. When John Austin says that the distinctions drawn in language, based on the experience of many generations, are more likely to be soundly based than ones dreamed up by a philosopher in his armchair, he has of course only English in mind; no need for him to say so. But as regards knowledge, one of his favourite topics, English has historically fared badly; the main verbs in 'I ken John Peel' and 'I wot John Peel keeps hounds' are both obsolete, we have to make do with 'know'. Polish has 'znam' for 'I ken' and 'wiem' for 'I wot', and a third verb 'umiem' for 'I know how' (with an infinitive). No doubt Polish sometimes blurs distinctions that are easily made in English; my point is that a good knowledge of more than one language is valuable as a safeguard against some sorts of bad thinking.

I oppose the idea that thoughts are so shaped by a language as to be untranslatable. The Aristotelian tradition was preserved in Arabic, and Hungarian and Chinese logicians can communicate with ones who write English. Nor is it only in such abstract matters that an extreme difference in grammatical structure does not impede communication. Grammatically speaking, the difference between male and female is strongly marked in Polish, and not at all in Hungarian; but Polish and Hungarian mores do not correspondingly differ.

> Poland is not lost: Poland lives and will live. *Witajcie, Rodacy*!

In extreme old age Hobbes diverted himself by writing an autobiography in Latin elegiacs. I have in my time written autobiographical elegiacs, and earlier I quoted one line. (I never tried to cover my whole life this way, as Hobbes did.) But now with a further piece I conclude:

> Sexaginta annos complevi hucusque novemque,
> In Domino sperans, dum vocet ipse: Veni.

University of Warsaw, 1985

ACKNOWLEDGEMENT

I wish to express my very great gratitude to Dr Harry Lewis for the immense trouble he has taken over producing this volume.

REFERENCES

[1] Geach, P. T.: 1972, *Logic Matters*, Basil Blackwell, Oxford, pp. 23–27.
[2] Geach, P. T.: 1969, 'Form and Existence', in *God and the Soul*, Routledge and Kegan Paul, London, pp. 42–64.
[3] Geach, P. T.: 1963, 'Aquinas', in G. E. M. Anscombe and P. T. Geach, *Three Philosophers*, Basil Blackwell, Oxford, pp. 65–125.
[4] Anscombe, G. E. M.: 1981, 'The Intentionality of Sensation: A Grammatical Feature', in *The Collected Papers of G. E. M. Anscombe, Volume Two: Metaphysics and the Philosophy of Mind*, Basil Blackwell, Oxford, pp. 3–20.
[5] Anscombe, G. E. M.: 1981, 'The First Person', in *The Collected Papers of G. E. M. Anscombe, Volume Two: Metaphysics and the Philosophy of Mind*, Basil Blackwell, Oxford, pp. 21–36.
[6] Geach, P. T.: 1972, 'Designation and Truth', in *Logic Matters*, Basil Blackwell, Oxford, pp. 189–193. (First published in *Analysis* 8(6), 1947–48, 93–96.)
[7] See Geach, P. T.: 1972, 'Imperative and Deontic Logic', in *Logic Matters*, Basil Blackwell, Oxford, pp. 271–272.

WILLARD VAN ORMAN QUINE

SELECTED CORRESPONDENCE WITH GEACH

I had a letter from an unknown reader, Peter Geach, in February 1949. In the ensuing thirty years his letters to me have run to 147. They brim with fresh ideas on logic, philosophy, and language, they bristle with jibes at bad ideas, and they sparkle with curios from the mediaevals. They have been an unfailing source of stimulation, instruction, and entertainment. My bulging file of them is a happy prospect for posterity.

Meanwhile, at Burton Dreben's suggestion, I am offering a brief preview. Since, however, this purports to be my contribution to the present volume rather than Geach's, I have had to limit my selections from his letters to passages that elicited substantial responses from me. Consequently the bits from Geach, for all their merit, are not meant as the cream of his crop.

Geach to Quine, 4 April 1961

... I now feel very doubtful as to whether a belief that could be *expressed* by the believer in a first-person sentence $\ulcorner\varphi(\text{I myself})\urcorner$ could suitably be *reported* by predicating of him the predicate $\ulcorner$① believes that φ(①)$\urcorner$. But I have no doubt that $\ulcorner$—believes that $\varphi(\text{himself})\urcorner$ is a Shakespearean predicate, which if true of anything is true of it by any other name, as '—smells sweet' is of a rose.

The import of the distinction between direct and indirect reflexive pronouns comes out in the following example:

(1) Oedipus called down divine vengeance upon the murderer of Laius

can be so read that, assuming Oedipus to be the murderer of Laius, we know that (1) is true if

(2) Oedipus called down divine vengeance upon Oedipus

is also true. We may then say: Though he was not aware of it, Oedipus was calling down vengeance upon himself. Now whereas

Harry A. Lewis (ed.), Peter Geach: Philosophical Encounters, 27–43.

(3) Oedipus called down divine vengeance upon himself

may thus be read as equivalent to (2), we may *also* be making such a report of what Oedipus said as would be acceptable as true iff Oedipus said something to the effect 'May the immortal gods punish *me*!'. In the former case, 'himself' is replaceable by 'Oedipus' and is a *direct* reflexive; in the latter, 'himself' answers to whatever first-person pronoun Oedipus used, and is an *indirect* reflexive. But on *either* interpretation

(4) —called down divine vengeance upon himself

is a Shakespearean predicate, true of the murderer of Laius iff true of Oedipus.

Now let's consider the Ernest of your pp. 154ff [in *Word and Object*[1]] who is trying to shoot a lion. If we expand 'Ernest is trying to shoot a lion', in accordance with *one* possible interpretation, into:

(5) Ernest is trying to bring it about that he shall shoot a lion

the 'he' is surely an indirect reflexive pronoun, corresponding to the 'I' of Ernest's boast 'I shall shoot a lion' or wish 'may I shoot a lion!'; and this would differ from the reflexive pronoun of (your (11)):

(6) Ernest is trying-to-cause himself to shoot a lion

which you surely meant as a *direct* reflexive pronoun, replaceable here by 'Ernest'. It is to be sure harder to make up a story that would discriminate (5) and (6): but not too difficult. Let Ernest be the English adviser of an Oriental potentate: at his employer's wish, Ernest is carefully arranging a lion shoot for an unnamed honoured guest of the potentate—who is planning a pleasant surprise for Ernest: *he* is the honoured guest! In that case, (6) is true, if 'himself' is indeed direct-reflexive, and (5) is false. But just on that account (6) appears *not* to be an adequate account of 'Ernest is trying to shoot a lion'....

Quine to Geach, 28 June 1961

... The contrast that you bring out in connection with Ernest's lion-hunting by your fable about the Oriental potentate is a contrast that I was aware of, and would draw in very much your own words: 'Ernest is trying to bring it about that he shall shoot a lion' versus 'Ernest is trying to cause himself to shoot a lion'. In the former of the two sentences we can supplant 'he' *salva veritate* by 'Ernest' or other singular terms which are known or thought by Ernest to designate himself. In the second sentence we can supplant 'himself' by any singular term that designates Ernest, beknownst or unbeknownst to himself.

In an interim draft of *Word and Object* I pointed up this contrast by using 'the toughest novelist' where you use 'the honored guest'. For the fact is that Ernest is not the English adviser of an Oriental potentate, but an American novelist whose other name is Hemingway. My contemplated use of 'the toughest novelist' was in allusion to Hemingway's notorious parade of virile and bellicose behavior. This allusion I embellished with my sardonic fiction of his *not* knowing that he was the toughest novelist. It pained me to scrap those paragraphs, but I did so when I came to appreciate... that 'Ernest is trying to cause himself to shoot a lion' *must* prevail over 'Ernest is trying to bring it about that he shall shoot a lion' insofar as we are concerned to paraphrase 'Ernest is hunting lions' or even 'Ernest is trying to shoot a lion'; it must prevail on pain of giving 'Ernest' incompletely referential position in the latter two sentences. Since I was talking of lion-hunting, this reflection made 'Ernest is trying to bring it about that he shall shoot a lion' irrelevant and best omitted from an already complex and cluttered exposition. Such anyway was my somber judgment, much though it pained me to chuck the bit about the tough novelist....

Geach to Quine, 12 July 1961

... You hold that in

(1) Ernest is trying to bring it about that he shall shoot a lion

'Ernest' has an incompletely referential occurrence. Apparently your grounds are that the question whether 'he' can be supplanted by some designation α must be settled, not according as α does or does not designate Ernest, but according as α would or would not be accepted by Ernest as a designation of himself. I agree with the negative part of this; I'm not sure if I agree with the positive part; that is, I am not sure whether e.g. from (1) and

(2) Ernest knows or believes that 'the toughest American novelist' designates himself

we can infer

(3) Ernest is trying to bring it about that the toughest American novelist shall shoot a lion.

But I'll let that pass. In any event, it seems to me that the occurrence of 'Ernest' in (1) *is* purely referential. To be sure, if in (1) we replaced 'he' by a repetition of 'Ernest', the second occurrence of 'Ernest' would not be purely referential; and so if 'Ernest . . . he . . .' in (1) were going proxy for 'Ernest . . . Ernest . . .', we should have to say that the sole occurrence of 'Ernest' in (1) was not purely referential; just as in

(4) Ernest was called that because of the importance of being earnest

'Ernest . . . that . . .' goes proxy for 'Ernest . . . Ernest . . .', and thus the sole occurrence of 'Ernest' in (4) is not purely referential. But 'he' in (1) does not simply look back to the antecedent 'Ernest': it is an indirect reflexive pronoun, corresponding to an 'I' of *oratio recta* (reporting a speech put into Ernest's mouth). This may be brought out by a paraphrase of (1) employing *oratio recta*:

(5) Ernest's current endeavours are made in a hope expressible in some such words as: 'May I shoot a lion!'

Analogously, (3) would be paraphrased thus:

(6) Ernest's current endeavours are made in a hope expressible in some such words as: 'May the toughest American novelist shoot a lion!'

Whether (6) follows from (5) and (2) is a question that seems to me to have much the same difficulties as whether (3) follows from (1) and (2). But in any event, the occurrence of 'Ernest' in (5) is purely referential; so I'll maintain that the occurrence of 'Ernest' in (1) is purely referential too. Of course this means that we *cannot* regard

(7) — is trying to bring it about that he shall shoot a lion

as a predicate of the form ⌜ ① is trying to bring it about that $\varphi($ ① $)$⌝; and I am now inclined to think that there *is* no such form — that the idea I had of there being such a form involved a muddle between direct and indirect reflexive pronouns. . . .

Quine to Geach, 25 July 1961

You misunderstand me. I do not hold that 'Ernest' in (1) has incompletely referential occurrence. I hold (as you do) that the occurrence of 'Ernest' in (1) is referential and that of 'he' is not. My point was rather that 'Ernest' in 'Ernest is hunting lions' has incompletely referential occurrence *if* we construe 'Ernest is hunting lions' as (1). For, the 'Ernest' of 'Ernest is hunting lions' expands into the 'Ernest' *and* the 'he' of the putative expansion (1). This is why I elect:

Ernest is trying to get himself to shoot a lion

rather than (1) as expansion of 'Ernest is hunting lions'.

Geach to Quine, 31 July 1961

Thank you very much for your letter—only now I feel more than ever puzzled as to your criteria for purely referential occurrence of a proper name. For you say that although the single occurrence of 'Ernest' in each of the following propositions:

(1) Ernest is hunting lions
(2) Ernest is trying to bring it about that he shall shoot a lion

is a purely referential one, nevertheless the occurrence of 'Ernest' in (1) *would not be* purely referential if (2) were taken to be a paraphrase of (1); and this is because with such a paraphrase we should be representing the 'Ernest' of (1) by 'Ernest . . . he . . .' in (2). I do not see how this reason can have convinced you. You are well aware that the question which part of a paraphrase 'represents' which part of the proposition paraphrased is not one that has a clear and unequivocal answer in general; and though in particular cases an intuitive solution may be impellingly convincing, I'ld be willing here to back my intuition against yours. For if 'Ernest' is indeed used in a purely referential way in both (1) and (2), then there can be extracted from (1) and (2) the predicates

(3) ① is hunting lions

and

(4) ① is trying to bring it about that he (himself) shoots a lion

Now *my* hunch is that (4) is what represents (3) when we paraphrase (1) as (2). And if so then there is no question of representing 'Ernest' in (1) by 'Ernest . . . he . . .' in (2); 'Ernest' just represents 'Ernest'. . . .

Quine to Geach, 23 August 1961

The difficulty is not a difference in intuitions, I insist, but purely a failure of communication; and I am determined to set it right.

My appeal to 'what represents what in paraphrase' was not meant as any supplementary channel of evidence for anything. I had hoped it would suggest the Giorgione parallel ('Giorgione was so called' explained by 'Giorgione was called "Giorgione"'), as a means not of proving anything but of quickly indicating my objection to:

(1) ① is trying to bring it about that ① shoots a lion

as a paraphrase of '① is hunting lions'. Let me now abandon that expository line. I think you do agree that (1) will not do as a paraphrase of '① is hunting lions', because of the non-referential position of the second '①' in (1). Indeed, I should regard the very use of a circled numeral in this second position as meaningless.

This does not mean that I object to your construing 'Ernest is hunting lions' as 'Ernest is trying to bring it about that he (himself) shoots a lion', for in so doing you are construing '① is hunting lions' not as (1) but as:

(2) ① is trying to bring it about that he shoots a lion.

My only reason for preferring:

① is trying to get ① to shoot a lion

is that it avoids appealing to an irreducible category of reflexive reference, or, perhaps, analyzes it out. . . .

Geach to Quine, 28 August 1961

Very many thanks for your last letter, which does remove my remaining difficulties, and along with them any fear that I'd got you wrong about Leibniz's Law as the criterion for purely referential occurrence of a name. . . .

Geach to Quine, ca. 20 February 1969

You'll have had my letter about the contradiction resulting from '$3 \cdot 0 = 0$' if we use the *Methods of Logic* definition of '$\kappa \cdot \lambda$' [namely, as the size of a class that can be partitioned into κ parts of size λ]. [2] Clearly my argument will show that such a contradiction arises from '$\kappa \cdot 0 = 0$' if we take κ to be any natural number > 1. Intuitively this is so because *no* class (not even the null class) can be formed by putting together κ null classes, if $\kappa > 1$....

I've thought of a fairly simple way to remedy these troubles....

'$\kappa \cdot \lambda$' for '$\hat{\alpha}(\exists \mu)[\mu$ is a class of mutually exclusive **existent** classes. α **contains only** members of members of μ. μ has κ members. $(\beta)(\beta \epsilon \mu \; . \supset . \; \alpha \cap \beta \epsilon \lambda)]$'.

I have **underlined** the divergences from the wording of your definition....

Quine to Geach, 3 March 1969

I am dumb with amazement. This is the second time, at least, that you have pointed out the falsity of something that I had stated as obvious in *Methods of Logic*. The other one I remember was 'Everyone will have had a mother'. This one is 'A class has $m \cdot n$ members if and only if it is breakable into m parts of n members each'. Even this is false. There remains little to cling to, here below.

The remedy in your white letter of about February 20 is about the best I can think of. I can improve it slightly: instead of 'mutually exclusive existent classes' we can still say simply 'mutually exclusive classes'. For, automatically, at most one of them will be empty. What is crucial is that we not forbid extra members.

We can still come this close, then, to an intuitive statement: A class has $m \cdot n$ members if and only if it is exhausted by m disjoint classes and shares n members with each....

Geach to Quine, 10 November 1969

... Subject to my doubts and hopes regarding a possible thorough godly reformation of set theory from the bottom up, my heart is still a bit with the zigzag theory....

My heart still favours the zigzag theory against the limitation of size theory because the latter seems to draw its inspiration from a view of classes as built up out of their members. It is philosophically more congenial to me to regard a predicate that is *not* somehow twisty as determining *two* classes, the Ins and the Outs, on an equal footing. But the future of the zigzag theory looks a bit bleak, in view of the history of troubles about Nn and mathematical induction in NF and ML—particularly, the devastating result that sethood of Nn can be proved in ML only if ML is inconsistent! Is there, to your mind, *no* remedy for this except the adhokey-pokey of adding 'Nn ∊ V' as an axiom? . . .

Quine to Geach, 20 November 1969

. . . I share your preference for a zigzag theory. More basically, I see the following as the *desideratum cardinale sive primi ordinis*: a class is determined by any membership condition, for all values of any further free variables therein, provided that '∊' occurs in it only before free variables. This amounts to saying that a quantificational schema has an extension for each choice of extensions as interpretations of its predicate letters. This reifies virtual classes, insofar as they are not infected by a prior theory of real classes.

If the mention of quantificational schemata in the above formulation be liberalized to admit also identity, as I could wish, then all finite classes are assured. For they are determined by simple enumeration of members, these being represented by free variables. Already, then, ML is disqualified; for it lacks the finite classes of its ultimate classes. NF remains in the running, but it has its shortcomings over induction. (I do not mention the axiom of choice; see *Set Theory and Its Logic*, revised edition, pp. 296, 299.[3]) I'd like to see some way, better than NF, of fulfilling the *desideratum cardinale cum identitate*.

I am interested in your early dream of a notation that would show membership and not state it. In 1935 I was dreaming of a contextual definition of membership that would show the meaninglessness of type violations by a natural failure to reduce such formulas to primitive notations. The last paragraphs of 'Toward a calculus of concepts' touch on this, and 'A theory of classes presupposing no canons of type' goes to work on it. But it is no good; a vulnerable denotation theory is presupposed. . . .

Geach to Quine, 13 June 1971

I am moved to a mild protest at your remark . . . that 'predicable' and 'count noun' are idle variants on each other. For me, a count noun is always an actual expression, and a syntactically simple one at that; whereas a predicable is rather a rule for getting sentences or clauses; the expression quoted after the words 'the predicable' will not be a quoted piece of a sentence, but an indication of how sentences or clauses of a certain class are to be constructed. I said in *Reference and Generality* § 26[4] that 'John shaved John' and 'Peter shaved Peter' are sentences containing one and the same predicable, which is *not* to be identified with the only shared segment, the word 'shaved'. In specifying that predicable I might have written something like this:

> the predicable '—shaved—' (both blanks to be filled up the same way)

Of course *your* circled-numeral notation for what you used to call stencils, '① shaved ①', is neater; but I think my way of speaking is just as perspicuous (and in writing *Reference and Generality* I wanted to keep printing costs down!). . . .

Quine to Geach, 23 June 1971

I am pleased and relieved to be reminded of your distinction between count nouns and predicables. It is a familiar and important distinction, but I had forgotten that it went with these words. The predicable, as a function from nouns to sentences, is indeed a *construction* in the sense of my *Philosophy of Logic* [5], though not a grammatical construction (since it imports lexicon and not merely particles).

In my work in progress (*Roots of Reference*) I have not had to take up this proper notion of predicable. My early recourse [in the book] to the 'such that' construction enables me, as it happens, to say what I need to say by talking of complex general terms that *are* actual and even continuous expressions. I do not consider this an advantage of my present approach, but it is a side effect. I am glad now that it justifies me in banishing 'predicable' from the plethora of terminological alternatives.

I had already banished 'predicate' because of *Reference and Generality*, where you commendably reserve this word to describe a role in a sentence. I regretted not having taken the same line in *Philosophy of Logic*, where 'predicate' is so persistent a refrain. But then there are also the so-called predicates way back in *Methods of Logic*, with their circled numerals, and indeed the predicate letters and predicate calculus of still earlier decades. Can we call the predicate letters *predicable* letters? Can Neptune stem the something something tide? Discouraging reflections.

The plethora of interchangeables is now down to 'count noun' and 'general term' and my readily expendable 'individuative term', 'term of divided reference'. I get some cheer from reading that count nouns, for you, are single words only. I now feel well justified in adhering to 'general term' in *Roots of Reference*. Granted, what behave as general terms (because as predicates) are sometimes mass terms rather than count nouns; but I think I am accommodating that. . . .

Geach to Quine, 13 July 1975

. . . I have been aware of my unredeemed pledge to send you comments on your book. But I do that herewith. . . .

Notes on *The Roots of Reference*[6]:

(p. 5) There is a point about causality that so far as I know comes out simply and plainly only in McTaggart (my father's supervisor, my philosophical hero from teenage years): A serious causal generalisation won't have the form 'Each *A* event is followed by a *B* event' but 'Each *A* event is followed by a *B* event in a relation *R* to it', e.g. not 'Each ingestion of a gram of KCN is followed by death' but '. . . by death of the body that performed the ingestion'.

How do you track energy through time? 'No entity without identity' *ought* to worry us here!

(p. 6) You here pick up my last point: but I'm *still* left unhappy. It is not just that in the *subatomic* realm identity criteria for parcels of energy go astray: even on a gross macroscopic level, are there *anything like* tagging methods for parcels of energy, as there are for parcels of matter?

(p. 8f) I applaud the point about 'by virtue of' and the insufficiency of using a non-material conditional. Likewise, and from long since (*Mental Acts*), your complaint about Ryle is congenial.

(p. 10) If a disposition *is* a physical state or mechanism, does this for you mean any more than: A body has such-and-such a disposition if such-and-such a physical state or mechanism is present? But the logical structure of $\ulcorner$① is disposed to $\varphi\urcorner$ isn't illuminated by finding a non-dispositional coextensive predicate for each φ.—And can *you* consistently gloss the '*is*' any other way?

(p. 11f) We need a logic for the theories we have, not for the 'finished theory of reality' that we have not. If the general dispositional idiom is here to stay, we need a logic for it, right now!

(p. 12) I've often thought you ought to worry more about probability. Can one do e.g. physics, or a lot of the application of statistics, without a sort of quantifying-in that you'll object to?

(p. 18) If 'disposed to' is a predicate-forming operator on predicates, then 'second-order disposition' is likely to be intelligible. But what *can* the term mean in your framework?

You don't notice the logical peculiarity of this so-called polyadicity involved in similarity: viz that there are argument places in which may stand a single proper name or a list of names equally well, *salva congruitate*. '*A*, *B*, and *C* (all) fit together' is a simpler example: this argument place will take any list with $2 + n$ items in it, and the result is not reducible to some conjunction of dyadic fittings-together. Think of the topological example of the three rings, all of which together are indissolubly joined, but no two of which remain joined if the third is removed.

(p. 19) 'Why does this College produce so many first-class men?' 'The College has a very tough policy of flunking out the inferior students.' Explanations by natural selection seem to me as unsatisfactory as this one. The high ability of many of the students who *aren't* flunked out and their coming to *this* College have just *not* been explained.

(p. 29f) I find this account of learning terribly unrealistic; and the story of the red shawl and the yellow rose is surely mere fantasy. As for 'preposterous'—the story is preposterous indeed.

(p. 38) The top line of this is a bit dark to me. If someone visibly reacts to what he sees, I may be said to see him seeing it; but a lot of the time I can see no such thing. And by the way—how rarely 'Red.' (the observation sentence, as opposed to the *Begriffswort* 'red') comes in our discourse! or even in the discourse of very young children learning to talk!

(p. 59) There is in fact not *much* more use of demonstrative pronouns in Geach's examples than in most writers on logic. As an improvement on the Heraclitus example (which awkwardly uses a mass term) let me shift to another.—In England the members of the Heralds' College have, as such, names: Garter, Norroy, Bluemantle, etc. Lord Whoosis, visiting the Heralds' College to trace his pedigree, may see the same herald on two occasions (say Bluemantle) but not the same man—the former Bluemantle has been promoted, or died—and again may see the same man on two occasions but not the same herald—the man who was Bluemantle, and is Garter. It's easy to see how the argument pattern on p. 150f. of *Reference and Generality* could be adapted to this case; and the peculiarities of a mass term like 'water' no longer come in. I think grammarians would parse 'Garter', 'Bluemantle', etc. as proper names: rightly, to my mind—with the associated criterion of identity 'same herald'. (This material was much tossed about in discussion between myself, Elizabeth, and a clever graduate student of hers, Harold Noonan.)

(p. 71) I've often thought that colour words dispose one to Platonic participation theory rather than to universals *in rebus* shared integrally by many things: 'red', 'white', etc. gravitate toward substantival status, and what we vulgarly call red is an object with a certain degree of imitation of Red. It is none the worse if the object also imitates White or Blue i.e. is pale red or purplish red.

(p. 72) 'Colour' is naturally often understood as chromatic colour (= coloury colour!) excluding white gray and black. In that case 'Colour(ed).' is a perfectly good observation sentence.

(p. 80) Present-day mores may make us wonder even about bachelor and unmarried (male) man. If men who haven't been married become women by grace of surgery and drugs, are they still bachelors? (It surely has nothing analytic about it whether this sex change is feasible!) Plainly 'bachelor' has a certain *Porosität* in face of such questions.

(p. 88) Alas, as you no doubt would agree, it's not only for characters in fiction or theology that identity troubles arise. A lot of what we blithely take to be understanding of our fellows' discourse depends on our apparent ability to sort out the occasions when a proper name is *meant* to be used for repeated reference to the same object—we appear able to do this even between equiform proper names—and we think we can do this without ourselves ever having attached the name to a present object. But *what* are we claiming to identify, and how?

(p. 91) Your 'dualistic position' seems to me a strong one.

(p. 96f) I am awestruck of your neo-Russellian paradox! Just because the substitution operator makes sense for any grammatical category, my hunch is that

(*x* not(*x* vicè *x*)) vicè (*x* not(*x* vicè *x*))

could have a near-vernacular transcription that was not so ungrammatical *in appearance* as the string at the top of p. 97. After all, by *that* sort of transcription 'Do vicè (*x* how *x* you *x*)' would come out (if I've not blundered) as 'Do is a thing *x* such that how *x* you *x*'; but 'How do you do' is OK for all that.—But I've not yet seen how to do the trick.

(p. 98f) You shift a bit between 'unspecifi*ed*' and 'unspecifi*able*'. Is this innocent? And it's not at all absurd to suppose that *any given* apple or rabbit would be named for immediate purposes of discourse.

(p. 102f) As you know, I cannot abide complacent, at-ease-in-Zion espousal of the now prevalent limitation-of-size style of set theory. And if we think (as we should!) of attributes, then the intuitive foundations of the limitation-of-size theory are entirely unattractive.—I think Russell was right in *Principles of Mathematics*, however, in thinking of *two* roots: linguistically speaking, predicates and lists—lists of proper names that can stand *salva congruitate* in *certain* contexts where proper names can stand singly (see my note on p. 18). For long I hoped the logic of lists would be a graspable clue in set theory; maybe, but I now fear not for me.

(pp. 105–11) Fascinating! I don't know what comment to make. (A nice point on p. 140f. in the same connection.)

(p. 114f) I prize this darkness-dispelling remark on the Skolem-Löwenheim theorem.

(p. 118f) I am with you about these bogus economies (numerals instead of numbers). I am uncertain whether modal notions ('inscrib*able* numerals') afford any comford: it might be a motive for looking into modal logic. But I don't forget Frege's quip about how the road through the possible leads in the end to the impossible—*un*inscribable numerals!

Addendum on Natural Selection and Induction: Remember Russell's chicken in *Problems of Philosophy*! At the very best natural selection could guarantee only that our inductions have not been *too* out of kilter with the world for us to live and bring up our children.

Quine to Geach 29 July 1975

I am grateful for your painstaking comments on *The Roots of Reference*. Here are my answers.

The point about causality that you cite from McTaggart shows unusual sensitivity on his part to the workings of bound variables. The insensitive rejoinder would be that the formula 'Each *A* event is followed by a *B* event in a relation *R* to it' is already covered by 'Each *A* event is followed by a *B* event', since the '*B*' in the latter formulation can be taken in particular as '*B* event in a relation *R* to it'. Such a rejungent would be guilty of breach of back-reference.

How to track energy through time? I think there are *prima facie* two ways: by correlating the changes in the velocity vectors of two bodies in the case of impact, and by correlating the changes of temperature of two neighboring substances in the case of heat transmission. In theory the latter case reduces to a statistical summation of the former case, at the molecular level.

In answer to your comments on dispositions, I would distinguish between ontological and logical concerns. In identifying the disposition with what some writers have called its physical basis, my motive is ontological: I am cleaving to a physicalistic ontology and saying that dispositionality is a trait only of our method of specification. What then of the logical side? There would be in my *austerely* regimented language of science no separable disposition operator; hence no logic of dispositions. The disposition operator would still be separable and freely applicable in the second-rate practical language, along with propositional attitudes and the like. (I am invoking here the 'double standard' of *Word and Object*, §45.[7]) So I do recognize a place for a logic of the disposition operator, just as I do for a logic of the propositional attitudes. My own writing even proceeds mostly in that second-rate practical language. I freely apply the disposition operator; in particular I get second-order dispositions. It is my second-rate way of referring, still *per accidens*, to hypothetical neural states or structures.

Now to your comment on probability. Probability is at its clearest, of course, where we divide the cardinality of one class of events by that of another to determine a frequency. When someone attributes probability to sentences, in ways not obviously reducible to such frequencies, I am at a loss to make sense of the attribution even apart from problems of quantifying in. This is why I haven't probed deeper. Of course the references to probability on page 12 have to do with disposition theories that I reject.

I was conscious of the point you raise about the indefinite polyadicity of similarity. One way of accommodating it logically is to treat the similarity predicates as a sheaf of predicates, each calling for a distinct number of arguments. A second way would be to treat it as a one-place predicate predicable of finite classes of stimulations. Your logic of lists would serve here. A third way would be to develop a logic of multigrade predicates (to revive a term that I once coined for Goodman to use).

'Why does this College produce so many first-class men?' The answer to the corresponding question regarding natural selection is mutation. Selection by survival of the fittest is indeed negative; random variation through chance mutation is what brings occasional traits worth selecting.

You are right about the want of realism in the examples of infantile learning in pp. 29f, 38. I have been intent on convenient schematism, to the neglect of naturalistic surfaces. A recent attack by G. A. Miller brought home to me that this is poor expository strategy.

You query the top line of page 38. My thought is merely that the parent sees that the child is facing the red thing open-eyed (or the child sees that the parent is doing so). I don't require either party to have incontrovertible knowledge that the other party is seeing the thing.

In my characteristic four-dimensional vein I would entify your Bluemantle as a spatially and perhaps temporally discontinuous object whose temporal segments are temporal segments of various men. I have taken this line with respect to the president of the United States. Contexts must of course be adjusted; we say that Bluemantle was born in 1908 to mean that the man whose present time slice is a time slice of Bluemantle was born in 1908.

Your suggestion of a Platonic participation theory, apropos of color, suggests that you think of red as distributed about a single reddest norm. I think rather of a plateau with sloping sides.

Your comment regarding 'colored' on my page 72 makes me wonder if you overlooked my hyphen. I was concerned with the functor '-colored' as in 'coffee-colored'.

Your remark about 'bachelor' illustrates the fact that analyticity, when defined (as I suggest) in terms of a community-wide uniformity in the learning of certain words, may change its extension with succeeding generations of learners. It also suggests the interesting point that in the midst of such transition a sentence that has not quite ceased to be analytic, by my criterion, may have already turned false.

There is a paper by Herbert Bohnert that bears on your problem of what we are claiming to identify, and how, when we sort out equiform names of persons known only by report. Bohnert adapts Ramsey's treatment of theoretical predicates, applying it rather to names. He construes proper names as existentially quantified variables. The device, like Ramsey's own, has the drawback of requiring indefinitely long scopes, except where we are prepared to drop a subject for good. But it has a certain appeal.

'Do something such that how it you it.' Yes, this way lies madness.

What strikes you on pp. 98f as vacillation between 'specified' and 'specifiable' will fall properly into place, I think, if you recognize that I am explaining substitutional quantification for a language with a closed, finished lexicon. Something is specifiable only if already designated by a string in the language, albeit long and forever unuttered. 'Specified' might suggest that somebody already used that string and performed the act; so I wrote 'specifiable'.

Charles Parsons has said tentatively kind words about inscribability, as have vaguer thinkers before him; possibly Hilbert. I lean rather to the line you quote from Frege—*per possibilia ad impossibilia*.

Your postscript on natural selection, induction, and Russell's chicken prompts me to stress that I see in natural selection no argument in support of induction, but only an explanation of the puzzling fact that induction *has* worked so well.

Harvard University

REFERENCES

[1] Quine, Willard Van Orman: 1960, *Word and Object*, The Technology Press of The Massachusetts Institute of Technology and John Wiley and Sons, New York and London, pp. 154–156.

[2] Quine, Willard Van Orman: 1950, *Methods of Logic*, Holt, Rinehart and Winston, New York, p. 234.

[3] Quine, Willard Van Orman: 1969, *Set Theory and Its Logic*, Revised Edition, The Belknap Press of Harvard University Press, Cambridge, Massachusetts, pp. 296, 299.

[4] Geach, Peter Thomas: 1962, *Reference and Generality*, Cornell University Press, Ithaca, New York, pp. 30–31.

[5] Quine, Willard Van Orman: 1970, *Philosophy of Logic*, Prentice-Hall, Englewood Cliffs, N. J.

[6] Quine, Willard Van Orman: 1973, *The Roots of Reference*, Open Court, La Salle, Illinois.

[7] Quine, Willard Van Orman: 1960, *Word and Object*, The Technology Press of The Massachusetts Institute of Technology and John Wiley and Sons, New York and London, pp. 216–221.

PETER GEACH

HISTORY OF PHILOSOPHY

In the preface to Wittgenstein's *Tractatus* there is a well-known sentence in which he refuses any obligation to assign the sources of his thought; so he has been thought 'ahistorical' and even culpably ignorant. Even in that preface, however, he owns his indebtedness to Russell and Frege; and though he read few philosophical works, he read certain works often and intensively: I remember on his bookshelves William James's *Principals of Psychology*, a German-language selection from Aquinas's *Summa Theologica*, Augustine's *Confessions*, and Frege's *Grundlagen*. So he is no precedent for anybody who would neglect the great thinkers of the past.

I have done a lot of teaching in the history of philosophy. Not having been at Oxford as a teacher of philosophy I have taught little of Plato (the *Theaetetus* and the *Sophist* only) and Aristotle not at all. On the other hand, I have often taken part in the teaching of medieval philosophy which formed a compulsory part of the courses for single-honours philosophy students at Leeds; and I once gave courses in medieval philosophy as a visiting lecturer at Oxford. The subject is shockingly neglected in English universities: I suppose this bad tradition arose partly from anti-Catholic sentiments, and partly from a foolish Humanist contempt for Latin that is not up to artificial Ciceronian standards: we see such an animus when Hume tells us in his *History* that in the middle ages students crowded to Oxford to hear lectures in 'bad Latin' on 'worse logic'. Of post-medieval authors I have lectured on Hobbes, Locke, Berkeley, Hume, Descartes, Spinoza, McTaggart, Russell (*The Principles of Mathematics*), and Wittgenstein. Kant, I always felt, is not for me to expound; still less Hegel.

What then has been my methodology in all this teaching? I have been much helped by the wise prescriptions of Richard Robinson in his book *Plato's Earlier Dialectic* [1]; I excluded from the collection *Logic Matters* an essay of mine on the Third Man which, I fear, sinned grossly against these canons. But there are two dangers about historical work. A man would not write a good history of physiology if he ascribed to Aristotle the doctrine that the blood circulates; he could

Harry A. Lewis (ed.), Peter Geach: Philosophical Encounters, 45–47.

not write a good one if for him it were a disputable matter of speculation whether the blood does circulate. I need hardly spell out the parallel in the history of logic. A man once countered an argument of mine with 'That's a very logical argument; but if you knew as much about the history of logic as I do, you wouldn't trust logical arguments'! I am confident that, being logically analphabetic, he knew *nothing* about the history of logic, by any decent standards: you might as well expect a man with no sense of smell to be a connoisseur of wine.

In the United States, and to some extent in other countries, there prevails a really scandalous situation in studies of medieval philosophy. People set up as experts in medieval philosophy, and win praise for their work, who are fairly unenlightened about formal logic and grossly ignorant of Latin grammar. They are learned in bibliography and palaeography, but these skills are unfortunately not regarded as strictly ancillary to knowledge of logic and of Latin. As a result, howlers abound, and work appears of a quality for which many a schoolboy has been flogged. Learned men come out with such Latin as *in corpo* and *ens per alio*; I have seen *ens*, despite *entia rationis* and so on, treated as a masculine noun and its accusative 'emended' to *entem*; one author rendered *mentior*, 'I lie', in one place as 'I am cheated' and in another as 'I am mentioned'. I could easily fill a page with similar examples. The culprits form a circle who, in Lewis Carroll's words

oil each other's little heads
With mutual flattery's golden slime,

and from time to time recruit pupils to the circle: books rarely get reviewed by people outside the circle. I am afraid that the status of medieval studies in universities is too insecure for those who do better work in the field to be willing to expose this racket; I as a reviewer have no scruples about exposing bad work if I have occasion to. The worst of it is that if an incompetent editor (or translator) has once published some little-known medieval text in Latin (or English version), work on that particular author may find no publisher for twenty or thirty years.

In another place I have discussed the predicament of a teacher of the history of philosophy, when he finds the work he is dealing with to be radically confused. Charitable construction of a philosopher's thesis as at worst false and refutable rather than confused beyond redemption, is I fear not always possible. How can one ever fairly report in indirect speech what a philosopher said, if what he said is a muddle? What is it to report, present to students, the *right* muddle? Certainly it would be no good to try to get one's own mind into a similar muddle; that would be like getting drunk to understand what a drunk man is saying! I have no theoretical view of this matter to expound; in practice I have done what I could, as I best saw things, from case to case.

The University of Leeds

REFERENCE

[1] Robinson, Richard: 1953, *Plato's Earlier Dialectic* (Second Edition), Oxford University Press, Oxford.

DESMOND PAUL HENRY

ABELARD AND MEDIEVAL MEREOLOGY*

I. ANTECEDENTS

It is a pity that the stock story of early medieval thought tends to concentrate on something called the 'universals controversy' and does so in a way which inappropriately subsumes the twists and turns of a highly complex situation under somewhat ill-fitting headings. Although a start has now been made on a saner account of the matter both in general[1] and insofar as it affects Abelard,[2] nevertheless the usual connotations of a term such as 'realism' when applied to the topic of universals render somewhat startling the realisation that one such theory attacked by Abelard was the polar opposite of any otherworldly Platonic-style theory, namely the '*collectio*' theory. It is yet a greater pity that in his attack on this theory[3] Abelard by no means does justice either to it or to his own wide-ranging account of part/whole relations. At the time of his attack his maturer thoughts (in the *Dialectica*) were still to come, yet some of the essentials of that later work are already to be found in his gloss on Boethius' *De Divisione*, a gloss dated as belonging to the end of his first teaching phase.[4]

In II below is presented a brief and inadequate characterisation of some of Abelard's theories and themes; in III these are applied to contemporary discussions which have a bearing on his own sad fate and on that of a certain cat to whom we have been genially introduced by Professor Geach.

IIA. MEREOLOGY AND SUB-MEREOLOGY

Abelard states clearly enough what was by his time the standard distinction between integral and distributive wholes:

The whole with respect to substance is one thing insofar as it embraces quantity (being then said to be an *integral* whole) and another in respect of the *distribution* of a shared essence (it being then a *universal* whole) as when a species is distributed among its individual members thus: 'This is one man, that is another...'. This whole evinces its universality by being predicated of its individual elements, whereas such does not hold at all in respect of an integral whole, since the predication is then of the aggregation of all

Harry A. Lewis (ed.), Peter Geach: Philosophical Encounters, 49–64.

the elements at once, e.g. 'This house' is predicated of these walls, this roof, and this foundation, taken all at once. That which is an integral whole cannot be a universal whole, since universality does not involve parts in a quantitative sense, but rather consists in the distribution of its common nature through a plurality of whose individual parts it is predicated. The integral whole, in contrast, involves only a singular predication in respect of itself, as in the case of Socrates, who is made up of these members.[5]

In thus making the contrast between the *integral* and the *distributive* ('*secundum diffusionem*') whole, Abelard is propounding a distinction corresponding to that made by Lesniewski between *collective* and *distributive* classes.[6] In general *a* is the *collective class* (or *integral whole* or *complete collection*) of *b*'s if and only if

(1) *a* exists,
(2) at least one *b* exists,
(3) every *b* is a proper or improper part of *a*, and
(4) every proper or improper part of *a* has a common proper or improper part with a *b*.

(This is a rough expression of Lesniewski's definition from his 1916 system, recorded as D2 in Sobocinski 1954–5). Abelard's examples appear to presuppose clauses (1), (2) and (3) of this characterisation, and there is no doubt that at times he had a good intuitive grasp of the latitude afforded by clause (4). This permits the countenancing of parts which are not themselves *b*'s, and among these may be parts which are not altogether outside one another, or discrete. His example of the collective class made up of three discrete elements but which also has at least *two* two-element parts, each having one element in common, goes in this direction:

Besides, we assert that this trio is made up of these three [discrete] elements, and any two of those elements yields a pair. Who then can deny that this trio is made up of [at least] two pairs? For indeed the common element along with *this* element makes up one pair < and with *that other* element another pair >, so that the trio contains within it [at least] two non-identical pairs, as *this* pair is not *that* pair, since the latter has a proper part which the former lacks. (For if *this* pair's make-up involves this element, and *that* pair's make-up does not involve this same element, then by a necessary logical consequence it is obvious that *that* pair is not *this* one).[7]

In spite of this argument Abelard does not always take advantage of this latitude, probably because of its apparent paradoxes and because the resultant possibly non-discrete parts need not have those readily available names in ordinary language which we and Abelard generally find it more convenient to handle ('foot', 'head', 'roof', and so on).

However, his mention of quantity when speaking of the integral whole ensures that questions concerning the contrasts between diminution and destruction or between mutilation and destruction (originally raised in chapter 27 Book Δ of Aristotle's *Metaphysica*) are dealt with by the appropriate special theories of the particular sorts of objects in question. Indeed, from a systematic point of view such particular theories involve what could be called 'sub-mereologies', i.e. elaborations on substitution-instances of general mereology. Examples of the ways in which such sub-mereological considerations interact with general considerations will be provided below.

IIB. 'PRINCIPAL PARTS': DESTRUCTIVISM AND THE *FRAGMENTUM SANGERMANENSE*

Arguments at the level of mereological instances often rely on unspoken presuppositions as to which parts are principal, and which subordinate, in a given type of whole. We take it that there is a sort of natural order in the way in which we figuratively carve up wholes, analogous to the 'natural' order of dividing a stretch of discourse into sentences, sentences into words, words into syllables, and syllables into letters: 'The natural order is to divide each thing into those parts whose bringing together immediately constitutes the whole'.[8] Abelard places on record his opinion that parts as thus described are 'principal parts',[9] yet he also amply recounts what he sees as its difficulties in the course of his discussion of two opposing opinions on this matter. There are, he says, those who look at the matter from the constitutive point of view (*secundum constitutionem*) as opposed to those who use criteria of destruction (*secundum destructionem*). The former call those parts 'principal' which are not parts of parts, but only parts of the whole, e.g. this wall, this roof, and this foundation in respect of this house. But this 'constitutive' view is, he says, refutable by the simple consideration that we can make a two-fold division of the object, and this can always operate at a point which makes a part which is allegedly *not* a part of a part into a part which *is* part of a part. In the house case, putting the foundation on one side of the division and walls and roof on the other, makes either wall or roof into a part of a part. And, had the head been posited as a principal part of a man, it can easily be rendered secondary (i.e. part of a part) by making a division of the foot from the rest of the body, of which the head then becomes a part. This examples again shows how a part usually accounted secondary, such as a foot, can equally well become primary, in the sense of not being part of a part,[10] and could be used against the 'natural' division of discourse: separation off of a single letter from the rest of the paragraph makes the sentences into parts of the other part which results.

(Abelard is here exploiting that latitude on part/whole relations permitted by the definition of collective class given above. His wall is here one of various non-discrete parts and corresponds to the element which he depicted as the common part of the two pairs in the example of the trio, cited above. The pebble which is part of the wall, and the wall which is equally a further part of the house (both of which figure in the next phase of his discussion) are likewise non-discrete.)

The next phase centres on the destructive criterion for the discrimination of principal parts. This accounts those parts principal which, on being destroyed, destroy the substance of the whole, e.g. the cutting off of the head immediately destroys the man. This substantial destruction is supposed to be distinguished from mere quantitative diminution, such as that which occurs when Socrates loses a fingernail. Yet Abelard recounts a criticism of this position which appears to obliterate this distinction by demanding exact specification of the whole which undergoes change. If, for example, that whole is *this* house (as opposed to just *a* house) then (according to the criticism) the distinction between substantial destruction and quantitative diminution lapses. *This house* is just the collective class of its parts. Comprised in this class is *this wall*, which in turn comprises *this pebble*, and destruction of the latter therefore destroys *this house*. It therefore looks as though for a whole thus specified *any* quantitative diminution, no matter how small, amounts to substantial destruction; the original identity no longer abides.[11] From the destructive point of view, therefore, given this sort of specification, all the parts appear to be equally 'principal parts',[12] and in consequence each tiny quantitative change generates a new house:

> But we do not say that if this pebble is removed then *a house* does not remain, but only that that house which was made up from that pebble and the rest of the parts does not remain: that particular house was made up of parts, and was referred to as 'this house' in terms of the composition of all those parts. Then the substance of *this house*, which is based on all the stuff pertaining to it, does not remain, and the house in question necessarily perishes substantially. No longer will one be able to say 'This house is' of the house that which previously was, that is, of the collection of this pebble and the rest of the parts together; this can now only be said concerning the collection of the rest of the

parts, a collection which is quite distinct from that which previously existed, the distinction between which is akin to that of a part to a whole whose quantity comprises that part: that object which is *now* said to be 'this house' will have been a part of the first house: *illud enim quod modo 'haec domus' dicitur, pars fuerit prioris domus.*[13]

It is here that the contemporary *Fragmentum Sangermanense*,[14] associated with Joscelin of Soissons, takes up the discussion in somewhat the same terms as those used by Abelard, in that it uses the example of Socrates, and puts the problem as to whether or not what is called the 'major identifiable object' (*natura magna*) which previously was a part of Socrates *becomes* Socrates when Socrates' fingernail is cut off. In other words, even as the earlier house is made up of the pebble and the *natura magna* of the whole-house-minus-the-pebble, so also the earlier Socrates is made up of the fingernail and the *natura magna* of the whole-Socrates-considered-apart-from-his-fingernail. And it would appear that in each case the *natura magna* is first a part of the whole; that whole is then destroyed by the removal of the small part (the pebble, the fingernail) and the *natura magna* which was earlier a part then *becomes* the whole. Thus it would appear that that which earlier was not Socrates becomes Socrates.[15] Correspondingly that which earlier was not *this house* becomes *this house*.

Historically speaking, that aspect of this state of the discussion which is to prove most durable is the one which sees occurring successive destructions and generations of various individual houses, or of various individual beings each called Socrates. Socrates in now an entity composed of many objects each of which is a Socrates.[16] According to this view, which we may label 'Destructivism' (both on account of its content and its context of occurrence in Abelard's discussion), 'Socrates' becomes a common noun naming many objects each of which is a Socrates. This is a view which accords with that of Woodger in his account of the 'time-stretches' of spatio-temporal objects.[17] Such stretches comprise all the spatial parts, constituting a given object during a given time. The distinction between time-stretches is effectuated so as to avoid the difficulty of attributing incompatible predicates across time to the same subject, i.e. to avoid the general form of the crises of identity encountered by Abelard in the case of 'this house'. In order to avoid our having to say that 'Nicosia House' both has a given pebble in its wall (at an earlier stage) and does not have that pebble (at a later stage) one contrives two time-stretches, each of which is a Nicosia House, the earlier with the pebble, the later without. Likewise one Socrates has a fingernail, but another later Socrates lacks the nail. Whatever may be the merits of such a view for certain theoretical purposes, it at least abrogates the senses of familiar terms. We used to think that houses were things which were built, extended, reduced and battered in various ways; now we have to hold that they, like objects in general, are things susceptible of only one of any two contraries; further, objects which are agents can perform only one of various incompatible actions, with corresponding equivocations as this network of revised talk is expanded. Sense has to be made, for instance, of an excessively juvenile ten-second time-stretch Socrates performing some recondite action requiring five years of training. Doubtless suitable revisions involving previous Socrateses may be contrived to accommodate such apparent oddities. Yet it is perhaps by no mere coincidence that in Descrates' universe which consists of completely independent successive time-slices no physical object acts upon any other physical object; we have merely what can be described as the passage of various fractions of the divinely maintained constant quantum of motion passing through the various 'objects'—hence Malebranche's criticism of this position.

However, the *Fragmentum Sangermanense* having joined with Abelard in stating a position which can push one into such paradoxes, avoids them (and hence avoids Destructivism) by viewing Socrates as the collective class of both his spatial and temporal parts: he can still remain the same Socrates, although his being may now involve fewer spatial parts than before: *Destructa una parte Socratis, ceterae quae remanent non fiunt Socrates, sed ex ipsis fit Socrates, et ita secundum hoc neque cogimur dicere quod Socrates pereat neque quod non-Socrates Socrates fiat; sed hoc tantum, salva essentia Socratis, consentimus ex istis, ut Socrates ex paucioribus existat.*[18]

Here we have the transition to what may (in view of its context) be labelled the '*Sangermanense*' thesis. It is akin to that of our contemporary Lesniewski[19] in that it does not abrogate our pre-theoretical conviction that individual houses, persons, and other objects persist in being across time in spite of changes. 'Socrates' thereby names Socrates from the beginning to the end of his life, being restored to its function as the proper name of the collective class of the series of Socratic time-stretches. The latter can still serve as the notional justification for ascribing contrary or incompatible predicates to the same object. As Aquinas observes in a kindred situation, Peter's walk to church may be one thing when considered according to its natural kind (*secundum genus naturae*) but many relative to the various changes of intention from good to bad and *vice versa* along the way (*secundum genus moris*).[20]

It is remarkable that Abelard had in fact most of the material required for the move to the *Sangermanense* view in his discussion of temporal wholes, had he but realised its full import.

Thus he observes:

I shall now take a look at the nature of certain parts and wholes which appear to have features imcompatible with those described above, even though they are admitted to be of the same kind insofar as they are said to be integral. These are temporal wholes, such as *this day*, which is said to be made up of *these* twelve hours, and to be their collective whole (*totum constitutivum*). But this whole is said to exist in a fashion opposed to that applicable to other integral wholes, in that

(i) any of its parts suffices to establish the whole, and
(ii) the destruction of the whole destroys all of its parts.

For if the first hour is, then the day is said to exist, whence also if the day did not exist its first hour would be said not to exist. We are unable to take the existence of wholes which lack abiding parts, and which are such as we encounter in stretches of discourse or temporal stretches, as involving all their parts taken together. Indeed, their parts never exist simultaneously, but rather follow one upon the other, so that it is only insofar as their parts exist that we mark out the being of their whole. And while we say that a stretch of speech exists only upon the utterance of its final letter, to which it cannot but owe its meaning, we nevertheless say that a day exists given the existence of any one of its parts, be it the first, third, sixth or ninth hour, or by the flowing by [reading *permanante* for *permanente*] of any one of the moments thereof. But if we grasp the meaning of 'day' or 'discourse' correctly, then at no time do we assert correctly and truly that a day or a discourse exists; indeed at no time do there exist at once all the twelve hours which are said to be a day, or the many utterances which are said to be a stretch of discourse. For even if several men simultaneously gave voice to many utterances, it could not be said to be a stretch of discourse unless they followed one upon the other in such a fashion as to complete the sense of the discourse. In any case, letters or utterances voiced by several men would not be likely to combine into the sense of a single piece of discourse, but rather to be coincidentally uttered by several individuals whose understanding could well be fixed on diverse things; nor can the meanings of the individual parts be distinguished when they are uttered together. Thus the parts of a day must be taken to follow one upon the other in the same whole, and not to be this first hour and that other which does not belong to the same whole, or which occurs at the same time in another day, but rather to be this hour which goes before the other following hours in respect of the same whole. Otherwise we could say that a whole day was completed in various individual hours, insofar as many first hours could be taken together to form a day, or many third hours, all drawn from diverse wholes . . .

Yet if we are to admit the truth of the matter it never happens that these things are really made up of parts. For no composite is contained in one of its parts, nor would that be a part which was not exceeded by the quantity of its whole; indeed, were there but one part then it would be absolutely identical with its whole. There are never many parts simultaneously in time, since there never exist many of them at the same time. Whence time is never made up of many parts, since neither the first nor the middle nor the last parts linger on; always only one of them is actually there. Nor can it ever properly be said that a day exists, although figuratively we might say that it exists through a part, that is, a part of the day exists. Yet even this is not properly a part, since it exists alone and is never compounded. For if 'day' is just the name of several diverse hours, how can we properly say that a day exists unless the many hours continue in being? Every composite is the same as all its parts taken together [*simul*]; hence whatsoever can be attributed to the whole is also attributable to all its parts taken together, and *vice versa* [and] to whatsoever the whole is attributable, all its parts taken together are attributable, and *vice versa*. If, therefore, we are to stress the truth in this matter such whole should not be admitted. Philosophers should rather deal with them as if they were dealing with wholes, assembling together into a pseudo-whole [*quasi unum*] things which were past and are to be, so that they may show their nature, as if from that which is and from that which is not something might be. Hence things which in fact are not wholes are for theoretical purposes viewed as if they were wholes.[21]

Provided that a few necessary adjustments are made, and trivial misunderstandings removed, then the concrete embodiments of Abelard's days as full-blown spatio-temporal objects, and his hours and temporal parts as time-stretches, yield the requisite filling out of the *Sangermanense* thesis. Thus some of his qualms concerning temporal wholes are in a way justified: in his original sense they are fictitious, but not for the reasons he gives, but simply because there is no such thing as time, but only things timed. His taking of '*simul*' as entailing literal contemporality of parts when a whole is identified as all its parts taken together (*simul*), and his presupposition that all predications can only be *ut nunc* (i.e. as of now) would, if accepted, render all spatio-temporal objects just as fictitious as his days, and hence can safely be disregarded. For the rest, he gives highly useful reminders of the manner in which it is only to the extent that time-stretches exist that we 'mark out' their spatio-temporal wholes. This then enables us to agree that the sensing of spatial parts of one or more time-stretches of a spatio-temporal object justifies our 'marking out' not only that object's existence, but also its roughness, its bulk, its being blue, or its being the author of Waverly, or whatever. In particular a later time-stretch of the same subject may have more or less spatial parts than did an earlier time-stretch. Whether diminution in terms of spatial parts amounts to mutilation or destruction is a matter to be settled by experts in the sub-mereology of the particular sort of thing in question.

Slightly more sophisticated are Abelard's worries concerning the theses numbered (i) and (ii) above, but yet again his own work suggests how those worries might be obviated. Thus we may note that he is troubled, in the case of (i), because he believes that this is the false converse of a thesis which he had propounded earlier in *Dialectica*,[22] i.e.

(iii) The existence of the whole entails the existence of any part thereof.

(It was the equipollent of (iii), whereby the non-existence of the part entails the non-existence of the whole, noted by him as true,[23] which was the basis of the 'destructive' paradoxes recorded above). Likewise (ii) appears to militate against a remark which he has earlier put forth concerning:

(iv) If something is denied of the whole, then it is denied of all the parts taken together (*simul*).[24]

That remark made what appears to be the obvious point that (iv) does not guarantee the non-existence of parts when the existence of the whole is denied. Thus the three house-parts (walls, roof, and foundations) may still exist, and may even be joined together (in a non-houselike way) although existence has been truly denied of the house.[25] In (ii), however, attention is called to a case in which it is clear that denial of the existence of a whole is incompatible with the continued existence, in any fashion, of the parts pertaining to that whole; thus (one might say) there is no 368th day in 1967, and its hours, unlike the house-parts, cannot have any sort of joint or disjoint existence.

In point of fact all of Abelard's qualms on these scores are quite groundless. His remark contains the germ of a way of making more exact the sense of his highly ambiguous theses. This renders true all of (i), (ii), (iii), and (iv), and hence no incompatibility arises. The crucial point is that mereological theses stated in ordinary language (as are those of Abelard) can be falsified, and even those which make use of functors and variables can mislead, if attention is not paid at the outset to that distinction which is embodied in Abelard's remark and which has been constantly stressed by Professor C. Lejewski in the course of his teaching, namely the distinction which holds between parts-of-*X* and *X*-parts. Thus the whole consisting of parts-of-*X* (e.g. Abelard's house) need not be identical with the whole consisting of *X*-parts (e.g. walls, roof, and foundations either disjoined or joined together in a non-houselike sort of way). In general the reasons for such non-identity will depend on the nature of the *X* in question. Thus the maker's parts-list for a car is based on technical information as to the vehicle's make-up and repair needs. This is a list of car-parts. However, the *parts-of-the-car* cannot be exhaustively listed, and their being parts-of-the-car is dictated by the purely mereological considerations outlined in IIA above. Plainly, therefore, being a car-part is not at all the same thing as being part-of-the-car, although among the parts of the car there may well be some which are identifiable as having past careers as car-parts. And after Abelard's house has tumbled down, there still are the house-parts which previously were parts of a house. In other words, although *X* may be *made up of X*-parts, the collective class of the *X*-parts need not be identical with *X* (considered as the collective class of parts-of-*X*). Exemplifications of this fact could be expressed in terms of a perfectly ordinary non-identity, fogged by the invention of a philosopher's '"is" of material constitution',[26] the nearest approach to which in unperverted speech would probably be the simple predication of an adjective (e.g. 'This desk is wooden', 'This rod is metallic').

A further consequence of the aforementioned distinction is that the whole of the *X*-parts can sometimes last longer than the whole of the parts-of-*X*. (Where there isn't any *X*, there aren't any parts-of-*X*, although there may be *X*-parts.) The way in which use of a variable which fogs the distinction can lead to confusion may be seen in the statement of what has been called a 'fundamental dilemma', i.e. 'Either 'concrete collection' is defined is such a way that concrete collection *X* has the same principle of individuation as the jug or it is not so defined'.[27] We now see that here are more than the two alternatives of a dilemma, since for '*X*' we can substitute either 'jug parts' or 'parts of the jug', whence obvious puzzles when the jug is broken into jug-parts. Only in the sense of the second substituend is the jug identical with the collective class of its parts.

Abelard has in many places stated the essentials of this distinction between *X*-parts and parts-of-*X*. The case of the house-parts apart from the house, mentioned above in his remark concerning (iv), already implies it. Slightly more incisive is the statement that on the destruction of the whole, although the parts may remain on their own account (*per se*) they no longer do so insofar as they are parts (*non . . . in proprietate partis*).[28] An even more definitive version figures in his discussion of a possible objection to Destructivism's thesis on principal parts:

But perhaps it might be objected: this pebble having been removed and now existing on its own outside the aggregated house (for after all the substance of a body cannot altogether be annihilated) *this house* which formerly was, still is; that is, this pebble and the things that were parts along with it still exist. Now indeed this last point cannot be denied. But it does not thence follow that *this house* exists. Those things which lack house-like composition cannot be said to be *this house*, and for *this house* to be, it is not sufficient that just its makings exist, otherwise even before the house was made the wood and stones could be described as a house; for this the composite requires a certain structure.[29]

The accent on the genitive construction in the Latin correlate of 'parts-of-*X*' tends somewhat in the same direction.[30] Thus Abelard's own material can be used in the solution of the difficulties which he sensed in connection with (i) and (ii) as opposed to (iii) and (iv). We have seen how, if there's a part-of-*X*, then there must be an *X* for it to be a part of; hence (i) is generally true, and need no longer be seen as the possibly false converse of (iii). Of course 'part thereof' in (iii) must then be taken in the 'part-of-*X*' sense, since otherwise it or its equipollent which states that the non-existence of the part entails the non-existence of the whole can easily be falsified. That equipollent can now be applied to Abelard's house without any of the paradoxes of Destructivism ensuing, provided that it is specified in the *Sangermanense* mode as the collective class of its spatial *and temporal* parts. In general, only *X* and parts-of-*X* are existentially interdependent in the way implied by (i) and (iii). The same goes for (ii) and (iv). Denial of the existence of Abelard's house entails denial of the existence of parts of that house, but not necessarily of that-house-parts. In spite of the numerous occasions on which Abelard seems to be making the requisite distinction, he still does not have a sufficient grip on it to avoid the troubles which follow upon its neglect. A further opportunity to see such troubles arises in his quite voluminous discussions of paradoxes in connection with increase.[31]

III. THE MISFORTUNES OF TIBBLES

We have now amassed sufficient material to deal with some of the tales of Tibbles presented by Professor Geach. The two with which I happen to be acquainted are presented for their superb puzzle-value, and we are of course only incidentally concerned with the points which their author may originally have had in mind when inventing them. The first, which I only have at second-hand, involves the cat Tibbles which is on the mat at time t, about which time somebody 'picked out, identified, and distinguished from Tibbles a different and rather peculiar animate entity — namely Tibbles minus Tibbles' tail. Let us suppose that he decided to call this entity 'Tib'. . . . Then both Tib and Tibbles were on the mat at time t_1.' The puzzle continues: suppose Tibbles loses his tail at time t_2; won't we then have two cats still sitting on the mat? 'Tib is certainly sitting there. In a way nothing happened to *him* at all. But so is Tibbles. For Tibbles lost his tail, survived this experience, and then at time t_3 was sitting on the mat.'[32]

In terms of the considerations raised above the puzzle may be construed as relying for its force on two main features, i.e. (a) the sub-mereology appropriate to cats; (b) a surreptitious double-think involving simultaneous application of both the Destructivist and the *Sangermanense* views. Thus (a), combined with the *Sangermanense* view of Tibbles as the collective class of his time-stretches, ensures the survival of Tibbles after the loss of his tail. In the same vein the other object is Tib. This is the correlate, constructed with reference to some of the visual and spatial aspects of (a), of that *natura magna* encountered in the earlier stages of the *Fragmentum* mentioned above, and which from being non-Socrates becomes Socrates. But although a later time-stretch of the *natura magna* Tib may happen to be spatially identical with the contemporal time-stretch of Tibbles, nevertheless Tib itself (as long as one remains non-Destructivist) is certainly not a cat. The collective class of *its* time-stretches by no means constitutes anything like a cat-career (birth, suckling, growth, and so forth). Indeed, it has never done any of the things with which cats must be credited if they are to be cats at all. Further, it was not, prior to the tail loss, sitting on the mat (as an incautious rendering of the puzzle might lead one to believe). At any rate such non-session on the part of the earlier Tib does follow if we adopt the common medieval maxim that actions are to be attributed to the *suppositum*, and not to the part.[33] It is on this basis that we may eliminate the earlier time-stretch of Tib from the class of seated things. Tibbles sat; its spatial parts did not. The maxim reminds us that although the ultimate justification for asserting that Tibbles the cat is sitting on the mat lies in the appropriate contact between certain spatial parts of his anatomy and the mat, we attribute the sessional act to the cat as a whole, and not to its rear (or any other) portion, except perhaps for the purpose of a joking *synecdoche*. In short, Tibbles the cat is one object; Tib the non-cat is another. The divergences between both their natures and their careers suffice to avoid confusion. Thus far, then, we have a position allowed by the *Sangermanense* view.

However, the Destructivist view, according to which each successive time-stretch of Tibbles-the-cat is a new Tibbles-the-cat, provides us with an entity of the latter sort which is in some sense on the mat at the end of the process described in the puzzle. Also, if we are to be consistent Destructivists we must credit ourselves with a new Tib. The old Tib was part of an old Tibbles, but the new Tib is spatially identical with a new Tibbles. It is at this point that the double-think occurs. Having thus reaped profit from the Destructivist swings we inconsistently attempt to simultaneously benefit from the *Sangermanense* roundabout; this would allow us to retrospectively prolong the new Tib-Tibbles back in time to a point antecedent to the tail loss. This inconsistent prolongation, combined with forgetfulness as to the non-cathood of the Tib segment of its product, generates the illusion that we are faced with two converging cats.

Considerations akin to those just invoked may also be brought to bear on the puzzle concerning the thousand-fold multiplication of the cat on the mat, a puzzle once again generated by Professor Geach.[34] Thus there seems to be a simple proof (with which, I may hasten to add, its author disagrees) that if the cat Tibbles is sitting on the mat, then there are at least 1001 cats sitting there:

> A normal cat has at least 1000 single hairs.... Let us suppose that we have firmly identified 1000 single hairs in our cat's fur, H_1, H_2, ... H_{1000}. Let C be the largest continuous mass of feline tissues located on the mat. Then for each hair H_n there is a proper part C_n of C, consisting of all of C except H_n; and every such part C_n is discriminable both from C itself and from every other such part of C. There are then 1000 such parts. Fuzzy as the concept *cat* may be, it is clear that C is a cat, and so is each part C_n of C, differing from C by just one single hair. So there is not just one cat called 'Tibbles' sitting on the mat; at least 1001 cats are sitting there. Of course there is a great deal of overlap and sharing of organs among those 1001 cats; but what has logic to say against that? It may easily happen on a small scale between Siamese twins....

On the assumption that we are talking about the cat Tibbles, and do not wish to go in for Destructivism, there is little need to go much beyond the simplest theses of general mereology to dispose of the puzzle; each C_n is a proper part-of-the-cat-Tibbles, and hence by the thesis already implied in IIA above (if *a* is a proper part of *b* then it's not that *a* is *b*) each is not the cat Tibbles, nor any other cat either, since 'otherness' on the cat-level (according to the sub-mereology of *cat*) requires discreteness. As parts-of-the-cat-Tibbles, however, they are distinct, though non-discrete, like the pairs in Abelard's trio (IIA above). Further, on the grounds invoked above to wipe out Tib's alleged session on the mat, one puts aside any attempt to infer validly that any one of those parts is sitting on the mat. Although the sub-mereology of *cat* served to eliminate the other cats, that same sub-mereology, with its presuppositions about 'principal parts', and hence as to what would constitute a *natura magna* in respect of a cat, is nevertheless at the same time intruding in somewhat the same fashion as before and infecting the argument with a distracting dimension, leading to an at least potential apparent multiplication of cats in the Destructivist style.

Alternatively, the mention of a 'continuous mass of feline tissue' as the whole cat in question could bring the discussion round to the level of inferences based on the makings of a continuous whole. In respect of such inferences Abelard speaks thus:

> It might look as though some parts are the same as the whole, namely those which are of the same nature as the whole, e.g. if you consider a bronze rod which can be divided into this and that lesser rod which likewise are of bronze. The rod then has parts like itself, as when one says, 'This lesser bronze rod is a bronze rod'. But it is wrong to take it that therefore the whole is predicated of the part. For *that rod made up of two rods combined* cannot be truly applied to either of them taken independently. But whatever it is that bronze rods have in common can be predicated of the whole and of the parts. . . .[35]

In other words, although both *C* and the proper parts of *C* may be *feline* (even as the tree and its parts are *wooden*, and all bronze rods *brazen*) none of the parts is the cat Tibbles, nor (on the sub-mereological grounds stated above) is it any other cat either.

Department of Philosophy,
University of Manchester

NOTES

* Some themes of the present paper are also discussed in Desmond Paul Henry, 'Abelard's Mereological Terminology', in E. P. Bos (ed.), *Medieval Semantics and Metaphysics* (Artistarium Supplementa II), Ingenium Publishers, Nijmegen, 1985, pp. 65–92.

[1] E.g. De Rijk (1962–7), Kretzmann (1967), Pinborg (1972), the *Cambridge History of Later Medieval Philosophy* and various of the *Cahiers de l'Institut du Moyen-age Grec et Latin* of the University of Copenhagen.
[2] Tweedale (1976).
[3] Geyer (1919, pp. 14–15).
[4] Dal Pra (1969, p. xxvii): 'fin dal primo periodo del suo insegnamento'.
[5] De Rijk (1966, pp. 546.24–547.5; *cf.* pp. 339.30.35, 574–576) and Dal Pra (1969, pp. 166, 193).
[6] See, for example, Sobocinski (1954–5), Luschei (1962 para. 4.9–4.10), Henry (1972, pp. 118–129).
[7] Minio-Paluello (1958, pp. 114.23–115.5).
[8] De Rijk (1966, p. 548.31.35).
[9] De Rijk (1966, p. 553.1.2).
[10] De Rijk (1966, p. 549.21.24).
[11] De Rijk (1966, p. 550.1.17).
[12] De Rijk (1966, p. 551.16.17).
[13] De Rijk (1966, p. 550.18.29).
[14] Cousin (1836, pp. 505–550).
[15] Cousin (1936, p. 511).
[16] Paul of Venice (1499, f.56va).
[17] Woodger (1951, pp. 202–205).
[18] Cousin (1836, p. 514).
[19] Lejewski (1957, p. 254).
[20] *Summa Theologica* I–II q. 20 art. 6c. The various senses in which this one/many contrast may be taken are dealt with in I–II q.17 art. 4c.
[21] De Rijk (1966, pp.553.8–554.35).
[22] De Rijk (1966, p. 343.34.35).
[23] De Rijk (1966, p. 346.31.34, p. 348.26.27).
[24] De Rijk (1966, p. 344.22.23).
[25] De Rijk (1966, pp. 344.33–345.8).
[26] As occurs in Wiggins (1968, pp. 80–81), Wiggins (1967, pp. 10–11).
[27] Wiggins (1967, p. 13).
[28] De Rijk (1966, p. 575.13.15).
[29] De Rijk (1966, pp. 550.33–551.2).
[30] De Rijk (1966, pp. 547.27–548.10).
[31] Some aspects of these are treated in Henry (1972, pp. 118–129).
[32] Wiggins (1968, p. 94). I may add that Professor Wiggins rejects the conclusion, but for reasons which appear not to coincide with the ones I adduce.
[33] *cf.* Paul of Venice (1499, f. 56vb).
[34] Kindly sent to me in the course of correspondence.
[35] De Rijk (1966, p. 576.12.20).

REFERENCES

Cousin, V.: 1836, *Ouvrages Inédits d'Abelard*, Imprimerie Royale, Paris.
Cousin, V.: 1859, *Petri Abaelardi Opera, Tomus Posterior*, Aug. Durand, Paris.

Dal Pra, M.: 1969, *Pietro Abelardo: Scritti di Logica*, La Nuova Italia Editrice, Firenze.
De Rijk, L. M.: 1962–7, *Logica Modernorum* (3 Volumes), Van Gorcum, Assen.
De Rijk, L. M. (ed.): 1966, *Abelard's Dialectica*, Van Gorcum, Assen.
Geyer, B. (ed.): 1919, *Abelard's Logica 'Ingredientibus'* (*Beitrage zur Geschichte der Philosophie und Theologie des Mittelalters* 21, Heft 1), Verlag der Aschendorffschen Verlagsbuchhandlung, Munster.
Häring, N. M.: 1965, *Life and Works of Clarembald of Arras*, Pontifical Institute of Medieval Studies: Studies and Texts 10, Toronto.
Henry, D. P.: 1972, *Medieval Logic and Metaphysics*, Hutchinson University Library, London.
Kretzmann, N.: 1966, *William of Sherwood's Introduction to Logic*, University of Minnesota Press, Minneapolis.
Kretzmann, N.: 1967, 'Article on "Semantics, History of"', in Paul Edwards (ed.), *The Encyclopedia of Philosophy*, Macmillan Publishing Co. and the Free Press, New York, Vol. 7., pp. 358–406.
Lejewski, C.: 1957, 'Proper Names', *Proceedings of the Aristotelian Society*, Supp. Vol. 31, 229–255.
Lejewski, C.: 1973, 'A Contribution to the Study of Extended Mereologies', *Notre Dame Journal of Formal Logic* **14**, 55–67.
Luschei, E. C.: 1962, *The Logical Systems of Lesniewski*, North Holland, Amsterdam.
Minio-Paluello, L.: 1958, *Twelfth-century Logic II: Abaelardiana Inedita*, Edizioni di Storia e Letteratura, Rome.
Paul of Venice: 1499, *Logica Magna, Pars Prima*, Venice.
Pinborg, J.: 1972, *Logik und Semantik im Mittelalter*, Fromann Holzboog, Stuttgart.
Sobocinski, B.: 1954–5, 'Studies in Lesniewski's Mereology', *V Rocznik Polskiego Towarzystwa Naukowego na Obczyznie*, Rok, London.
Tweedale, M. M.: 1976, *Abailard on Universals*, North Holland, Amsterdam.
Wiggins, D.: 1967, *Identity and Spatio-Temporal Continuity*, B. Blackwell, Oxford.
Wiggins, D.: 1968, 'On Being in the Same Place at the Same Time (with One Remark About Categories and Materialism)', *Philosophical Review* **78**, 90–95.
Woodger, J. H.: 1951, 'Science Without Properties', *British Journal for the Philosophy of Science* **2**, 193–216.

ANTHONY KENNY

FORM, EXISTENCE AND ESSENCE IN AQUINAS*

In the present century few philosophers in philosophy departments in British universities have made any very serious study of medieval philosophy: scholastic writers such as Aquinas have been read rather by historians and theologians. Those few who have taken a strictly philosophical interest in medieval thinkers owe a great debt to Peter Geach for his writings on Aquinas and on later medieval logicians, and most recently for his work as General Editor of the edition of Paul of Venice being prepared under the auspices of the British Academy.

My own debt to Geach in the area of medieval philosophy is particularly great. Before meeting him in Oxford I had been trained in Rome in scholastic philosophy as presented in seminary textbooks, but I had read little medieval philosophy at first hand. Indeed I regarded scholastic systems as dull and dead by comparison with the living excitement of analytic philosophy. It was above all from Geach that I learnt the importance of distinguishing between Aquinas and his twentieth-century admirers: I grew familiar with the pocket edition of the *Summa* which he kept always to hand, and came to envy his phenomenal memory for passages of the text. I learnt too that some of the most apparently crabbed and arcane topics of medieval disputation were, beneath the surface, verry closely linked to items of contemporary debate. One of the most stimulating essays in the history of philosophy I have ever read was Geach's 1955 Aristotelian Society lecture, 'Form and Existence'.[1]

'Form and Existence' contains a most valuable comparison between Frege's theory of functions and Aquinas' theory of forms. Just as Frege regarded a predicate, such as '. . . is a horse', as standing for a particular kind of function, namely a concept, so Aquinas held that a general term such as 'horse', standing in predicate position, referred to a form. The form which is referred to by the predicate which occurs in the sentence 'Socrates is wise' may be referred to also by the phrase 'the wisdom of Socrates'; but this latter expression must not be construed as 'wisdom, which belongs to Socrates', just as 'the square root of 4' does not mean 'the square root, which belongs to 4'. 'The wisdom of

Harry A. Lewis (ed.), Peter Geach: Philosophical Encounters, 65–75.

Socrates' refers, in Geach's terminology, to an *individualised* form; the expression which indicates the generic form, the form strictly so called, is not 'wisdom', nor 'the wisdom of Socrates' but 'the wisdom of. . . .'[2] 'Wisdom' *tout court* means nothing in heaven or earth; wisdom is always wisdom *of*: as Aquinas puts it, it is of something (*entis*) rather than itself some thing (*ens*). Against Plato's doctrine that the form signified by a general term is 'one over against many', Aquinas insisted that the question 'one or many' is itself only intelligible if we ask it in relation to a general term that signifies a form or nature.

Geach admits that the account which he gives of individualised forms does not accord in all respects with Aquinas' language: but it is a most interesting analysis in its own right, whether or not it is to be found in its worked out form in Aquinas' writings. Geach treats Aquinas as Aquinas treated Aristotle — improving his insights, tactfully masking his confusions, charitably resolving his ambiguities. This may exasperate historians, but it is the philosophically rewarding way to read a classic text. But in some cases, it seems to me, Geach's benign interpretation fathers on Aquinas ideas and arguments which are not his and of which he might do well to disown the paternity.

One such case is the distinction which Geach maintains between an individualised form and the *esse* of that individualised form. This distinction he attributes to Aquinas and defends by reference to Aquinas' writings. The distinction, it seems to me, is not to be found in Aquinas in the terms in which it is propounded by Geach. It is, however, related to a distinction which is undoubtedly to be found in Aquinas: the distinction, in every creature, between essence and existence. The distinction between individualised form and *esse* is related to this distinction, but it is not the same distinction, since the essence of a human being, such as Socrates, is something really distinct both from Socrates and from his substantial form or soul. It includes matter, flesh, bone, as well as the soul of Socrates, though it does not include any particular bit of matter, flesh or bone. But Geach's arguments for the distinction between individualised forms and their *esse* would, if valid, provide reasons for a parallel distinction between individualised essences and the existence corresponding to them.

The distinction between individualised forms and *esse* was first set out and argued for in 'Form and Existence' in 1955; the arguments were set out at greater length, and some objections to them were countered, in Geach's contribution on Aquinas in the Anscombe-

Geach volume *Three Philosophers* in 1961.[3] In what follows, I quote 'Form and Existence' from the text reprinted in my anthology *Aquinas: A Collection of Critical Essays*.[4]

In order to show that any individualised form is really distinct from the corresponding *esse* Geach presents three arguments drawn from Aquinas' writings. The first may be called the argument from the shareability of natures; the second, the argument from intensive magnitude, and the third, the argument from intentional existence. None of the arguments, in my view, succeeds in establishing Geach's distinction between *esse* and form or Aquinas' distinction between essence and existence.

The first argument is thus stated by Aquinas in a well-known passage of the *Summa Theologiae*[5] to show that God is in no genus.

> All members of a genus share the quiddity or essence of the genus which is predicated of them in the category of substance. But they differ in respect of *esse*, for a horse's *esse* is not the same as a man's, and this man's *esse* is not the same as that man's. So wherever something belongs to a genus its quiddity or essence differs from its *esse*. In God, however, as has been shown, there is no such difference. Clearly, therefore, God is not a species within a genus.

Aquinas' language here does not suggest that he is talking about an *individualised* form or essence. If horse and rider can be said to share animality, and Peter and Paul to share humanity, then on the face of it, the animality and humanity in question are not individualised but general. If Aquinas was talking about individualised humanities, he could say not just that this man's *esse* differs from that man's, but also, with equal right, that this man's humanity is distinct from that man's. What is being distinguished from *esse* in this passage appears to be not the individualised form, Peter's humanity, but rather Humanity as such.

Geach paraphrases Aquinas' argument in such a way as to avoid this Platonic implication.

> If x is F and y is F, then in respect of F-ness x and y are so far alike; the F-ness of x will indeed be a different individualised form from the F-ness of y, but they will be, as F-nesses, alike. But when x *is* and y also *is*, the *esse* of x and the *esse* of y are in general different as such.[6]

Geach illustrates the sense in which the *esses* of two individual substances are in general distinct by contrasting the normal situation with an imaginary one in which several substances share a single *esse*.

In the fairy-tale, all the human members of the family and the family cat shared a single life, that is, a single *esse (vivere viventibus est esse)*; and when the betrothed of the youngest daughter took a pot-shot at the cat, its death was the death of the whole family. In actual families, animality is common to all the members of the family, including the cat, but *esse* is not, and so killing the cat has no such consequence.[7]

It is difficult to see why Geach describes the lot of the fairy-tale family by saying that they all shared a single *esse* or life, rather than by saying that their several lives and *esses* were bound together by a common destiny such that they would all end together. When father, mother and daughter simultaneously die, the Registrar General even of fairyland will surely add three deaths, not one, to his statistics. We can, if we like, say that they all share a common death, like the passengers in an aeroplane crash: but they share a common death only in the same sense in which they share a common humanity. If we want a rule for individuating and counting deaths, surely the most natural rule is to say that there are as many deaths as there are individuals that die. In just the same way, if we insist on counting humanities, it seems that we must say that there are as many humanities as there are individuals that are human.

It is hard to know whether Aquinas would give the same reason as Geach for saying that this man's *esse* is not the same as that man's. Passages parallel to Ia,3,5[8] do not elaborate. In the *Summa against the Gentiles* we read:

Whatever belongs to a genus differs in respect of *esse* from other things in the same genus; otherwise a genus would not be predicable of many things.[9]

In the fairy-tale family there are many animals: in so far as *esse* is a predicate, it is predicable of each of them; it is the *esse* of *many animals*. There is no distinction between *esse* and generic predicates to be found along these lines.

In *Three Philosophers* Geach presents a more developed form of the same argument.

In view of Aquinas' doctrine of universals, it is hard to see the force of his saying that two men or two animals 'share in a quiddity or essence'. So far as his words go, he might well have been taken to argue that since the *esses* are different while the quiddity is the same, the quiddity of each man or animal must differ from the *esse*—only this cannot be his mind, since for him the humanity of this man is not identical with the humanity of that man, and the animality of a man is even unlike that of a horse. We should rather, I think, construe his argument thus; 'while the quiddities (the animalities) of two animals are certainly different, this difference arises from the side not of quiddity but of *esse*;

were there not difference of *esse*, there could be but one individualized animality in two animals even of different species; as there is but one individualized animality in the two kidneys of a given animal, or again in his kidney and liver, differently organized as these are.'[10]

Several things are puzzling about this intriguing suggestion: it is a pity that Geach did not develop it at greater length. To say that if two animals did not have a different *esse* they could not have a different individualized animality, just as two kidneys of a given animal have only one individualized animality, is to suggest that the two kidneys have a single *esse*. But one kidney may die or be removed while the other kidney continues to function; yet hitherto Geach has taken the possibility of *X*'s ceasing to exist while *Y* continues to exist as establishing that *X* and *Y* have different *esse*s. Moreover, in what sense is there an individualized animality *in* a kidney? In the passages just quoted from Aquinas it is clear that he has in mind a principle according to which *F*-ness is in a thing if and only if '*F*' is predicable of the thing. But 'animal' is not predicable of a kidney: a kidney is not an animal. If there is such a thing as kidneity, there are presumably two kidneities, and not just one, in a pair of kidneys.

To be sure, the functioning of an animal's kidneys is part of its functioning as an animal; and on Aquinas' view the whole animal functioning of the human being Peter is organised by a single form, Peter's substantial form or intellectual soul.[11] In this, quite different, sense one might speak of animality as being 'in' a kidney (for on Aquinas' view the animality of Peter is one and the same form as the humanity of Peter, though of course animality differs from humanity). It might seem therefore that we could establish a distinction between form and *esse* by appealing to the case of kidney transplantation. If Peter's kidney is transplanted into the body of Paul, do we not have a case of a single kidney, with a single *esse*, informed first by the animality of Peter and then by the animality of Paul? I am inclined to think that Aquinas would regard the transplanted kidney as being only by a figure of speech the same kidney before and after the transplant: as being a piece of matter informed first by Peter's soul (which organizes the matter into one kidney) and then by Paul's soul (which organizes the matter into another, different kidney). But whether or not this is the correct way to explain the topic on Thomist principles, the transplantation does not provide an instance of a difference between an individualised form and the *esse of that same form*. If there is such a thing as a form of kidneity, that lasts just as long as the kidney lasts; while if we are talking about the individualized humanities of Peter and Paul, then it is the *esse* of Peter and Paul's souls, not the *esse* of any shared kidneys that is relevant to the problem of the real distinction. Neither in its original nor in its modified form does Geach's first argument give us any grounds for distinguishing between an individualised form and that by which the individualised form *is* or continues in existence.

A more promising argument is drawn from the consideration of intensive magnitude.

It may be that *x* is *F* and *y* is *F*, and that they have the same specific *F*-ness, but yet the *F*-ness of *x* is more intense than that of *y*. Moreover, the *F*-ness of *x* may become more or less intense; and increase of *F*-ness plainly resembles a thing's coming to be *F*, whereas a decrease of *F*-ness resembles a thing's altogether ceasing to be *F*. Now difference between *x* and *y* as regards intensity of *F*-ness is not difference precisely as regards *F*-ness (especially as we may suppose *x* and *y* to have the same specific *F*-ness); it is rather, I wish to say, difference as regards the existence of *F*-ness—the *F*-ness of *x* exists more than the *F*-ness of *y*. So also a change in mere intensity is a change as regards existence; increase in the intensity of *x*'s *F*-ness resembles the coming to be of *x*'s *F*-ness, both being additions of existence; decrease in the intensity of *x*'s *F*-ness resembles the ceasing of *x*'s *F*-ness, both being subtractions of existence. Here again, there is a real distinction between the *F*-ness of *x* and the *esse* of this individualised form; while the *F*-ness as such remains unchanged, its existence may vary in degree.[12]

Geach gives no reference to Aquinas for this argument; but there are to be found in the *Summa Theologiae* two lengthy treatments of intensive magnitude, one in connection with forms and dispositions in general,[13] one in connection with the particular case of the theological virtue of charity.[14] The theory presented in those two passages seems to disagree in some important respects with the argument presented by Geach.

Aquinas says of charity that its *esse*, like that of any accident, is to inhere in a substance. ('*Accidentis esse est inesse*' is a common slogan in his writings: where 'F' is an accidental predicate, the existence of a particular F-ness is simply something's being F.) Hence, he goes on to say,

for charity to increase in respect of its essence is one and the same thing as its inhering to a greater degree in its possessor. It increases essentially not by beginning or ceasing to be in its possessor, but because it begins to be to a greater degree in its possessor.[15]

He is not saying—as Geach would have him say—that the charity as such remains unchanged while the *esse* varies in degree; he is sayig rather that the change in the essence of the charity is the very same thing as a change in the degree of its existence in its possessor. And so with all forms which admit of variation in intensity:

This is the mode of increase proper to every form which is intensified, because the *esse* of such a form consists in its entirely in its inhering in its possessor.[16]

Whether or not Geach's argument corresponds to Aquinas' mode of speaking in passages such as these, it is an interesting one which deserves consideration in its own right. One difficulty which arises is this: if the temperature of water in a kettle rises from 98° Centigrade to 99° Centigrade, how are we to decide whether this consists in the departure of one form (the form of *being 98° hot*) and its replacement by another (the form of *being 99° hot*), or whether it is simply an increase of *esse* in one and the same form, hotness? Geach recognizes that there are difficulties in treating an increase of heat as an instance of the intensification of a form; but he goes on to say

So far as I can see, *sound* is a good example. A louder and a softer sound may be qualitatively identical; and a sudden increase of loudness resembles a sound's suddenly starting, a sudden decrease of loudness its suddenly stopping.[17]

But in a similar manner, when the sound of A flat on the flute increases during a crescendo from n decibels to $n+2$ decibels, we can inquire whether this is the intensification of a single form or the replacement of one form by another.

The reply would no doubt be that wherever we have the possibility of *continuous* increase or decrease in the intensity of a quality it is more intelligible to regard the change as a variation in *esse* than as the rapid succession of an infinite series of infinitesimally differing forms. In *Three Philosophers* Geach remarks in this context that Hume was wrong to draw the conclusion that all the degrees of a given quality are perfectly distinct from one another like different colours or tones.[18]

In this case, then, there seems to be good reason to distinguish between an individual form and the *esse* of that individual form. But as Geach admits, the argument lacks generality: a shape, or a relation like fatherhood, or a substantial form, does not admit of differences in intensity. This point is made, with respect to various forms essences, by Aquinas himself in a number of places. For instance:

The criterion for assigning something to a species must be hard and fast and precise. There are therefore two things which may make it impossible for a form to be possessed in differing degrees. First, the form may be what qualifies its possessor for membership of a species. No substantial form can be possessed in different degrees. That is why Aristotle says that just as number does not admit of more or less, neither does substance in the sense of species (i.e. there is no *more* nor *less* in the possession of specific form).... Secondly, it may be that precision is part of the notion of a particular form, so that if anything possesses that form, it must possess it altogether or not at all. That is why specific numerical terms cannot have 'more' or 'less' added to them: each specific term or number is made up of indivisible units. The same is true of specific terms assigning numerical measurements to continuous quantities ('two cubits long', 'three cubits long') and of relational terms ('double', 'treble' etc.) and of terms assigning shape ('triangular', 'quadrilateral' etc.).[19]

Thus Geach's second argument, if valid, establishes a distinction between an individualised form and its *esse* only in the case of those forms that admit of being possessed in varying degrees. In other cases it cannot establish the distinction: in particular, it cannot be used to establish any distinction between the individualised essence of a substance and that substance's *esse*.

The third argument offered by Geach, however, is one of unrestricted generality. It is based on the distinction between a form as thought of, and a form as existing in nature; and since whatever can exist can be thought of, if the argument works in one case it will work in all cases.

Aquinas' doctrine of the intentional existence of forms may be looked on as an answer to the question: when I think of *X*, what is it that makes my thought be a thought of *X*? When I think of redness, what makes my thought be a thought *of redness*? When I think of a horse, what makes my thought be a thought *of a horse* and not of a cow? Aquinas replies that what makes my thought of a horse a thought *of a horse* is the very same thing as makes a real horse a horse: namely, the form of *horse*. The form exists, individualised and enmattered, in the real horse; it exists, immaterial and universal, in my mind. In the one case it has *esse naturale*, existence in nature; in the other case *esse intentionale*, existence in the mind. Geach puts the matter thus:

What makes a sensation or thought of an *X* to be *of an X* is that it is an individual occurrence of that very form or nature which occurs in an *X*—it is thus that our mind 'reaches right up to the reality'; what makes it to be a *sensation* or *thought* of an *X* rather than an actual *X* or an actual *X*-ness is that *X*-ness here occurs in the special way called *esse intentionale* and not in the 'ordinary' way called *esse naturale*.[20]

This is admirable as an exposition of the doctrine of *esse intentionale*: but it provides no reason whatever for concluding to a distinction between a particular individualised form and the *esse* of that form. For it is no part of Aquinas' doctrine that there is one same individualised form of horse which occurs in a particular horse, say Eclipse, with *esse naturale* and occurs also in my mind with *esse intentionale*. What we have are two different individualisations of the same form, not two different existences of the same individualised form.

Geach writes:

> When Plato thinks of redness, what exists in Plato is not a certain *relation* to redness or red things but *is* redness, is an individual occurrence of the very same form of which another individual occurrence is the redness of this rose.[21]

There is an equivocation in the sense of 'individual occurrence' here. The occurrence of redness in a particular rose is an individual occurrence because it is an occurrence of redness in a particular rose: it is the redness of this rose. The occurrence of redness when Plato thinks of redness is not individual by being the thought of the redness of any particular thing, but by being a thought thought by a particular thinker, namely Plato. It was a constant doctrine of Aquinas that thought, as such, is not directly of individual things at all—neither of individual forms like redness nor of individual substances.[22] When I think of Socrates there is no form of Socrateity having intentional inexistence in my mind: there is only the universal form of humanity, which I can use to think of Socrates only by placing it within a context of sensory imagery (*phantasmata*). The individual humanity of Socrates has *esse naturale* in Socrates but it does not have *esse intentionale* in my mind or anyone's mind; the universal, humanity, has *esse intentionale* in my mind but it does not have *esse naturale* in Socrates or in any human being. In neither case do we have one same individualised form with two different modes of *esse*: but that was what was needed to establish Geach's thesis in its universal form.

I conclude that neither Geach nor Aquinas has shown that there is as a general rule a real distinction between an individualised form and its *esse*. In the particular case of forms which admit of continuous variation in intensity the distinction may be made out: but this gives no reason for accepting a real distinction between substantial forms where there is no question of variation in intensity. Most importantly, Geach's arguments lend no support to the celebrated Thomistic doctrine of the real distinction between essence and existence.

In my view, Aquinas' teaching on essence and existence rests on a confusion between generic and individual essences. Aquinas thought that in all creatures essence and existence were distinct, whereas in

God the two are identical. If we understand 'essence' in the generic sense (corresponding to a Fregean function such as denoted by '. . . is God', '. . . is a human being') then the doctrine of the real distinction between essence and existence is the truism that whether or not a concept is instantiated is quite another matter from what are its constituent characteristics; but the doctrine that the two are identical in God is the nonsense that the question 'What kind of thing is God?' is to be answered by saying 'There is one'. On the other hand, if we understand 'essence' in the individual sense in which we can speak of the individualised humanity possessed by Socrates and by Socrates alone, then the doctrine of the real distinction in creatures becomes obscure and groundless. As Aquinas often insists, for a human being to exist is for it to go on being a human being; Peter's existing is the very same thing as Peter's continuing to possess his essence; if he ceases to exist he ceases to be a human being, and his individualised essence passes out of the nature of things.

To expose in detail the confusion in the doctrine of essence and existence would take us beyond the limits of a short paper. I have tried to do so elsewhere. The confusion which I have tried to expose is one of which I would never have become aware had I had not been taught, by Geach's 'Form and Existence', to keep a sharp eye open for the distinction between generic and individualised forms.

Balliol College,
University of Oxford

NOTES

* The substance of the present essay is also published in Anthony Kenny, *The Heritage of Wisdom*: Essays on the History of Philosophy, Basil Blackwell, Oxford, 1987.

1 Geach (1955).

2 *Cf.* Aquinas: *Summa Theologiae* Ia 3, 2 ad 4; Ia 50, 2.

3 Anscombe and Geach (1961).

4 Kenny (ed., 1969).

5 Aquinas: *Summa Theologiae* Ia 3, 5 ('*Tertio quia omnia . . . in genere sicut species.*)

6 In Kenny (ed., 1969, p. 49).

7 In Kenny (ed., 1969, p. 50).

8 *E.g. De Ente*, 6; *De Potentia* 7, 3.

9 Aquinas: *Summa contra Gentiles* I, 25.

10 Anscombe and Geach (1961, p. 94).

11 Aquinas, *Summa Theologiae* Ia 76, 3.

[12] In Kenny (ed., 1969, p. 51).
[13] Aquinas: *Summa Theologiae* Ia IIae 52.
[14] *Op. cit.*IIa IIae 24.
[15] *Op. cit.* IIa IIae 24, 5 ad 3.
[16] *Op. cit.* IIa IIae 24, 5c.
[17] In Kenny (ed., 1969, p. 51).
[18] Anscombe and Geach (1961, p. 93).
[19] Aquinas: *Summa Theologiae* Ia IIae 52, 1c.
[20] Anscombe and Geach (1961, p. 95).
[21] Anscombe and Geach (1961, p. 95).

REFERENCES

Anscombe, G. E. M. and P. T. Geach: 1961, *Three Philosophers*, Cornell University Press, Ithaca, New York.

Geach, P. T.: 1955, 'Form and Existence', *Proceedings of the Aristotelian Society* **55**, 251–272.

Kenny, A. (ed.): 1969, *Aquinas: A Collection of Critical Essays*, Macmillan, London.

BOGUSLAW WOLNIEWICZ

ON THE DISCONTINUITY OF WITTGENSTEIN'S PHILOSOPHY

1

How is Wittgenstein's later philosophy related to his earlier one? Despite much discussion there is little clarity on that. So let us give it again some consideration.

The *Tractatus* is a masterpiece of rare power and ravishing beauty. Its content is a profound and highly coherent philosophy of language, based upon a radically new kind of metaphysics: the metaphysics of facts and situations. (Meinong, with his notion of the 'objective' of a proposition, apparently was moving in the same direction. But he never came near asking himself any of the two crucial questions: (1) When, if ever, are the objectives of different propositions identical? (2) What, if any, is the relation of the objective of a compound proposition to the objectives of its components?) Moreover, the *Tractatus* anticipated many of the later developments of logical semantics, especially those commencing around 1950 and connected with its algebraization. The kernel of its message may be put down as follows.

2

The fundamental problem of the *Tractatus*, as of all philosophy, concerns the relationship of thought and reality. This relationship is mediated by language, and so it may be decomposed into the relative product of two relations: one between thought and language, the other between language and reality. Let us mark the latter by 'φ', the former by 'ψ'.

It has been maintained that according to the *Tractatus* the projective relation φ between language and reality has to be an isomorphism. This, however, is not borne out by the text. To satisfy the conditions laid down by Wittgenstein it is enough for φ to be a *homomorphism*, and this already makes a lot of a difference. In the first place, we are confronted now with two delicate questions: (1) Which is the direction of that homomorphism: from language to reality, or the other way

Harry A. Lewis (ed.), Peter Geach: Philosophical Encounters, 77–81.

round? (2) Is it a homomorphism *onto*, or merely one *into*? Neither of these questions has a trivial answer in the context of the *Tractatus*.

We assume here that the relation φ is a homomorphism on the language L onto the reality R, i.e., that $\varphi: L \rightarrow R$. Thus reality is a homomorphic image of language. But language is the totality of propositions, and the reference of meaningful propositions are possible *situations*. Consequently, reality is not the world, but the *logical space*; i.e., it is not the totality of facts, but the totality of possibilities. Thus language is more capacious than the world, and the number of propositions is greater than even that of situations.

The simplest non-trivial homomorphism of that kind is the well-known Fregean one. Language is mapped under it onto the set of the two classic truth-values, and the corresponding two-element Boolean algebra is then the logical space. Thus for Frege there are just two possible situations: the True and the False. This is so because his only stipulation with regard to the reference of propositions is that *contradictory* propositions cannot have the same reference. In the *Tractatus*, however, it is stipulated further that logically *independent* propositions cannot have the same reference either. This move is the gist of its logical atomism, transforming the Fregean homomorphism $\varphi: L \rightarrow \{1,0\}$ into the composition of two other ones: φ' on L onto logical space, and φ'' on logical space onto the set of truth-values.[1]

3

The aim of the *Tractatus* was to trace the boundaries of clear thought:

> Philosophy . . . should trace the unthinkable from within by means of the thinkable. By presenting clearly what may be expressed it will point to the inexpressible.[2]

The positivistically-minded members of the Vienna Circle deemed to recognize in these words their own 'demarcation problem', together with their own hostility towards 'metaphysics' and their cult of 'science'. It was a monumental misunderstanding. To Wittgenstein the metaphysical is indeed the inexpressible, but this is not to mean that it is regarded as some kind of delusion or hoax. On the contrary, the hoax is the idea of a 'scientific philosophy'.

In the *Tractatus* the tracing of the boundaries of the inexpressible was to be accomplished at one stroke. Logical space R fills the realm of the expressible E completely, i.e., we have $E = R$. Consequently, the homomorphism φ' is onto the expressible, and what is left, evidently, is only the inexpressible. This grandiose project, however, was soon to encounter grave technical difficulties, and then Wittgenstein simply dropped it. This was rash. Not all the difficulties were quite as insuperable as they might have seemed, and the *Tractatus* left room for manoeuvre. It might have been helpful, for instance, to weaken the homomorphism φ' to one *into* the expressible. Then instead of the one language L we could consider a whole series of languages $L_0, L_1, \ldots,$ and a corresponding series of logical spaces $R_0, R_1, \ldots$. The realm of the inexpressible would be approximated by the latter 'from within', starting from what is expressible in the language L_0 at hand. Certainly, the series of logical spaces need not be monotonic, and in advance there would be no telling whether what is inexpressible at a given stage L_1 is absolutely or only relatively so. Thus the final tracing of the boundaries of the inexpressible would recede to infinity, but for theory this could hardly count as an objection.

4

Such, in rough outline, is the doctrine of the *Tractatus*, with one of its vistas. What, on the other hand, is the doctrine of the *Philosophical Investigations*? There is none. The book is a formless accumulation of ambiguous hints and loose remarks regarding what it calls the 'uses of language'. It may be doubted, however, whether it really deals with language at all. And though this might have seemed a preposterous thing to say, still there are some good reasons to say so.

The basic reason is this: the 'language-games' discussed in Wittgenstein's later writings are neither languages, nor fragments of languages, nor even simplified models of such. They are rough and ready imitations of sign-systems of a quite different kind, namely of prelinguistic ones. Karl Bühler[3] has called them 'one-class systems', to distinguish them from languages which are 'two-class systems'. A one-class system is not a language, but a *code*, and its signs are *signals*. These are neither sentences, nor words. This distinction makes sense only with regard to those sign-systems which in addition to a *phonetics* (i.e., a description of the semiotically relevant perceptible features of the sign-vehicles), and to a *semantics* (i.e., a set of correlations between the signs and their reference), are provided also with a *syntax*. The rules of syntactic construction induce the differentiation of signs into 'single words' and 'whole sentences', thereby making the system of signs a two-class one. The semantics of a code is direct; it may be represented by a table or chart correlating the signals with certain situations. (These situations form a possibility space, which is the ontology of the code.) The semantics of a language is indirect. Instead of a table we have here a vocabulary correlating certain words with *objects*, and—due to syntax—the correlation of sentences with the situations they represent takes care of itself. In this way a two-class system acquires the properly linguistic function of *representation*, in addition—and in contradistinction—to the two animal functions served by the one-class systems: *manifestation* of emotions and *stimulation* of actions.

The potentialities of one-class systems to control animal and human behaviour are easily underrated. This gives rise to the idea that they are in fact primitive languages. A code need not be finite: discrete (or digital) codes have a finite number of signals, but continuous (or analogue) codes do not. Moreover, a signal need not be a simple sign. Out of two (or more) codes C_1, C_2 a product-code may be formed, and the signals of such a code $C_1 \times C_2$ are couples of signals of the two simpler ones. At first sight such complex signals may look like sentences, but the resemblance is spurious. It well may be that human languages have evolved genetically out of codes. There is, however, no more point to regard codes—or 'language-games'—as primitive languages, than there is to regard fish as primitive mammals.

5

Philosophy of language oscillates between the two poles of metaphysics and psychologism; i.e., between stressing the relation φ, or the relation ψ. The main representative of the former trend is Frege, and the *Tractatus* follows closely in his wake. The *Philosophical Investigations*, on the other hand, are the expression of a dramatic departure from Frege and towards psychologism. Language is now viewed by Wittgenstein not in its logical capacity as the medium of thought and the bearer of truth-value, but instrumentally as a means of communication. Its relation to reality fades out, and its connexion with feeling, thinking, and doing comes massively to the fore.

Psychologism and positivism go hand in hand. Metaphysics, though not condemned as a vice as in pure positivism, is viewed as a malaise to be cured. Thus Wittgenstein's later philosophy turns into a kind of psychotherapy, exerting itself to bring metaphysics to a stop. What for? This is never made clear, as if it were a matter of course. But it is not.

University of Warsaw

NOTES

[1] For details *cf.* Wolniewicz, B.: 1979, 'A Wittgensteinian Semantics for Propositions', in Diamond, C. (ed.) *Intention and Intentionality: Essays in Honour of Professor G E M Anscombe*, Harvester Press, Brighton.

[2] Wittgenstein, L., *Tractatus Logico-Philosophicus*, 4.113–4.114.

[3] See Bühler, Karl: 1965, *Sprachtheorie*, Gustav Fischer Verlag, Stuttgart (original edition 1934).

GEORG HENRIK VON WRIGHT

POSSIBILITY, PLENITUDE AND DETERMINISM (WITH SOME COMMENTS ON ANCIENT AUTHORS)

1

In his Commentary to Aristotle's *De interpretatione* Boethius says that 'Diodorus possibile esse determinat, quod verum aut est aut erit'.[1] Diodorus's view was well known and much debated in Antiquity. Modern writers often refer to it as his 'definition' of the concept of possibility.

Diodorus was also famous for his κυριεύων or Master Argument. This seems to have been an attempt to prove that everything which is possible either is or will be true.

There is some conflict between the view that Diodorus had both *defined* possibility as that which is or will be true and had wanted to *prove* this. To give an authentic reconstruction of Diodorus's thoughts on the basis of existing sources is hardly possible.[2]

A concept answering to Diodorus's 'definition' I shall call Diodorean possibility or, for short, *D*-possibility. I shall here discuss it and its relation to some other concepts of possibility and to some metaphysical ideas concerning the constitution of reality.

The questions to be discussed have been very much debated in recent literature. It is difficult to write about them without becoming entangled in scholarly disputes. I shall try to avoid this as much as I can.

2

The 'things' to which *D*-possibility is attributed are what I have elsewhere called *generic* propositions. Such propositions are expressed by grammatically complete indicative sentences, e.g. 'it is raining' or 'Socrates is sitting', but they can be pronounced true or false only when coupled with an 'occasion' for their truth or falsehood. By occasion I shall here (exclusively) mean a location t in time. As a

Harry A. Lewis (ed.), Peter Geach: Philosophical Encounters, 83–97.

schematic representation for a sentence expressing a generic proposition I shall use the letter *p*.

Generic propositions which are contingent may be ('turn out') true on some occasions and false on others. Such propositions are therefore sometimes said to have a *variable* truth-value. An *individual* proposition, e.g. the proposition that *p* at *t*, is true or false *timelessly* (atemporally). Timeless truth must not be confused with omnitemporal truth.

The attribution of *D*-possibility to individual propositions makes no sense, because to say that it is or will be true that *p* at *t* is not to say anything over and above that *p* at *t*.

It should be noted that the proposition that a certain (generic) proposition is *D*-possible is itself *generic*. Something which is *D*-possible today may come true tomorrow but never again. Then this thing is henceforth not *D*-possible, i.e. *D*-impossible. One and the same proposition may thus change its modal status from *D*-possible to *D*-impossible. The converse change cannot happen. If it is now the case that it will never be the case that *p*, then it is for ever the case that it will never be the case that *p*. *D*-possibilities can thus become extinguished but cannot be created.

If something happens over and over again for ever, then this thing is permanently or for ever *D*-possible.

D-possibility can be called a notion of *prospective* possibility. It refers to that which is or will be (true). One could reshape the notion a little and make it even 'more prospective' by omitting reference to that which already is (true). For some purposes this reshaping would be useful.

3

One can supplement *D*-possibility with a retrospective counterpart. One can also introduce a generalized 'omnitemporal' notion of possibility according to which something is possible if, and only if, it was, is, or will be true. Whereas propositions to the effect that something is prospectively or is retrospectively possible are *generic*, a proposition to the effect that something is in the omnitemporal sense possible ('happens sometime') has a determined truth-value.

The modal logic of Diodorean possibility (and of its retrospective counterpart) is the system known as S4.3. The modal logic of 'omnitemporal' possibility is S5.

Since *D*-possibility is defined in terms of truth and time, it can be called an *extensional* or *reductive* notion of possibility. One could also say that it is no 'genuine' modal notion at all. The same holds for its retrospective counterpart and for generalized 'omnitemporal' possibility.

A reductivist concept of possibility of the kind exemplified by *D*-possibility is sometimes called *statistical*.[3] In its most general form, the statistical conception equates an attribution of possibility with an *existential* statement. On the statistical view, roughly speaking, that a certain thing possibly happens ('can happen') means that such things *sometimes* happen — and that a certain thing possibly is such and such means that *some* (such) things are such and such.

4

Boethius also reports[4] that Diodorus defined notions of necessity and impossibility corresponding to his notion of possibility. Something is *D*-necessary if, and only if, it is and will always be true—and *D*-impossible if, and only if, it is and always remains false. We may add to this a definition of *D*-contingency meaning the (now or) sometimes true and (now or) sometimes false.

There is no evidence that Diodorus himself had acknowledged the existence of any other kind of possibility beside the kind nowadays commonly named after him. But he admitted a kind of necessity other than *D*-necessity. This is obvious from the first 'premiss' involved in his Master Argument. It says that 'everything which is past and true is necessary'.

This is a retrospective notion of necessity. But it is not the retrospective counterpart of *D*-necessity. That something is in that sense necessary means that it (now) is and always was true. Diodorus certainly did not wish to maintain that every past truth was of this sort. His thesis must be understood as an attribution of necessity to *individual* propositions true of the past. What the thesis amounts to is that, if it is a *fait accompli* that *p* at *t*, then it is from that time on necessary that *p* at *t*.

The idea that *facts* are, in this sense, necessary is also reflected in Aristotle's well-known words in *De interpretatione* 9 that everything which is is, when it is, necessary. In the language of the Schoolmen: 'Omne quod est quando est necesse est esse'. Or, applied to that which is past and true: 'Quod fuit non potest non fuisse'.

This idea about the necessity of past truth can be interestingly related to ideas about potentially and causal efficacy—but I shall not discuss these aspects of it here.

I shall coin for this idea the name *necessity of truth as fact* or *F*-necessity. If it is true that not-*p* at some time *t* not later than the present time, then it is *F*-impossible that *p* at *t*. On the received view of the interrelatedness of the modal notions, it is *F-possible* that *p* at *t* if, and only if, it is *not* the case that not-*p* at *t*, i.e. if it *is* the case that *p* at *t*. Thus *F*-possibility and *F*-necessity coincide. There is, in other words, no such thing as *F-contingency*.

The modal logic of *F*-necessity is a system in which there is 'no room' for contingency. Every proposition is either necessary or impossible. I have elsewhere described this system and baptized it the System *R*.[5]

F-necessity too is a *reductive* modal notion, although in a somewhat different sense from the one in which *D*-possibility is reductive.

5

Assume that *p* at *t*. Then, at *t*, it is *D*-possible that *p* and *F*-necessary that *p* at *t*. We may, however, wish to say that it is a *contingent fact* that *p* at *t*, i.e. that it *could have been* the case that not-*p* at *t*. Or, we may wish to deny this and maintain that it was *necessary* that *p* at *t* and could *not* have been the case that not-*p* at *t*. How shall we understand those statements about contingency and necessity respectively?

We cannot clarify them in terms of *F*-modal status, since there simply is nothing which is *F*-contingent. But nor can we do this in terms of *D*-modal status. That it is *D*-necessary at *t* that *p* means that it is true at *t* and will for ever remain true that *p*. That the same proposition is *D*-contingent at *t* means that since it, on our assumption, is true at *t* it will be false at some future time.

One might wish to suggest the following here: Saying that it is contingent that *p* at *t* means that we *did not know*, before *t*, whether it was going to be true at *t* that *p*. Denying this again would mean that we *knew*, before *t*, that it was going to be true at *t* that *p*. I shall call these notions *E*-contingency and *E*-necessity, respectively.

Like the *D*- and *F*-modalities, the *epistemic* or *E*-modalities, too, are reductive. They define modal status in the terms of our state of knowledge.

An answer in epistemic terms to our question need not be reductivist, however. One might also have suggested that the contingent character of the true proposition that p at t consisted in the fact that we *could not have known*, before t, whether it was going to be true at t that p. This answer in terms of *possible knowledge* retains a genuinely modal component. I shall not here inquire further into the nature of this idea of possibility, nor coin for it a special term.

Instead I shall consider another non-reductivist attempt to cope with our question. The suggestion then is that to say that, although at t it is true that p this need not be so, is not to make a statement about our knowlege nor about what is the case at other times, but is to maintain the existence of an *alternative possibility* at that very moment t. This is what the contingent character of the fact that p at t amounts to. Had there not existed such a 'real alternative' to this fact, it would have been necessary that p at t.

The non-reductive modal ideas which we are now considering I shall call '*real*' or *R*-modalities. *Is* there such a thing as real contingency (possibility, necessity)? If so, how shall we understand it? These are 'deep' questions in the sense that the answers given to them by different philosophers constitute *deep* divides between their philosophical viewpoints.

6

There is a metaphysical idea which has, relatively recently it seems,[6] become known under the name Principle of Plenitude or the *plenum* principle. According to it, roughly speaking, everything which is possible will, at some time or other, also be actual. Every possibility will materialize or realize itself with time. Nothing which is possible will for ever remain 'merely possible'.

These formulations cover a multitude of variants of the principle. In order to give to it a more precise content one must specify the kind of 'things' the possibility of which is being contemplated. *One* way of doing this is to say that they are (all) generic propositions which have a *truth-value when 'coupled' with a time t*. To maintain that the Principle of Plenitude holds good for them is to maintain that, if such a proposition is possible at t, then it is *D*-possible, i.e. either is true at t or will be true later.

As noted above (Section 2), *D*-possibilities may become extinguished, get lost. What is *D*-possible at t may not be this at a later time t'. Conversely, D-necessities may be created, but not lost. What is, at t, D-contingent, may at t', be D-necessary.

By postulating that such changes in modal status are excluded, one can give to the Principle of Plenitude a stronger content. Then, if something is ever D-possible, it is always D-possible. Once it has happened, it will repeat itself over and over again. This version of the principle entails (but is not entailed by) another famous metaphysical idea, *viz*. the idea of an Eternal Return (*ewige Wiederkunft*).

7

By virtue of his 'Master Argument' and his 'definition' of possibility, Diodorus may be said to have adhered to the *plenum* principle. He is also reputed to have championed *determinism*.

If the only accepted concept of possibility is that of D-possibility, then one must reject as 'meaningless' the idea that a proposition which is true at a certain time t could have been false at that very time. Nothing which happens can have a real alternative to itself at the moment of its happening.

This rejection of real alternatives is *one* sense of 'determinism'—but a rather 'slim' and uninteresting one, it would seem. We do not know exactly in what Diodorus's deterministic view consisted. But it is hard to imagine that it had consisted *only* in this.

If one acknowledges an idea of R-modality, then one might wish to *deny as being false* any proposition to the effect that a certain proposition is, at a certain time, R-contingent. All generic propositions are, when true, R-necessary. This would give to determinism a somewhat 'thicker' content than mere rejecting R-modality as meaningless.

But the difference would hardly be great. What *is* the difference between regarding the notion of R-contingency as senseless and maintaining that *no* proposition is, as a matter of fact, R-contingent? It would seem that the difference is nil unless we say something *more* to elucidate the meaning of R-modality.

8

Whether Aristotle adhered to the *plenum* principle is very much a matter of dispute. What is certain however, is that he rejected determinism. About this he was quite explicit. In an often quoted passage in

De interpretatione 9 he says that 'not everything is or comes to be of necessity.

Whether Aristotle was altogether successful in rejecting determinism is another matter. The answer depends upon how we understand the idea of *being or coming to be of necessity* as contemplated by Aristotle. To this, as far as I can see, we have no very safe clue.

Coming to be is a process in time. Something, e.g. that p, which was not there at a certain time t' is there at a later time t. Then it has 'come to be'. By saying that this happened 'of necessity' one might be saying that some time before this took place, say at t', it was already necessary or 'settled' that is was going to be that p at t. It was, in other words, *antecedently necessary* or *predetermined* (at t') that p at t.

We need not stop here to consider *how* such predetermination is possible, if at all. But we shall distinguish two forms of such predetermination—a weaker and a stronger form. I shall say that it is in the *weaker* sense predetermined that p at t if, and only if, at *some* time t', earlier than t, it is necessary that p at t. And I shall say that it is in the *stronger* sense predetermined that p at t if, and only if, at *all* times t', earlier than t, it is necessary that p at t.

Obviously, if it is necessary at t' that it will be the case that p at t, then it is also necessary at any time *between* t and t that p at t. Once an individual proposition has 'acquired' the status of being predetermined, it cannot 'lose' this status. This seems to be a logical feature of predetermination.

I now propose the following elucidation of the previously introduced notion of R-necessity: It is R-necessary at t that p if, and only if, it was antecedently necessary that p at t. One can distinguish a weaker and stronger form of R-necessity depending upon the weak or strong nature of the antecedent necessity in question.

The notions of R-possibility and R-contingency are defined in the received manner in terms of the concept of antecedent necessity or predetermination.

By the Principle of (Universal) Determinism we shall here understand the thesis that all true propositions are antecedently necessary. They are thus, when true, R-necessary (in addition to being F-necessary).

Was the determinism which Diodorus professed and Aristotle wished to deny a determinism definable in terms of antecedent necessity or predetermination? There is hardly enough evidence to permit a definite answer to the question. But there certainly is *some* evidence for an affirmative answer. Thus, for example, it seems to me obvious that the type of determinism which in the famous 'sea-battle'-passage Aristotle is anxious *not* to be forced to acknowledge is that the future is predetermined.

9

Let it be the case that p at t. Let it also be the case that at t' it was not antecedently necessary that p at t. The proposition that p at t is, in other words, antecedently contingent at t'. This means that 'the history of the world' which resulted in that p at t *could*, at t', have taken a different turn resulting in not-p at t. And since it could have done this at t', then also at any time before t'—but maybe not at any time between t' and t. Perhaps from some time t'' on, nearer to t than t', it was antecedently necessary that p at t.

Thus whether something which is the case is this contingently or necessarily may depend on the temporal perspective in which its being or coming to be is viewed. Something which, in a shorter perspective, is necessary, predetermined or inevitable may be contingent when viewed from a point further back in time. This is in perfect agreement with the way we often view things. Once the avalanche was 'let loose', the disaster was inevitable, but had people started to move away a little earlier it would not have happened. So, that X was killed then was, in a way, necessary (unavoidable), in another way contingent (might not have happened).

If it is antecedently contingent at t' whether p at t, then so it is at any time t'' between t' and t. It follows from this that if everything which ever is or happens is antecedently necessary—no matter whether in a shorter or longer perspective—then no proposition is at any time R-contingent. If, at t, it is R-possible that p, then this is also R-necessary, and hence it is true that p at t. And if, at some time t', it is antecedently possible that p at t, then it will also be true that p at t. The antecedently possible and the R-possible is also D-possible. And the D-possible is, *ipso facto*, either R-possible or antecedently possible.

Thus, if Universal Determinism is true, then anything which is or was antecedently possible will, in due time, come or have come true. In this sense Universal Determinism may be said to be a warrant of the Principle of Plenitude.

The converse relation, however, does not hold. It is perfectly conceivable that every generic proposition which, at a given time *t*′, is antecedently possible will at some later time *t* be true but that some or even all of these propositions, when true, are *R*-contingently true. Some (but not all) of the propositions, moreover, may at the time when they come true by *R*-necessity ('come true of necessity') but at some other time be *R*-contingently false.

10

The concept of *R*-possibility (necessity, contingency) which we introduced in Section 5 can be termed a *synchronic* modality in the following sense: To say that it is *R*-possible at *t* that *p* and to say that it is *R*-possible that *p* at *t* mean the same. The attribution of *R*-modal status to a proposition refers to the proposition individuated at the time of the attribution.

That it is *R*-possible at *t* that *p* shall mean, we said, that at some earlier time *t*′ it is (was) antecedently possible that *p* at *t*. This notion of antecedent possibility (necessity, contingency) is a *diachronic* modal notion. One could say, using pictorial language, that an attribution of antecedent possibility *projects* from a point *t*′ in time *a spectrum of possibilities* onto the time screen at *t*. This spectrum may shrink as the point of projection approaches the screen and broaden as we move the point further back. The projective rays are 'possible histories' or world-developments beginning in *t*′. We can illustrate this in a figure:

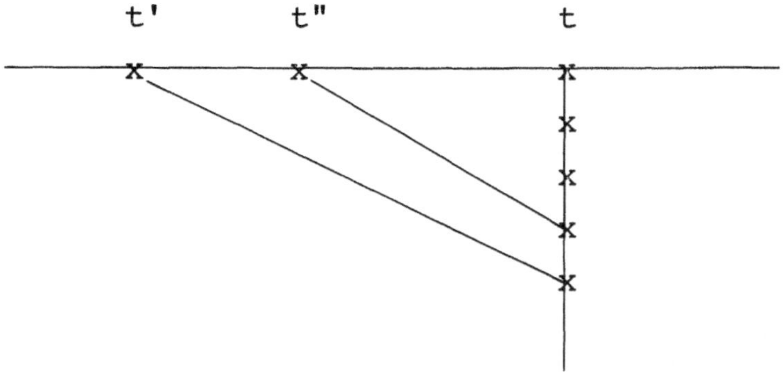

The crosses on the vertical axis indicate the 'points' at which the histories 'hit' the 'screen'. They correspond to a possible (total) state of the world, i.e. a conjunction of generic propositions and/or their negations instantiated at *t*. These conjunctions and their individual conjuncts are antecedently possible at the point of projection and are *R*-possible at *t*.

I have here used the diachronic modalities to elucidate the meaning of the synchronic modalities. This, in a sense, is arbitrary. Both types are equally 'basic' and irreducible to non-modal concepts. I shall therefore not introduce a new letter to mark the diachronic modalities, but distinguish the two aspects of 'real' modality as R_d- and R_s-modalities, respectively.

The diachronic modal status of the proposition that *p* at *t* may change with the choice of the point of projection *t'*. If that *p* at *t* is R_d-possible at *t'*, this is also R_d-possible at any time before, but not necessarily at any time after *t'*. By duality, if that *p* at *t* is R_d-necessary at *t'*, this is also R_d-necessary at any time after *t'* and before *t*, but not necessarily at any time before *t'*.

Something which takes place at *t* may have been R_d-contingent at *t'* but become R_d-necessary at later *t''*. One and the same true individual proposition can thus both be diachronically necessary and synchronically contingent. (These facts about modal status are important when one is discussing causality. They will not, however, be further discussed here.)

11

From what has already been said, it easily follows that the modal logic of *diachronic* *R*-modality is the system known as S4.

If it is true that *p* at *t*, then any time before *t* it was R_d-possible that *p* at *t*.

If, and only if, at some time before *t* it is R_d-possible that *p or q* at *t*, then at that time it is also R_d-possible that *p* at *t* or R_d-possible that *q* at *t*.

These are trivial observations. More crucial is the behaviour of iterated diachronic *R*-modalities. It follows from the prospective character of R_d-modality that an iterated modal operator must refer to an earlier time than the modal operator(s) within its scope. That it is R_d-possible that it is R_d-possible that *p* at *t* therefore means that at some time *t''* it is possible that at some time *t'* it is possible that it is true that *p* at *t*, where *t''* is earlier than *t'* which is earlier than *t*.

From the fact that the spectrum of R_d-possibility projected onto time *t* may become narrower, but cannot become broader, as the point in time from which the projection is made approaches *t*, it follows that if, at *t''*, it is R_d-possible that, at *t'*, it is R_d-possible that *p* at *t*, then it is also R_d-possible at t'' that *p* at *t*.

For the very same reason concerning the size of the spectra it is *not* (necessarily) the case that if, at *t''*, it is R_d-possible that, at *t'*, it is R_d-impossible that *p* at *t*, it is (already) at *t''* R_d-impossible that *p* at *t*. The spectrum of possibilities may have shrunk from *t''* to *t'* so as to make things which were still possible at *t''* impossible at *t'*.

12

Before we can decide which is the modal logic of *synchronic R*-modality some further elucidations of this notion are needed.

By a set of possible histories up to *t* we shall mean all the alternative ways in which the world from its actual state at some time *t'* in the past might have developed up to time *t*. To each member of such a set there answers a total state of the world which is R_s-possible at *t*. The total state we define as a conjunction of generic propositions which are true or false when temporally specified. (Possible further determinations of the notion of 'generic proposition' can for the present purposes be ignored.)

That the proposition that *p* at *t* is *true in a* given *possible history* means that it is a conjunct in the total state of the world in that history at *t*. That the proposition that *p* is R_s-possible at *t* in a given history again shall mean that the proposition that *p* is a conjunct in the total state of the world in *some* possible history at *t*. That the proposition in question is R_s-impossible at *t* in a given history means, accordingly, that the proposition that not-*p* is a conjunct in the total state of the world in *all* possible histories at *t*.

Trivially, if it is true that *p* at *t*, then it is R_s-possible at *t* that *p*; and if, and only if, it is R_s-possible at *t* that *p or q*, then it is, at *t*, R_s-possible that *p or* R_s-possible that *q*.

Now assume that it is, at *t*, R_s-possible that it is R_s-impossible that *p*. Then the proposition that it is R_s-impossible that *p* is a conjunct in the total state of the world in *some* possible history at *t*. From this it follows that the proposition that not-*p* is a conjunct in the total state of the world in *all* possible histories at *t* and from this again that the proposition that it is R_s-impossible that p is also a conjunct in the total state of the world in *all* possible histories, and hence in the actual history too, at t. Herewith it has been proved that if, at t, it is R_s-possible that it is R_s-impossible that p, then, at t, it is also R_s-impossible that p.

This completes the proof that the modal logic of synchronic real possibility (modality) is the system S5.

13

As indicated above (Section 8), the question whether Aristotle adhered to the Principle of Plenitude has been a matter of dispute. Arthur Lovejoy, who coined the name for the principle and subjected its place in the history of thought to a thorough scrutiny, answered the question in the negative.[7] Jaakko Hintikka has argued forcibly for an affirmative answer[8]—but his position on the question has also been contested.

I cannot claim an opinion of my own on the issue. Perhaps it cannot be definitely settled on the basis of existing evidence. I shall illustrate the complications of the problem by two passages from Aristotle's writings. The first seems to favour a negative answer to our question. But whether it really does this will be seen to be debatable. The second seems explicitly to affirm the debated principle. But this impression may nevertheless be elusive.

14

The first passage is in *De interpretatione* 19a 13–17. In the Oxford translation it reads: 'It is possible that this coat may be cut in half, and yet it may not be cut in half, but wear out first. In the same way, it is possible that it should not be cut in half; unless this were so, it would not be possible that it should wear out first. So it is therefore with all other events which possess this kind of potentiality.'

What is being maintained here can be accommodated to match our terminology, as follows: Assume that the coat will not be cut. Then, since it is a perishable thing, it will wear out at a certain time t. From now on, it is antecedently possible that the coat is cut at some time between now and t. But that it will be cut at t or after is impossible because then it has worn out. Aristotle admits the possibility that the coat will not be cut before *t*. If this possibility materializes then the possibility that the coat will be cut will never materialize.

According to Alexander,[9] Diodorus held that it is possible for me to be at Corinth, if I am at Corinth or I am going to be at Corinth; should I die without having visited the place, it *would not have been* possible for me to be there. Aristotle would presumably have agreed with Diodorus that, if I never went to Corinth, the proposition that I was there is, 'in retrospect', impossible. This is *F*-impossibility. But would Aristotle also have denied that the proposition that I was going to be there is impossible, or denied that it would have been possible for me to be there—as apparently Diodorus would have done? He would then also have been committed to saying about the coat which was never cut but wore out first that it would not have been possible for it to be cut. And this seems to be contrary to the passage in *De interpretatione*. But exactly how he would have reacted to the question we cannot tell, *one* reason for this being that he did not clearly distinguish the various temporal perspectives—prospective, antecedent, retrospective—in which modal status may be attributed to something.

Let it be agreed that the coat which was never cut might have been cut. This seems good common sense. But how do we *know* this? An answer could be: It was a *coat* and some coats are cut, others wear out first. Or: the coat was made of *cloth* and some pieces of cloth get cut, others never. Perhaps the Principle of Plenitude holds for the potentialities which we associate with *kinds* of thing or with the *materials* of which things are made, but not for the potentialities which we attribute to *individual* things of a given kind or a given material. We might then say that the attribution of potentiality to an individual is secondary to its primary attribution to a universal. And perhaps one could defend the view that Aristotle adhered to the *plenum* principle for primary attributions of potentiality and also defend the view that for such attributions the principle is true. But this argumentation would require a more precise determination than we can give here of the distinction between 'primary' and 'secondary' attributions of possibility and potentiality.

15

In *Metaphysica* 1047b4 we read: 'it cannot be true to say "this is capable of being but will never be"'. This looks like a clear affirmation of the Principle of Plenitude. But in the context immediately following and immediately preceding the meaning of the statement assumes a different appearance.

Aristotle invites us to imagine a person who says of something which is impossible and therefore neither is nor will be—his example is a common measure of the side and the diagonal of the square—that it *can* be even though it never *will*. This person, Aristotle notes, would be saying something false. The source of his mistake is that he overlooks the distinction between things which will never be because they are impossible and things which, although possible, will never be. From the admission that certain things may but never will be, one must not falsely conclude that anything which never will be yet may be. It seems, in other words, that Aristotle is anxious here to uphold a distinction between the impossible and the never existing—and not to identify the two. If this impression is correct, the quoted sentence cannot be given in support of the opinion that Aristotle had accepted the *plenum* principle. The context speaks for the contrary.

This impression of what Aristotle had in mind is confirmed by what the text says a little earlier (1047a24–26) in elucidation of the notion of capacity. 'A thing is capable of doing something', it says, 'if there will be nothing impossible in its having the actuality of that of which it is said to have the capacity. I mean, for instance, if a thing is capable of sitting and it is open to it to sit, there will be nothing impossible in its actually sitting'. This sounds very much like an affirmation of *R*-contingencies. Because does it not entitle us to say that a man who is not, at a certain time, sitting *could* nevertheless *have been* sitting then? To be capable of sitting does not require that one actually is sitting but only that no impossibility is involved in the assumption that one is. Admitting real contingencies is compatible with adhering to the *plenum* principle. However, the evidence provided by the quoted passage is not decisive on the question whether the possible can for ever remain unactualized. Aristotle might have thought that although the capacity of sitting exists independently of whether the one who has this capacity is sitting or not at a given time, its existence still requires that, if he is not sitting then, he will necessarily be sitting at some other time. But the passage which we discussed from *De interpretatione* about the capacity of a coat to be cut speaks against this interpretation of Aristotle's view.

Academy of Finland

NOTES

[1] Boethius, *Commentarii in Librum Aristotelis* Περι 'Ερμηνειας, C. Meiser (ed.): 1877, ed. secunda, Leipzig, p. 234.

[2] The literature dealing with this fascinating topic is vast and has grown rapidly in recent decades. My own modest contribution to it is a paper 'The "Master Argument" of Diodorus', in E. Saarinen *et al.* (eds.), *Essays in Honour of Jaakko Hintikka*, D. Reidel, Dordrecht and Boston, 1979, pp. 297–308.

[3] I do not know the history of this term. Hintikka uses it in the same, or a closely similar, sense to the one in which I use it here. The earliest occurrence of the term known to me is with Oskar Becker in his book *Untersuchungen über den Modalkalkül* (Meisenheim/Glan 1952). Becker, however, does not mean by it a reductionist notion of modality. By the statistical interpretation of modality he means the view that possibility means truth in *some possible* world, necessity truth in *all possible* worlds, and impossibility truth (falsehood) in *no* (all) *possible* world(s).

[4] Ibid.

[5] *Cf.* my paper 'Some Observations on Modal Logic and Philosophical Systems', in R. E. Olson and A. M. Paul (eds.), *Contemporary Philosophy in Scandinavia*, Johns Hopkins University Press, Baltimore, 1972.

[6] By Arthur O. Lovejoy in his classic study of the role of this idea in the history of thought, *The Great Chain of Being*, Harvard University Press, Cambridge, Mass., 1936. See pp. 52, 337f. (Harper Torchbook edition).

[7] *Op. cit.*, pp. 55f. and 338.

[8] See Jaakko Hintikka, *Time and Necessity: Studies in Aristotle's Theory of Modality*, Oxford University Press, Oxford, 1973; and Jaakko Hintikka in collaboration with Unto Remes and Simo Knuuttila, *Aristotle on Modality and Determinism* (*Acta Philosophica Fennica* 29, no. 1), North-Holland, Amsterdam, 1977.

[9] Alexander Aphrodisias, *Commentarium in Aristotelis Analyt. Prior. Librum I*, in M. Wallies (ed.), *Commentaria in Aristotelem Graeca* II 1, G. Reimer, Berlin, 1883.

PETER GEACH

LOGIC

In two of my articles ('On Teaching Logic' and 'Why Logic Matters')[1] I have expressed my general views about the necessity of more widespread logical training and the methodology of logic teaching. Many educated people in England are logically analphabetic, and many even rationalize their contentment with such ignorance by the pretext that one can very well conduct one's affairs sensibly and competently without any logical knowledge.

Nobody, however, can be complacent about the risk of inconsistency. For what is wrong with inconsistency is not, as Strawson supposes, that contradicting oneself is like doing an act and then at once undoing it; nor, as he also says, is a criticism of somebody's discourse as inconsistent a criticism of a fault internal to the discourse, not concerned with its relation to the world. Discourse cannot have *only* this logical fault without being also faulty in its relation to the world: that is the trouble about inconsistency. A military plan, for example, is *bound*, if inconsistent, to lead to botched execution of orders: on one account, Marshal Grouchy received inconsistent orders at a crucial moment before Waterloo. An inconsistent theory about the world is bound to be factually false at some point. An inconsistent moral code is bound to contain some morally objectionable prescription. And so on. Only fiction is an exception; inconsistency is all right in fiction if the author gets away with it—if it is not so blatant as to disturb the reader; but this is because, as Frege puts it, in fiction the author is only presenting thoughts and is not concerned with their truth-value. (In fact careful investigation detects inconsistencies in great novels like Tolstoy's or George Eliot's; but what of it?) The exception arises just because here the relation of thought to the world, as describing what there is or presenting what shall be, is no longer in question; so the exception proves the rule about why inconsistency matters.

Now inconsistency is usually not a matter of head-on collision between two sentences: an inconsistent set of three or more members may have no inconsistent subset. (Astonishingly, a much used textbook says, to the contrary, that usually inconsistency of a plural set of

Harry A. Lewis (ed.), Peter Geach: Philosophical Encounters, 99–104.

short sentences *is* a matter of an inconsistent pair. The author has some standing as a mathematician; what then, as Wittgenstein said in a similar case, is the good of mathematics?) The only way to detect such inconsistency is to derive from the set a flatly inconsistent pair, or to show that the set 'is on all fours', shares a logical form, with a set thus provably inconsistent. We thus come to the heart of logic: the question of what follows from what.

Here again, a logically uncultivated person will think be can just *see* what follows from what. But men's logical vision is likely to be myopic if unaided by logical spectacles. Not only do people often fail to see, given premises A and B, what conclusion may be drawn; if C is offered as a conclusion from A and B, and the question is whether C follows, people's answers may be right only by chance. Let A be the premise 'All Spartans are equally brave', and B the premise 'Not all hoplites are equally brave'. Consider now the two conclusions C1 'Not all Spartans are hoplites' and C2 'Not all hoplites are Spartans'; which of these follows from the premises? Obviously four answers might be given: that C1 follows but not C2, that C2 follows but not C1, that both follow, that neither follows. When I have been trying to interest a general audience, or a group of students beginning logic, in the art I practice, I have found that roughly equal numbers of those present voted for each of the four answers. (When I set the question in an examination, I found that if, in spite of my warnings, students had recourse to some 'traditional' textbook and talked about undistributed middle and all that, they were actually *more* likely to give one of the three wrong answers; to my mind that is not surprising.)

So much for logic in practice, *logica utens* as the medievals called it. Logic is unproductive like book-keeping, but without sound accountancy a productive business may smash. Or, as the Stoics said, logic is like the shell of an egg; you cannot eat a shell, but an egg would be unmanageable without it. Some parts of philosophy are closer to human concerns than logic is; but in philosophy, unlike ordinary life, one may fail to see that reasonings must have gone astray because they led to patently absurd conclusions—nor is it always easy to see what is patently absurd. Philosophy (theology as well) needs the best logical tools available.

The forging and sharpening of logical tools is the business of theoretical logic, *logica docens*; without a good tradition here, *logica utens* is bound to suffer as well. I have not myself been a very creative theoretical logician; I have just a few results to my credit in modal logic and set theory, bricks for better men's buildings. But I believe I have helped to train a few men who will be better formal logicians than I.

I have followed in the steps of Neville Keynes and Sir William Hamilton in insisting that the theory of quantifiers shall not neglect two expressions which belong to the same syntactical category: 'most' and 'only'. (The blank in '—*S*s are *P*' may be filled up *salva congruitate* with 'all', 'some', 'most', or 'only'.) Of 'most' I have written elsewhere in this volume. Aquinas already assigned 'only' to the same category as 'all' and 'no(ne)', viz. of showing how a predicate latches onto a subject, *ordo praedicati ad subjectum*. (A slick review of *Reference and Generality* said that I got a great deal out of four Latin words! Of course what Aquinas says runs to more than four words, but the reviewer didn't bother to check my references.) I have rather disliked the existing jargon about quantifiers. The very word (due to Hamilton) expresses the clearly wrong view that quantifiers answer the question *how much* of the class of *S*s—all, some, most, or none of it—a predicate *P* is being affirmed of; obviously this will not do for 'only'. But here usage is too strong to be displaced. On the other hand I have some hope that people will pick up my suggestion of using the adjectives 'major' and 'minor' instead of the standard pair 'universal' and 'existential' ('particular'); I am imitating Polish logicians' use of 'duży' and 'mały', literally 'great' and 'little'; such a pair are both handily brief words and manifestly noncommittal philosophically.

I have defended the old application of the term 'name' to *some* uses of general terms, while admitting that other uses are irreducibly predicative. It is amazing what success Russell and Frege have had in promoting the contrary position that only singular terms for objects are names: this is clear from the fact that Strawson, no friend of Russell's philosophically, considers the treatment of general terms, in some contexts, as names to be a 'proposal' of—Peter Geach. Historically, the restriction of the term 'name' to singular terms was a proposal on the part of Russell and Frege: a proposal to my mind ill defended. (The young Frege in *Begriffsschrift* uses general terms, at least mass terms, as possible interpretations of *name*-letters; presumably, then, he held *at that time* the traditional view that 'gold' or 'water' is a shared name of different samples of a given material; it would be absurd to read into him Quine's view that 'gold' is a proper name of one single scattered object.) I have come in for many attacks over the matter; I have also been attacked for an unargued *acceptance* of the Russell-Frege 'dogma'; these latter attacks, persisted in despite my denials, I can only call impudently mendacious.

For many years I have waged war against the continued teaching of traditional logic, that confused mass of fragments of medieval logic which for about 300 years was the only formal logic to be found in textbooks. I kept it out of my own teaching, and would warn students against reliance on contemporary textbooks in the same line. My attacks may seem to fit in ill with my continuing admiration for Neville Keynes. To be brief, I may say that to my mind Keynes, like his contemporaries, was to some extent in a bog, but he was struggling to get out of it; the difficulties he honestly raises against the doctrine of distribution were the source of my own rejection of it; my campaign has met with some success.

Readers may be amused with a very brief refutation of the doctrine of distribution. Contrary to what is often said, there is a consistent reading of categoricals which validates all the laws of the 'square of opposition' and all the traditional syllogisms; as Moody and others have pointed out, medieval logicians already had this system. Now in this system '*SaS*' is not always true, being false if '*S*' is an empty general term; but if '*SiP*' is true, '*S*' cannot be an empty term; so the form '*SiP*, ergo *SaS*' is valid. This decisively shows that there is nothing logically sound in the illicit process rule, the rule against inferring a conclusion about every *S* when '*S*' has no distributed occurrence in any premise. It should here not be forgotten that this rule was applied in traditional logic to decide on the validity of single-premise arguments, e.g. to show that the conversions '*SaP*, ergo *PaS*' and '*SoP*, ergo *PoS*' are formally invalid patterns. Of course they can be shown invalid, but *non tali auxilio*.

Another campaign of mine has occupied many pages of published articles: since I began the work which crystallised out as *Reference and Generality*, I have been attacking the muddled doctrine of referring expressions. The expressions to which a common role of 'referring' is ascribed are a miscellany, clearly different among themselves in their syntax and semantics; if the first step in analysis is supposed to be: spotting referentially used expressions and assigning to each its reference; then subsequent steps will only get you into deeper and deeper mud. This campaign has been less successful, but I do not despair.

I end with a few words about the relation of my work on the philosophy of logic to ordinary language philosophy and to serious linguistics.

A native speaker is on the face of it authoritative about usage in his own vernacular; but that does not make him an authority about whether an argument stated in his native language is valid or invalid. Any vernacular has logically uninteresting quirks, idiotisms of idiom as Arthur Prior would say, and logicians do well not to respect these quirks in their theorizing: I may instance the sequence-of-tenses idiom in English ('James Smith thought π was (!) equal to 25:8') and the defectiveness of English modal verbs (the need to say 'ought to have shaved yesterday' instead of 'had ought to shave yesterday', or 'can come tomorrow' instead of 'will can come tomorrow'). Recognition of such quirks is much helped by mastery of some other language which does not share a given idiomatic quirk. (Some quirks, however, are reasonably widespread. A native speaker of Hebrew told me that the *Euthydemus* puzzle about 'your dog' and 'your father' can be reproduced in Hebrew, even though the possessive construction in Hebrew differs from the Indoeuropean pattern.) Arguments based on such quirks are just to be ignored. Of course one *may* thus be blind to a logically non-trivial point; but in general one is justified in refusing to heed the cavils of of ordinary-language philosophers. Like the traditional way of arguing the hind legs off a donkey (because it has *only four* legs), their arguments will not cross linguistic frontiers.

It has gratified me over the years that professional linguists have taken seriously my examples in *Reference and Generality* and elsewhere, and have even adopted from me some terms of art (e.g. 'pronoun of laziness'). Perhaps ungrateful, I must confess to scepticism about what some linguists assert. Learning a foreign language quite correctly is difficult; a few months' field study surely cannot be enough for the confident assertions sometimes based on it. I remember a Dutch linguist assuring me that I had been wrong to distinguish 'He must be ill' and 'He ought to be ill'; I have had a U.S. citizen correct me on a point of English people's etiquette; claims to have mastered idioms and polite usages in a very alien language learned from scratch are therefore inevitably a bit suspect in my eyes. Some claims strike me as mere traveller's tales; I do not believe in tribes who never use proper names, or who have *no* ways of marking the subject and the object of a transitive verb. And it is merely laughable when we are told, in a would-be impressive style, of a feature in some remote language which a moment's thought shows to be shared with some more familiar language—'In Eskimo languages the word for "snow" takes tense inflections'! I am of course sure that philosophers, logicians in particular, have much to learn from a serious study of language; but takeover bids by linguists are to be resisted. I remember one statistical study sent to me by a man who claimed to have thus made out the genuine logic of quantifiers, in contrast to the errors of logicians; naturally it went straight into the wastepaper basket.

A native speaker's confidence that he knows his way around *logically* in his own language is likely to be ill-founded. Of course we often do well to accept as plainly valid some argument for which as yet we lack a technique of formalization. But a logician is not like a gardener trimming a hedge in a formal garden where every path is familar; rather, he is like an explorer clearing paths in a jungle which grows apace.

> Still her old empire to regain she tries,
> For, born a goddess, DULNESS never dies.

The University of Leeds

NOTE

[1] Geach, P. T., 'On Teaching Logic', *Philosophy* **54**, 1979, pp. 5–17; 'Why Logic Matters', in H. D. Lewis (ed.), *Contemporary British Philosophy (Fourth Series)*, George Allen and Unwin, London, 1976, pp. 86–99.

J. E. J. ALTHAM

PLURAL AND PLEONETETIC QUANTIFICATION

INTRODUCTION

This essay has three parts. In the first, I make some general remarks about relational quantifiers, and, using 'there are more φs than χs' as an example, I provide an application to syntax. A relational quantifier is one that binds one variable in each of an ordered n-tuple of open sentences, where $n \geq 2$, but all my examples will concern cases where $n = 2$. The *plural* quantifiers 'nearly every φ is χ' and 'many φs are χ' are also best treated as relational. What is said in this part will be familiar to some readers, but knowledge of it is perhaps sufficiently patchy to justify including it. In the second part I make some remarks about domains of discourse appropriate for systems formalising reasoning with relational quantifiers. The matter is less straightforward than in the case of the standard, one-place quantifiers. This discussion involves mention of principles of inference that impose requirements of proportionality on plural quantifiers. The third and last part is devoted to the quantifier 'most φs are χ', which is relational and evidently proportional. It is, to use Geach's word, a *pleonetetic* quantifier. The close relationship between numerical principles and principles governing plural and pleonetetic quantifiers — a relationship already illustrated in the first two parts — raises the question how far the basic principles of plural and pleonetetic logic should ideally be from obviously numerical principles. I give some principles for monadic pleonotetic logic, which fall into two groups. In one group all the principles *look* straightforwardly logical — at least to my eye. The principles in the other group are *formulated* using only logical notions, but they are most naturally *grasped* via their relationship with numerical principles. I conclude, after providing examples of how to prove things with these principles, by leaving open the question whether a line can satisfactorily be drawn between logical and arithmetical reasoning.

Finally in this introduction I must acknowledge large debts to Professor Geach. The debts are general and specific. Some specific ones are acknowledged in what follows. The general one is for the

Harry A. Lewis (ed.), Peter Geach: Philosophical Encounters, 105–119.

stimulus of his published writings and many letters, for his encouragement and his example.

I

The *standard* quantifiers are the existential and universal ones, and are extremely familiar. A standard quantifier binds one variable in one open sentence. In Frege's hierarchy of levels, it is a second-level concept-expression, that is to say, an expression with one gap in it and such that when that gap is filled with a first-level concept-expression, the result is a sentence. These descriptions, and especially the Fregean one, suggest generalisations of the notion of a quantifier. Only one kind of generalisation, however, is relevant here. That there is a category of second-level expressions each of which has *one* gap suggests the idea of categories of second-level expressions each of which has n gaps. There would be such a category for each n. Expressions of this kind are naturally called n-place relational quantifiers; each of the n gaps requires to be filled by a first-level concept-expression. An n-place relational quantifier binds one variable in each of an ordered n-tuple of open sentences. For such a quantifier the notation '$(Qx) \langle \varphi_1(x); \varphi_2(x); \ldots; \varphi_n(x) \rangle$' seems appropriate, adapting the notation for ordered n-tuples; the semi-colons are merely convenient punctuation-marks distinguishing one gap from another.

It is a defect of most treatments of elementary logic that only the one-place quantifiers find a place in them. Logic is studied for theoretical understanding, and also as an instrument to discipline the deductive intellect. Both purposes would be better served if a wider variety of quantifiers received more general attention, and it is especially surprising that the latter purpose, of intellectual discipline, has not been better served. Relational quantifiers exist in natural language, people make inferences from propositions containing them, and the principles of *valid* inference where these quantifiers are involved are in many cases by no means obvious. An instrument to help us make the most of our premisses in this area, and to preserve us from fallacies, would therefore seem desirable.

Frege famously analysed statements of numerical identity in terms of the relation *being like-numbered* with. The number of φs is the same as the number of χs just in case the concept φ is like-numbered with the concept χ. This relation is evidently a second-level one, since it relates first-level concepts. There is a quantifier that expresses this relation; to say that the concept φ is like-numbered with the concept χ we can simply say 'there are exactly as many φs as χs'. This two-place relational quantifier could be introduced into first-order logic. A possible notation would be '$(=x)\ \langle\varphi(x);\chi(x)\rangle$'. Other two-place relational quantifiers are 'there are more φs than χs', which can be written '$(>x)\langle\varphi(x);\chi(x)\rangle$', and 'there are at least as many φs as χs', or '$(\geq x)\langle\varphi(x);\chi(x)\rangle$'.

Somebody might say that the introduction of such quantifiers into first-order logic was inappropriate or even improper, on the ground that the notions they represent are numerical. That objection fails. Given that the language has an identity-sign, the explicitly numerical quantifiers 'there are *n* φs' are definable in first-order logic. Besides, Frege successfully argued that his analysis of numerical identity is not circular, since the account of *being like-numbered with* does not depend on the concept of number. One might similarly show that in order to determine that there are more φs than χs, it is not necessary to count either. It might then be rejoined that even if 'there are exactly as many φs as χs' is not essentially numerical, it should rather belong to second-order predicate logic, on the ground that the analysis of that relation is in terms of the existence of a one-one relation, and so quantifies over relations. This rejoinder is, however, not decisive. The semantics of *standard* first-order logic makes free use of set-theoretical notions: that the semantics of certain relational quantifiers would quantify over relations should not disqualify these quantifiers, which syntactically belong with first-order logic.

The workings of relational quantifiers are more clearly seen if we give a few examples of how to represent sentences of natural language, using the notation already sketched. In particular, there seems some syntactical insight to be gained by considering sentences containing both standard one-place quantifiers and relational ones. To start with a simple case, consider

(1) More people are hungry than have full bellies.

Let the universe of discourse be restricted to people, and form the two predicables '—is hungry' and '—has a full belly'. The quantifier relates the concepts these express, in that order, and (1) is written

(1)′ $(>x)\ \langle H(x);\ Fb(x)\rangle$.

(Here and in subsequent examples I use obvious abbreviations of the predicables). Consider next

(2) More people have written a poem than have composed a symphony.

Isolate first '—has written a poem' and '—has composed a symphony'. The former is then written as '$(\exists y)$ (y is a poem and—has written y)', the latter as '$(\exists z)$ (z is a symphony and—has composed z)'. (2) itself then comes out as

(2)′ $(>x)\langle(\exists y)\,(P(y)\ \&\ W(x,y));\ (\exists z)\,(S(z)\ \&\ C(x,z))\rangle$.

The relational quantifier can thus clearly have a one-place one within its scope; it can also have relational quantifiers within its scope, as in the example

(3) There are more drivers who have caused more accidents than they have suffered than drivers who have suffered more accidents than they have caused.

Confining the universe to drivers (3) goes over into symbols as

(3)′ $(>x)\,(>y)\langle A(y)\ \&\ C(x,y);\ A(y)\ \&\ S(x,y)\rangle;$
$(>z)\langle A(z)\ \&\ S(x,z);\ A(z)\ \&\ C(x,z)\rangle$.

A relational quantifier can have another within its scope since the application of a relational quantifier to a pair of formulas results in a *single* formula, which can then be one of a further pair for a second application of a relational quantifier. Thus also a standard quantifier can have a relational quantifier within its scope, as in

(4) There is a book that more people have bought than have read;

which, in symbols, comes out as

(4)′ $(\exists x)\,(Bk(x)\ \&\ (>y)\langle P(y)\ \&\ Bt(y,x);\ P(y)\ \&\ R(y,x)\rangle)$.

Here the scope of '$(>y)$' constitutes a single formula with a free variable.

Note on the other hand that the following formula is ill-formed:

(5) $(>y)\,(\exists x)\,\langle Fyx;Gyx\rangle$.

The reason why (5) is ill-formed is as follows: in constructing (5), we start with the open sentences 'Fyx' and 'Gyx'. The next stage would have to be the application of the existential quantifier, but this is impossible unless 'Fyx' and 'Gyx' are first combined into a *single* formula (since '$(\exists x)$' is a *one*-place quantifier). This combination, however, is only effected by the application, in (5), of '$(>y)$', an application which is supposed to come *later* than that of '$(\exists x)$'. (5) is therefore not coherent. These reflections can be applied to natural language. First compare

(6) John has bought a book and Mary has read one.

and

(7) John has bought a book and Mary has read it.

The difference made by the terminal 'one', contrasted with 'it', is this. In (6), we have two existential quantifiers, each with narrow scope; (6) comes out as

(6)′ $(\exists b)\ (\mathrm{Bt}(j,b))\ \&\ (\exists b)\ (\mathrm{R}(m,b))$.

In (7), we have one quantifier, whose scope is the whole, i.e.

(7)′ $(\exists b)\ (\mathrm{Bt}(j,b)\ \&\ \mathrm{R}(m,b))$.

Now consider

(8) More people have bought a book than have read one,

and

(9) More people have bought a book than have read it.

The contrast between (8) and (9) is not the same as that between (6) and (7). (8) goes into

(8)′ $(>y)\langle(\exists x)\ (\mathrm{Bk}(x)\ \&\ \mathrm{Bt}(y,x));\ (\exists z)\ (\mathrm{Bk}(z)\ \&\ \mathrm{R}(y,z))\rangle$.

(9), however, is much less straightforward. First of all, it sounds odd. Secondly, (7) can be rewritten as the equivalent

(7)″ John has bought some book or other and Mary has read that book.

Rewriting (9) similarly yields the still odder

(9)′ More people have bought some book or other than have read that book.

Since different people may have bought different books, there need be no such thing as that book, as (9)′ purports to claim. The fact is that (9) is odd because its structure suggests a reading in which there is one existential quantifier whose scope would *not* include 'more than', but *would* include each of the pair '—has bought...' and '—has read...'. But this would make it like (5) in the respect in which that is not coherent. For another, more obviously incoherent example, take

(10) More people have seen a portrait of the Queen than have painted it.

It now seems to me that the *plural* quantifiers, 'many', 'few', and 'nearly every', are best treated as relational. It is clear, to start with, that *how* many φs have to be χ for it to be true that many φs are χ varies with φ, as was recognised in the last chapter of [1]. But the treatment there was unsatisfactory, since the variation was accommodated by varying the sense of the plural quantifier itself, whereas what we want is a quantifier with a single sense that will itself induce the variation with different fillers for φ. A suggestion that has this effect is that in saying that many φs are χ we are in effect comparing the φs that are χ with the φs that are not, and saying that the former do not fall far short of the latter. Similarly, nearly every φ is χ if and only if there are far more φs that are χ than φs that are not χ. These suggestions, which were made in [2], clearly make the plural quantifiers relational.

II

In what follows, the universe of discourse is treated as consisting of the *A*s. That is, '*A*' is a predicable such that anything is in the universe of discourse if and only if it is an *A*. I am going to raise questions about the size of the universe. The usual assumption in elementary logic is that the universe of discourse is merely non-empty, that there is at least one *A*. There is no space to argue the point here, but I proceed on the hypothesis that there is nothing sacrosanct about the assumption of non-emptiness. Depending on our purposes, we could use a logic valid even in an empty domain, or we could make some other requirement.

Consider first

(11) If every *A* is φ, then many *A*s are φ.

If the universe of As is merely non-empty, then I take it that there will be cases of (11) that are not true, since I want to stick to the claim that if there is only one A, it is false that many As are anything whatever. Note next that if the universe is allowed to be empty, formulas of the form '$(\exists x)$ $(\varphi(x) \vee -\varphi(x))$' are not universally valid, and that one main reason in standard logic for insisting on non-emptiness is to obtain the additional richness of principles that comes if there are theorems of the form '$(\exists x)\varphi$'. Similarly, if the universe is merely non-empty, there are no theorems such as 'Many $(\varphi \vee -\varphi)$s are $(\chi \vee -\chi)$'. To avoid such impoverishment, the assumption could be made that the logic is to be applied only in universes that have many members. (If *this* notion of 'many' is itself to be relational, it can be considered as relating the universe of As to some wider background universe of discourse.) But now consider

(12) If nearly every A is φ, then many As are φ.

If the universe is required to be no larger than what minimally suffices for there to be many As, (12) will not be universally valid. To validate (12) it must be assumed that the logic is to be applied only in universes with at least $2n-1$ members, where n is the least number of As that counts as many. I pointed this out on p. 66 of [1]. In that publication, unfortunately, I did not realise that this process of further determining the universe of discourse can be continued. For consider the principle

(13) If nearly every A is φ, and nearly every A is χ, then many As are both φ and χ.

This is an appealing principle. For example, if in some College nearly every Fellow is lazy, and also nearly every Fellow is malicious, then many Fellows are both lazy and malicious. (We cannot of course strengthen the consequent to 'nearly every Fellow is both lazy and malicious'.) Admitting (13) has, however, curious consequences. I need first to point out that I am here retaining the interdefinability of 'nearly every' and 'many', argued for in [1]. Thus 'nearly every A is φ' may be defined as 'not many As are not φ'. Now the combination of this principle with the semantic rule that many As are φ if and only if at least n As are φ (for suitable n), together with the assumption that the universe contains at least $2n-1$ members, does not suffice to yield the validity of (13). If, for example, there are exactly $2n-1$ Fellows, it could be that nearly every one of them was lazy, nearly every one of them malicious, and yet only one of them both lazy and malicious. E.g. take $n=3$, and let the Fellows who are not lazy be none of them the same as those who are not malicious, and let there be two of each these classes.

Now nearly every A is φ if not many As are not φ, iff at most $n-1$ As are not-φ. So if the antecedent of (13) is true, at most $n-1$ As are not-φ and at most $n-1$ are not-χ. We know that the universe of As contains at least $2n-1$ members, so there could be no overlap between not-φ and not-χ. So, to make sure that where the antecedent is true the consequent is also, we must make sure that when $n-1$ that are not-φ, and a different $n-1$ that are not-χ, have been subtracted, there are still at least n As left. But that amounts to the requirement that the universe contain at least $3n-2$ As. This result may seem bizarre. Even allowing that the size of universe it is appropriate to assume may vary with the logic, it seems strange that we should arrive at so apparently arbitrary a formula as $3n-2$ as the minimum size of universe. Yet the result has been reached by a sequence of natural stages.

The result has a further consequence, pointing out which may reduce somewhat the appearance of strangeness. Taken together with our other assumptions (13) imposes a proportional requirement. Its effect is that a necessary condition of the truth of 'nearly every A is φ' is that over two-thirds of the As be φ. Correspondingly, it is a sufficient condition for 'many As are φ' that at least a third of the As are φ. For suppose we could have that nearly every A is φ and not more than two-thirds of the As are φ. Then at least a third of the As are not φ. Similarly for χ, yielding that at least a third of the As are not χ. In an interpretation in which nothing that was not φ was also not χ, at least two-thirds of the As would be either not φ or not χ. By supposition that nearly every A is φ, at most $n-1$ As are not φ. Combining that with the parallel supposition for χ, we obtain that at most $2n-2$ As are either not φ or not χ. But this, we said, is at least two-thirds of the As. So there must be not more than $3n-3$. But by hypothesis about the size of the universe, there are at least $3n-2$ As. So unless more than two-thirds of the As are φ wherever 'nearly every A is φ' is true, (13) will not be valid.

(13) is merely the first member of the series of principles

(14) If nearly every A is φ_1, nearly every A is φ_2, ... nearly every A is φ_m, then many As are φ_1 & φ_2 & ... φ_m.

The higher the value of m for which a principle of this type is admitted, the higher the proportion of As have to be φ for it to be true that nearly every A is φ, and the smaller the proportion that suffices for many As to be φ. In fact, in admitting a principle in the series (14) for a particular m, we are imposing the requirement that if nearly every A is φ, then more than $m/m+1$ of the As are φ. Further, for a given minimum number n as the least that counts as many As, an argument similar to the one given for the case $m=2$ yields the result that the universe of discourse should contain at least $m(n-1)+n$ members.

Clearly there is no definite answer to the question which of the principles in the series (14) are logically valid. The sense of 'nearly every' and 'many' in natural language is not sufficiently sharp. On the other hand, a formal logic for plural quantifiers does have to be sharp, in the sense that for any m, a formulation of that logic must either definitely include or definitely exclude the mth principle in the series (14). So there cannot be an exact match between the quantifiers of natural language and their formal counterparts. Concepts may be vague without that precluding determinate relations holding among them, and indeed the vagueness of plural quantifiers does not preclude *some* determinate principles from holding. But these quantifiers do have a dimension of vagueness that prevents their logic from being everywhere determinate.

III

Principles of the form (14) have a very logical *look* to them, and yet they impose proportional requirements, whereas one might naturally think that issues of proportion are arithmetical. Other considerations also in the first two parts of this essay suggest that the line between what is properly logical and what is not is none too clear. For further illustration, I now go on to consider some principles for the logic of the quantifier 'most', or *pleonetetic* logic. I shall be considering only the monadic case. 'Most' is directly a proportional quantifier in the finite case, and in the general case most φs are χ if the cardinal number of the φs that are χ is greater than the cardinal of the φs that are not χ, or, in brief, if there are more $\varphi\chi$s than φ not χs. Using '*aut*' for exclusive disjunction, one might think that conversely 'more than' could be defined in terms of 'most': one might think that there are more φs than χs if most φ *aut* χs are φs. However, as Geach pointed out to me, this latter equivalence does not always hold in an infinite domain. He mentioned also another consideration that obstructs the definition of 'more than' in terms of 'most'. If we say that there are more φs than χs, we do not presuppose that there is any containing universe embracing both φs and χs, whereas we do make this presupposition in the case of 'most'. Thus it is perfectly in order to say

(15) There are more coins in John's pocket than ideas in his head, but many would declare the following not merely false, but nonsensical:

(16) Most of the coins in John's pocket are ideas in his head.

For these reasons 'most' will be treated here as a distinct quantifier, although the truth-condition of 'most φs are χ' will remain that there are more $\varphi\chi$s than φ not χs. The truth-condition marks out 'most' as being a relational quantifier. I shall not need a full-blown symbolism in what follows, but shall merely abbreviate e.g. 'most smokers cough' as 'Most Ss are C'. (For the dual quantifier whose sense is that of 'not most φs are not χ', I much like Geach's way of putting it, namely 'every other φ is χ').

In Ch. 13 of his [3] Geach shows how to test arguments involving 'most' for validity. Very briefly, he assigned letters to the cells of a Carroll diagram, and then used the algebra of inequalities. This method is effective, but although the interpretation of the algebra is no different from that of the sentences in the argument, so that semantically the method is enlightening, it is not itself a logical method. The *explicit* concepts in it are not easily thought of as logical. So one might look for a method that *is* logical, or more nearly logical. However, it is not entirely clear what is involved here. For we can *imitate* the algebraic method using only logical terms (assuming of course that 'most' *is* a logical term). I now explain this imitation.

First, we put premisses into a normal form corresponding to the break-up of the universe of discourse into cells of a diagram. Thus 'most As are B' is turned into 'most AB v AB' are AB v $A'B$', if only these two terms are involved. (Henceforth I use a prime to negate a term). If a third term 'C' is involved, we write instead 'most ABC v ABC' v $AB'C$ v $AB'C'$ are ABC v ABC' v $A'BC$ v $A'BC'$.' Similarly for four or more terms. Next, here are some principles allowing the addition or deletion of disjuncts in the subject or predicate-term of 'most'-sentences. Notice that the cells corresponding to the disjuncts in sentences in normal form are disjoint.

(P1) Most φs are χ; every χ is ψ; *ergo*, most φs are ψ.

This permits the move from 'most φs are χ' to 'most φs are $\chi \vee \psi$', i.e. adding a disjunct in the predicate.

(P2) Most φs are χ; every ψ is χ; *ergo*, most $(\varphi \vee \psi)$s are χ.
(P3) Most φs are χ; every ψ is φ; *ergo*, most $(\varphi \vee \psi)$s are χ.
(P4) Most φs are χ; *ergo*, most $(\varphi \vee \psi)$s are $(\chi \vee \psi)$.
(P5) Most φs are $\chi \vee \psi$; no φ is ψ; *ergo*, most φs are χ.
(P6) Most $(\varphi \vee \chi)$s are ψ; no χ is ψ; *ergo*, most φs are ψ.

All these principles have a solidly logical feel about them. The same cannot be said with equal confidence about the following:

(P7) Most $(\varphi \vee \chi)$s are φ; most $(\chi \vee \psi)$s are χ; the φs χs and ψs are pairwise disjoint; *ergo*, most $(\varphi \vee \psi)$s are φ.

(P7) is a *transitivity* principle. Thus if the φs, χs and ψs are pairwise disjoint, the first two premisses are true if there are more φs than χs and more χs than ψs, and the conclusion if there are more φs than ψs. In this way a principle of the algebra of inequalities is imitated in purely logical terminology. Now consider

(P8) Most $(\varphi \vee \chi)$s are φ; most $(\chi \vee \psi \vee \zeta)$s are $(\chi \vee \psi)$; the φs, χs, ψs and ζs are pairwise disjoint; *ergo*, most $(\varphi \vee \psi \vee \zeta)$s are $\varphi \vee \psi$.

Given the third premiss, the first two say in effect that there are more φs than χs, and more $(\chi \vee \psi)$s than ζs; the conclusion that there are more $(\varphi \vee \psi)$s than ζs. Thus (P8) is a kind of *substitution* principle, corresponding to the principle that if $a > b$, and $b+c > d$, then $a+c > d$.

Finally, we have

(P9) Most $(\varphi \vee \chi)$s are φ; no ψ is χ; *ergo*, not (most $(\varphi \vee \chi)$s are χ).

This corresponds to the principle that if $a > b$, then not $b > a$.

Here now are some sentences exemplifying the last three of these principles. In Cambridge, nobody is a Fellow of more than one College (Honorary Fellows don't count), so the Colleges provide us with pairwise disjoint sets of Fellows. Then, as an example of (P7), we have

(17) Most of those who are either Fellows of Trinity or of St John's are Fellows of Trinity; most of those who are either Fellows of St John's or of Caius are Fellows of St John's; nobody is a Fellow of any two of these Colleges; *ergo*, most of those who are either Fellows of Trinity or of Caius are Fellows of Trinity.

Here is an example of (P8):

(18) Most of those who are either Liberals or National Frontists are Liberals; most of those who are either National Frontists or Tories or Labourites are National Frontists or Tories; nobody belongs to two of these parties; *ergo*, most of those who are either Liberals or Tories or Labourites are Liberals or Tories.

For (P9):

(19) Most of those who either drink or abstain drink; no drinker abstains; *ergo*, it is not the case that most who either drink or abstain abstain.

It is now time to see some of the principles in action. Geach has provided me with a considerable number of unobvious pleonetetic principles of inference. Here is an example of one of them.

(20) Most smokers are male sedentary workers; most sedentary workers are women who don't smoke; *ergo*, most women don't smoke.

Abbreviate the premisses and conclusion to yield

(20)′ Most As are BC'; most Bs are $A'C$; *ergo*, most Cs are A'.

For ease of reading, I give each cell a name as follows.

a ABC	e $A'BC$
b ABC'	f $A'BC'$
c $AB'C$	g $A'B'C$
d $AB'C'$	h $A'B'C'$

The premisses are then, in formal form

1. Most $(a \vee b \vee c \vee d)$s are $b \vee f$.
2. Most $(a \vee b \vee e \vee f)$s are $e \vee g$.

(P5) allows deletion in the predicate, so from 2 we get

3. Most (a v b v e v f)s are e.

(P6) allows deletion in the subject-term, to yield from 3

4. Most (b v e)s are e.

Similar operations on 1 yield

5. Most (a v b v c)s are b.

Applying (P7) to 4 and 5 we get

6. Most (a v c v e)s are e.

Thence, by (P4)

7. Most (a v c v e v g)s are e v g.

The table may now be used to translate 7 back into

8. Most Cs are $A'C$, i.e. most Cs are A'.

Here is another, more complicated example. It is essentially the problem of the Rev. Jeremial Prodnose, of pp. 63–4 of [3], but slightly simplified.

(21) Most smokers drink; most drinkers use bad language; most users of bad language smoke. Most who smoke, and drink if they use bad language, have neither of the latter bad habits. Most who drink, and smoke if they use bad language, have neither of the latter bad habits. And most who use bad language, and smoke if they drink, have neither of the latter bad habits. We show that these six propositions are an inconsistent hexad. The premisses are written as follows:

(21)′ Most As are B; most Bs are C; most Cs are A; most $A(B \equiv C)$s are $B'C'$. Most $B(A \equiv C)$s are $A'C'$. Most $C(A \equiv B)$s are $A'B'$.

First transform the last three into

most (ABC v $AB'C'$)s are $AB'C'$; most (ABC v $A'BC'$)s are $A'BC'$; most (ABC v $A'B'C$)s are $A'B'C$.

Using the same lettering for the cells as before, the premisses go into normal form as

1. Most ($a \vee b \vee c \vee d$)s are $a \vee b \vee e \vee f$.
2. Most ($a \vee b \vee e \vee f$)s are $a \vee c \vee e \vee g$.
3. Most ($a \vee c \vee e \vee g$)s are $a \vee b \vee c \vee d$.
4. Most ($a \vee d$)s are d.
5. Most ($a \vee f$)s are f.
6. Most ($a \vee g$)s are g.

(P5) allows deletions in the predicates of 1–3, yielding

7. Most ($a \vee b \vee c \vee d$)s are $a \vee b$.
8. Most ($a \vee b \vee e \vee f$)s are $a \vee e$.
9. Most ($a \vee c \vee e \vee g$)s are $a \vee c$.

(P8) is now the principle to use. By this, from 9 and 4 we get

10. Most ($d \vee c \vee e \vee g$)s are $d \vee c$.

By the same principle, from 8 and 6 we get

11. Most ($g \vee b \vee e \vee f$)s are $g \vee e$.

Now from 10 and 11, by (P7):

12. Most ($d \vee c \vee b \vee f$)s are $d \vee c$.

Then, from 5 and 7 we get

13. Most ($f \vee b \vee c \vee d$)s are $f \vee b$.

But from 12, by (P9), we get

14. Not (most ($d \vee c \vee b \vee f$)s are $b \vee f$).

13 and 14 are evidently each other's contradictory.

These illustrative proofs indicate, as I think the reader will agree, that *this* way of doing pleonetetic logic is less perspicuous than testing by explicitly algebraic means. It is really just the two principles (P7) and (P8) whose application it is not easy to get used to. But perhaps this is only a matter of a lack of familiarity. If we were brought up on principles like these as we are brought up on standard logic, would we see them as sufficiently distinct to regard them as at best dubiously logical, because of their close relationship to algebraic principles?

Gonville and Caius College,
University of Cambridge

REFERENCES

[1] Altham, J. E. J.: 1971, *The Logic of Plurality*, Methuen, London.
[2] Altham, J. E. J. and Neil Tennant: 1975, 'Sortal Quantification', in Edward L. Keenan (ed.), *Formal Semantics of Natural Language*, Cambridge University Press, Cambridge.
[3] Geach, P. T.: 1976, *Reason and Argument*, Basil Blackwell, Oxford.

G.E.M. ANSCOMBE

ON A QUEER PATTERN OF ARGUMENT

The pattern to which my title refers is:

1° If *p*, then *q*.
2° If *r*, then not (if *p* then *q*).
3° If not *p* then *r*.
∴ *p* and *q*.

We get 'not *r*' from the first two premises and then '*p*' from 'not *r*' and the third; with the first one again this gives us the conclusion.

ARGUMENT A

Now consider the following instance of the pattern; I will call it argument A.

1′ If that tree falls down, it will block the road for a long time.
2′ That's not true if there's a tree-clearing machine working.
3′ If the tree doesn't fall, there will be a tree-clearing machine working.
∴ The tree will fall and the road will be blocked for a long time.

The argument is formally valid and its premises are mutually compatible. They can all be true together just in the case where the tree falls and blocks the road for a long time. How is it then, that there is obviously something screwy about this argument?

The answer seems to arise from a consideration of the subject-matter. In order actually to 'come to the conclusion' you would have to hold the premises true. But, one would say: how can you be confident of 1′, if 2′ is true and is also *as* pertinent as the truth of 3′ makes it? How can you believe that the tree will block the road for a long time if it falls, if you grant that this won't be true if there's a tree-clearing machine operating? You might, indeed, if you were quite certain that there was no question of there being a tree-clearing machine. But you accept that if the tree doesn't fall, there will be one. So, unless you

Harry A. Lewis (ed.), Peter Geach: Philosophical Encounters, 121–135.

already believe that the tree will fall, how can you be sure that there won't be a machine working? Of course, if you are sure that the tree will fall, we can understand how you might be in a position to accept the truth of those other propositions too. You may have some reason to think that if the tree does fall, no tree-clearing machine will be working. So you judge that none will be working, and hence that (2′) casts no doubt on (1′). But in that case, the validity of argument A notwithstanding, it by no means tells us the real grounds of your conclusion. You believe that the tree will fall, and that if it falls the road *will* be blocked. Those are the real grounds for believing that it will fall and the road be blocked. You *assent* indeed to those other two propositions, and from them and the hypothetical that you do have as a real ground, the same conclusion follows. But the premises as given could not be the grounds of belief in the conclusion, even though they are all assented to and the conclusion does follow from them.

(There is an exception we must make in order to be quite accurate. If one took the three hypotheticals as true on someone's authority, one could use them as premises and believe the conclusion on the strength of them. Here the premises really would be one's grounds. But a question arises about the authority. If it were simply someone making judgements on the situation, the argument of the last paragraph would apply to it.)

If all this is right, we have here a rather interesting case of a set of propositions which entail a conclusion but are impossible grounds for coming to that conclusion, and that not in any trivial or obvious fashion. (Trivial cases are easy to construct; any circular argument is one, for example.)

The peculiarity of our case is that there doesn't seem to be any difficulty about reasonably judging any of the three premises to be true without having already judged the conclusion or part of it to be true. The difficulty lies in combining them in knowledge, or in a reasonable judgment, unless part of the conclusion is part of the ground for accepting the combination. One wants to say: that you can get this conclusion out of these three propositions is ground for doubting the conjunction of them! But the reason is not that the conclusion is itself false, let alone absurd. It is a perfectly possible proposition, and is objected to only as a conclusion from perfectly possible propositions, which are mutually compatible and from which it does follow!

The reason for objecting seems not to be anything peculiar to our example, but to reside in a fairly general character of the first two premises when the first one does not state any 'necessary truth', i.e. when q does not *follow* from p. The general character seems to be this: one can usually accept instances of 1° and 2° together, when q doesn't follow from p, only when one is independently confident that not r. The circumstances are rare, in which one thinks that if p then q, and just because of that, it being clear that if r, then not (if p then q), one can deduce that not r. Notice that when one judges 'If p then q' it is easy to invent propositions 'If r, then not (if p then q)' which one will accept without the slightest perturbation of one's belief that if p then q. E.g. in the example in hand, make the second proposition: 'If an angel will dive down and pick up any tree that falls, it's not true that if that tree falls it will block the road.' Or 'If the tree will evaporate as soon as it falls...' Our actual example is disquieting just because 'If there's a tree-clearing machine it's not true that...', though compatible with the first proposition, nevertheless seems to cast doubt on it. Is this a matter of form or content? One answers 'content', because of the compatibility. On the other hand, the *point* of producing a remark of that form, if someone has said 'If p then q', *is* in general to cast doubt on the hypothetical. Whether it actually does so is not just a matter of the content, but of the particular circumstances. There are circumstances in which it would be as silly to say 'Not if there's a tree-clearing machine working' as to say 'Not if there's an angel...' It might be wondered how we can discuss the matter outside any particular context; but we have a perfectly possible third premise, and if it is accepted as true, that at once means that the second premise *is* highly pertinent. It couldn't be true that if the tree didn't fall there'd be a tree-clearing machine, unless tree-clearing machines were a practical likelihood. (Of course our hypotheticals are not material hypotheticals.)

So the point is that the truth *and pertinence* of the second proposition must throw doubt on the first one except for someone with special information that there *won't* be a tree-clearing machine at work. And this point we can generalise. If we judge that if p, then q, and a true proposition 'if r, then not (if p then q)' is presented to our attention, that is liable to cast doubt on our first judgment. What casts doubt will not be the truth alone of the second proposition, but its truth combined with the impossibility of just dismissing it because we don't have to take its antecedent seriously.

There are large numbers of hypothetical judgments that are like this.

It is an interesting and important observation, that there is a whole class of judgments such that when we make them we are not implicitly dismissing as false everything that would falsify them. In contrast, when I make a categorical statement with appropriate confidence, it is very often the case that I can straightway rule out as false what would falsify it—just because I know that *it* is true.

Then we have perhaps discovered the special character of (theoretical) hypotheticals whose consequents don't follow logically from their antecedents. We might call this character 'essential defeasibility'. This will be the reason why, even though 'not *r*' follows from 'if *p* then *q* and if *r*, then not (if *p* then *q*)', still it may be highly unreasonable to deduce 'not *r*' from that conjunction.

But is this character quite general? Is there no such thing as an indefeasible sort of 'if *p* then *q*', where '*q*' doesn't follow from '*p*'? Or, rather, a sort of case where the judgment 'if *p* then *q*' has it in it to be indefeasible?—I mean, if one has judged it soundly, it could itself be used to reject '*r*' in 'if *r*, then not (if *p* then *q*)'. If this is possible, we can have a case where we can come to the conclusion '*p* and *q*' from the three hypothetical premises 1°, 2°, 3°.

ARGUMENT B

I believe that there are such cases. Here is one: we have a bank of coloured lights operated by an apparatus of circuits, switches, etc. behind them. There are, among others, a red and a green light and there is a green-inhibitor, which, when it is on, prevents the green from coming on. The main way it does this, is by changing a circuit which otherwise runs from red through green. Now let us suppose that we have our pattern of argument in this material. I will call it argument B:

1″ If the red light is on, so is the green.
2″ If the green-inhibitor is on, it's not true that if the red light is on, so is the green.
3″ If the red light is not on, the green-inhibitor is on.
∴ The red light is on, and so is the green.

The whole apparatus must either be actually working, with current turned on, lights flashing, etc., or else the description 'The red light is on' means 'The arrangement is such that, given current, the red light shines', etc. It doesn't matter which way we take it, so long as we stick to one way at a time.

1″ is said because the red-green circuit is intact—and the bulbs with their attachments in good working order.

2″ is said because the green-inhibitor is such as to break the red-green circuit, when turned on. It switches the path of any possible current so as to by-pass green.

3″ is said because the red switch, in its off position for red, is so connected with the green-inhibitor as to turn it on.

The conclusion follows, whichever way we take it. Either the display is working, and then the conclusion is that the red and green lights are actually shining, or, if it is not currently working, present adjustment of the mechanism is such as to make it shew red and green lights if there is current.

Here, then, is a case where an argument in this pattern is undisturbing. The reason is that there is no difficulty about concluding from 1″ and 2″ together that the green-inhibitor is off. And the reason for that is that 1″ is true because of a particular set-up, so that we could be satisfied that it was true, enough to respond to 2″ by saying 'Well, the green-inhibitor is off, then'.

At this point I ought to confess that I have found electrical engineers and other experts recalcitrant to the argument. They have not been willing to grant that 'red on, green on, and green-inhibitor off' is the only situation consistent with the three hypotheticals. But I think the reason is that the hypotheticals are not standard descriptions of machines. In particular, the second one is apt to puzzle an engineer. If we make an array shewing the off-on possibilities for red, green and the green inhibitors, we cannot say that any row verifies any of the hypotheticals; we can say that some rows are incompatible with the first and the third, but can't say even as much as that about the second. We may strike out, as incompatible with the conjunction of all three, any row that is incompatible with any one of them. Putting a tick for 'on', a cross for 'off', 'F' for 'incompatible', and 'PT' for 'possibly true', we can construct a compatibility-table as follows:

				Propositions		
	Red	Green	Green-Inhibitor	1″	2″	3″
a	√	√	√	PT	PT	PT
b	×	√	√	PT	PT	PT
c	√	×	√	F	PT	PT

	Red	Green	Green-Inhibitor	Propositions 1″	2″	3″
d	×	×	√	PT	PT	PT
e	√	√	×	PT	PT	PT
f	×	√	×	PT	PT	F
g	√	×	×	F	PT	PT
h	×	×	×	PT	PT	F

Only four lines (c, f, g, h) are ruled out as incompatible with one or other of the propositions. But we can delete the first two lines on the ground that from the data about the green-inhibitor, we know that it can't be on with green on too. This leaves us with d and e as compatible with each of our hypotheticals, and the engineer is apt to say: 'So, you see, there are two states compatible with these specifications, and you are wrong to say that red and green must be on. We can have red and green off and the green-inhibitor on.' He says this because—except for the exclusion of *green and green-inhibitor both on*—he demands that the matter be decided by what is excluded and what allowed by the hypothetical propositions considered severally. Of course he can accommodate the point about green and the green-inhibitor to his way of looking at the matter: it's really a fourth hypothetical, 'If the green-inhibitor is on, green is off', which ought to be put into the specifications. This will lead to the exclusion of the first two rows in the same way as the others are excluded. He makes nothing of the second proposition: 'If the green-inhibitor is on, it is not true that if red is on green is on.' By the methods we have just been applying, it has no consequences, and he is apt to conclude that it imposes no constraints on the construction or state of the apparatus.

However, we ought to have excluded all the first four rows. The reason is that although all but one of those rows are compatible with each of the three hypotheticals, *none* of them is compatible with the first two *taken together*; for this conjunction requires that the green-inhibitor be off. Thus row d is excluded after all, and row e alone is compatible with the three hypotheticals taken together.

So argument B is a case where the conclusion may be reached as a genuine deduction from those premises. The thing that is most characteristic, relative to this possibility, is that each of the hypothetical propositions is based on, is even descriptive of, a *quite particular* set-up. The more an hypothetical proposition approaches to being in effect a description of an arrangement or of the properties of an object or region, the more it will be possible to use it, if true, to reject the antecedent of a related proposition of the form 2°. One may wonder how a 'generic' proposition[1] of the form 2° could be known, in a case where its antecedent is regularly false. Our example gives a way. The green-inhibitor is not operative; it remains in the off position but its construction shews us how it is apt to operate. And, of course, it merely remains in the off position so long as the first hypothetical is true. Switching red off will at once activate the green inhibitor and make the first hypothetical false. But, again, we may know that the first hypothetical is then false, not because sometimes red is on and green not, but because the red-green circuit is broken. Not that the matter *has* to stand so, for we have not stipulated whether the *status quo* will be restored by merely switching red on again; if it is not, then the falsehood of 'If red is on green is on' will appear also in our now getting red without green. But if switching red on does restore the *status quo*, we shall not have red sometimes on without green, even when 'if red is on green is on' is false. To repeat, these hypotheticals are not truth-functional. And, further, we have a case of non-collapse of 'if p, then if p then q' to 'if p, then q'.

ARGUMENT C

This is perhaps best embedded in a story. It runs:

1‴ If NN. decides at four o'clock to leave at five, he will leave at five.

2‴ If the doors are locked at four fifty, it's not true that if NN. decides at four to leave at five, he will leave at five.

3‴ If NN. does not decide at four o'clock to leave at five, the doors will be locked at four fifty.

∴ NN. will decide at four to leave at five, and will do so.

I will make NN. Napoleon's son, the King of Rome; the time shall be when he was living at Schönbrunn. I will suppose (this is mere imagination on my part) that the usual practice at the palace is to have guards, who are supposed to prevent the prince from leaving, posted at all exits, except when they are locked up, and that they are regularly locked up at four fifty in the afternoon. The locking makes it impossible for the prince to get out so long as they are all locked up. However, on this particular day an Arch-duchess has been visiting. She is due to make her farewells inside the palace at four o'clock, and this is the moment at which, there is some suggestion, the prince might decide to leave with her. At four forty five she will descend with her lady-in-waiting, her jewel boxes, her maids, her guards; she will leave the palace and be ensconced in her carriage to the accompaniment of ceremonious salutes; at five she will drive away.

The prince has a tutor, perhaps new to the task, with some sense of his charge's princely dignity and power of decision; we may suppose that our first proposition: 'If the Prince decides at four that he will leave at five, then he will leave at five' is this tutor's indignant response to someone whom he hears laughing at the suggestion that the young man is thinking of leaving with the Arch-duchess. The second statement 'if the doors are locked at four fifty, it's not true that if the Prince decides at four to leave at five, he will leave at five' is made by a courtier who knows the usual practice with respect to the locking of the doors, and also knows that the prince will not be able (for some reason) to come down to get out of the palace *before* four fifty. The final remark is a contribution of the Emperor's secretary: he knows that the Emperor has given orders that if the prince makes no decision at four to depart with the Arch-duchess, the doors are to be locked as usual at four fifty—though nothing has been said about what is to be done if the prince does make such an arrangement.

The tutor, we will suppose, does not doubt the truth of either contribution from the other two; but, convinced of the royal dignity of his pupil, he reiterates that if the prince arranges—decides—at four to leave at five, he will do so. The argument then leads to the conclusion that the prince will decide to leave and will leave. But this is even madder than the argument about the tree.

It is not much to the point to remark that the tutor is wrong in his belief in the first proposition. *He* believes it; if he believes the other two observations, he ought to be able to draw the conclusion which follows from the three propositions. We may ask how he can believe the second proposition if he is so sure of the first; but he might do so because he knows enough of the palace practice to know that once the doors have been locked at their regular time, it is a complicated procedure to get them unlocked; it takes too long; he knows the Arch-duchess is unconditionally to set off at five, and the business of getting a door unlocked would delay the prince's exit from the palace far too long, if he tries to emerge after the doors have been locked. Clearly the tutor's willingness to accept the second proposition without being shaken in his belief in the first is determined by a further belief, namely:

X If NN. decides at four o'clock to leave at five, the doors will not be locked at four fifty.

But while the addition of this to the tutor's beliefs makes us understand why he is not bothered by the second proposition, it doesn't make any difference to the reasonableness of drawing the conclusion.

Notice that the third proposition doesn't work (as in our construe of the tree case) to ensure the pertinence of the possibility that the doors will be locked. In the tree case, we said 'How can one believe the first proposition if the second is as pertinent as the truth of the third shows it to be?' But the Emperor's order would be rather reassuring than otherwise to the tutor; to be sure, the doors may be locked, but his only information on whether they will be or not is that they will be if the prince doesn't decide to leave—and their being locked in that case doesn't matter.

Except that the tutor can apparently *derive* from the first two propositions that the doors will *not* be locked! Our question now must be: how can be hold both propositions, when their conjunction implies a proposition, whose falsehood he is perfectly prepared to entertain—so long as it's false only if his pupil does not arrange to leave? There must be something wrong with, or misinterpreted about, his belief in the first two propositions, if they are to license him to conclude that the doors absolutely will *not* be locked.

He holds these three propositions:

1''' If the prince decides at four to leave at five, he will leave at five.

2''' If the doors are locked at four fifty, it's not true that if the prince decides at four to leave at five he will leave at five.

X If the prince decides at four to leave at five, the doors will *not* be locked at four fifty.

and his belief in the third one, X, explains his equanimity in face of the suggestion that the doors *will* be locked. But what use is it to say that? It still follows from the first two, taken together, that the doors will not be locked at four fifty; and it still seems unreasonable to believe that on the ground of the first two. All the more because by taking this conclusion together with the original third proposition, 3‴, the tutor will be able to deduce the prince's actual choice and action. *That* is completely preposterous.

Since it is preposterous to draw the conclusion, the tutor, if he recognizes this, cannot after all believe the first two propositions together. Suppose he firmly adheres to the first one. What is he to say about the second? He must say that *it* is true only if the prince does not decide to leave. That is, he withholds assent from it and replaces it by:

(i) 'If the Prince does decide at four to leave at five, then it is *not* true that if the doors are locked at four fifty, it is not true that if he decides at four to leave at five he will leave at five.'

This might be called a somewhat mystical conviction, but it enables him to escape the absurd deduction. Can he grant that the second proposition is true *if* (not just *only if*) the prince does *not* decide to leave?

I.e. can he say:

(ii) 'If the Prince does not decide at four to leave at five, then, if the doors are locked at four fifty it is *not* true that if he decides at four to leave at five he will leave at five'?

He surely can't grant that. He insists that if the prince decides at four to leave at five, he will leave at five; to grant that in the case where he does not so decide this is false if the doors are locked would be very strange. Why? Because by proposition 3‴ the doors *are* locked if he does not make that decision—so, if he does not make that decision it is *not* still true that if he does make it he will leave; but it *is* true (*ex hypothesi*) that if he does make it he will leave; and so by *modus tollens* he *must* make the decision after all, and once more we have a preposterous deduction. Thus the tutor must assert (i) and refuse assent to the original second proposition (2‴), if he is to adhere to the first one and not be drawn into a fantastic deduction that the doors will not be locked (and so, via the third proposition, that the prince will decide to leave and will leave). If he cannot but grant the second proposition and is nevertheless unwilling to infer that the doors will not be locked, he must allow the second proposition to cast doubt upon the first. That is, he must do so in the circumstances, because it is evidently quite on the cards for the doors to be locked.

ARGUMENT D

Argument A (about the tree) might occur, crazily, as follows: the three propositions are statements made by three different people talking together, and a fourth straightway says he believes them all and draws the conclusion! This is crazy because, in the absence of any indication from the first one, he should not be assumed still to believe what he said at first. It is not of course ruled out that he still does think the same. For it may be a mere fragment of his thought—he may be already convinced that the tree will fall *and* will block the road. Then he has given vent to this thought only in a partial form when he says 'If that tree falls, it will block the road.' In this way one who is too timid to assert what he knows or believes, especially about the future, may come out with such an 'If...'

This leads on to the fourth embodiment of the pattern, argument D. It might be produced by a less evasive and tortuous Johannes de Silentio picturing Isaac in the interval in which he has realised that *he* is the intended sacrifice, and before Abraham's hand is stayed. Isaac reasons:

1'''' If God has promised my father that he will be the father of a great nation through me, then my father will be.

2'''' If my father kills me, it's not true that if God has promised him he will be the father of a great nation through me, then he will be.
(Therefore he is not going to kill me.)

3'''' If God has not promised my father that he will be the father of a great nation through me, my father *is* going to kill me.

∴ God has promised that to my father and it will be fulfilled.

This argument differs from all the others in that in the first proposition the consequent necessarily follows from the antecedent. The question is, why should Isaac believe the second proposition? Many children have been murdered by their parents. Upon the belief 'What God has promised, he will perform' neither this fact nor the present threat of such a death has any bearing at all. 1'''' 'If God has promised my father that through me he will be the father of a great nation, then he will be', is true in the way the generality is. We must suppose Isaac's thought to be, not a mere statement of something obvious—and irrelevant—but only the tip of an iceberg. In reality, Isaac's belief is that God has promised Abraham that. That is what he will have been brought up to believe. So the first proposition has for him the significance rather of '*Since* God has promised my father that, my father will etc.', and not only the significance of the necessary hypothetical. This would also explain why Isaac should think of the second proposition. There has been this promise. If his father kills him, the necessary hypothetical itself, the irrefragable 'What God promises, he will perform' will be falsified. The third proposition states something that must have been dreadfully obvious to Isaac—if his belief in the existence of that promise is wrong, it is all up with him. It *looks* as if it were all up with him. But it cannot be, because God performs his promises, and if Isaac is killed God doesn't perform his promises. So it isn't true, it cannot be true, that this is his end. If that's not true, then since clearly $3''''$ *is* true its antecedent is false. Of course we made Isaac assume that already, in explaining why he might believe $2''''$; but if, believing $1''''$ and $2''''$, he manages to believe he will not be killed, the evident truth of $3''''$ strongly reinforces his belief in the promise.

Note that there is an alternative possibility for Isaac, which does not involve his drawing the conclusion that Abraham will not kill him. He might reject $2''''$ and replace it by:

> If God has made that promise, it is not true that if my father kills me it is not true that if God has promised that through me my father will be the father of a great nation, he will be.

This is parallel to the proposition (i) by which we described the tutor as replacing $2''''$ in argument C. I suggested it was a somewhat mystical conviction: the tutor is so sure of his pupil's royal dignity and independence, that he brushes aside the evidence on the consequences of the door's being locked—if the prince once makes that decision, he'll carry it out no matter what. The suggestion was pretty absurd and was only made to evade the preposterous conclusion. In fact of course the tutor should abandon the first proposition rather than the second as soon as he sees the consequences of accepting all three. (Still, there is something about this case that remains obscure and intensely puzzling.)

The parallel proposition for Isaac is not in the same position as the tutor's proposition (i). It is essentially the thought that even if Abraham kills him, God can raise him up and still fulfil the promise. And, as with the original proposition 2'''', Isaac's only reason for believing it would be his belief in the promise. (There is not in his case the odd problem of avoiding a preposterous conclusion out of the three hypotheticals.) Of course no germane conclusion now follows from the three propositions. Proposition 3'''', which suggests the doubt that must threaten Isaac, would now be amplified as 'If God has not made that promise, my father is going to kill me and that *will* be the end of me.' But the propositions have fallen apart and are not combined in any argument.

We can easily parallel argument D in more familiar matter. Let the propositions be: 'If they've published the award, they'll give me the money', 'Not if there's no money in the kitty' and 'If they haven't published the award, there's no money in the kitty.' Instead of Isaac, we have an anxious young woman who has *heard* that they've published the award, but who has some rather good evidence that there'll be no money in the kitty. There are differences: the first proposition not being necessarily true, it's not too puzzling why she should accept the second; she might after all take it as casting doubt on the first. So we have to put in that the standing of the 'they' is such as to make it seem out of the question that they should publish the award and not give the money. If that *is* out of the question, then why should she accept the second proposition—unless (and here is a close parallel with Isaac) she believes they have published? No money in the kitty means no payment will be made—just as no Isaac means no descent of a nation through him. Harmless, *unless they have published*—but she *thinks* they have. If she's really convinced of that and of 'their' probity, she may reject the threatening evidence of no money in the kitty. Perhaps she quite sees that (in the circumstances) if they *haven't* published, it means there's no money in the kitty; for not merely has she heard that they've published, but (we will suppose) they ought to have done so—*unless* they have no money to pay the award. So if she is sure about their having published, and firm in her confidence in them, she infers that there must be money there, and her belief that they have published is reinforced, for only if there isn't money will they *not* have done so.

In these cases the argument does not function to establish the conclusion p & q, for it operates because the person deploying it already believes that p. But it's not like the tree case, where the second and third proposition did no work that we could see, once we found out that the believer in the first proposition believed that *p*. In argument D the logical relations of the three propositions give us a structure of confidence, probable anxiety and doubt, clinging to or abandoning conviction, which is worth examining. A man who believes an instance of *p, if p then q, and if r then not (if p then q)* may have to bet on *r*'s *not* being the case, though any evidence that antecedently seems germane to whether *r* is the case or not speaks *for r*. That is, the things you would enquire into in order to find out whether *r* held all suggest it does hold. He may have to make some grave decision (our young woman can easily be made an instance of this): he may even have to bet his life. And yet the grounds for believing the proposition 'if *p* then *q*' may not themselves tend to make *r* improbable. As we saw in examining the engineers' objections to argument B, the propositions which severally allow a case may exclude it when combined, and the grounds for believing the separate propositions may have very little to do with one another.

CONCLUSION

In discussing this pattern of argument, I have raised a problem rather than clarified anything. It is quite easy to go on constructing further arguments on the same pattern. E.g. 'If that dog's got rabies, that man is a goner; that's not so if the man takes such-and-such a course of injections; if the dog's not got rabies, the man will take those injections: *ergo*, the dog's got rabies and the man is a goner'! And so on. The arguments are often quite crazy. But aren't they valid? And aren't both premises and conclusions quite possible—given, for each case, a suitable story? (The third premise usually needs a story. But stories are easily supplied.) It is possible that minute scrutiny of the case would always yield reason to say the premises would not be jointly believed. But why does one know in advance that one wants to find something of the kind? Am I wrong—against the engineers—to think argument B all right? Is there usually some lurking ambiguity? For the *form* is surely valid. And the use of such an argument to reinforce conviction in threatening circumstances, as in the D arguments, is as sound as the convictions themselves.

Several people have suggested that what would really be meant by the second premises would be, schematically, 'If *r*, then if *p* then not *q*'. But that is too strong. 'Not (if *p* then *q*)' means that *p*'s being the case doesn't involve *q*'s being the case, not that it involves *q*'s not being the case.

University of Cambridge

NOTE

[1] Here, contrary to the practice of Russell, Church, etc., I speak of a proposition as capable of being now true, now false: von Wright calls a proposition so conceived 'generic'.

JAAKKO HINTIKKA

GEACH AND THE METHODOLOGY OF THE LOGICAL STUDY OF NATURAL LANGUAGE

In the many-splendored field of the logic and semantics of natural languages, Peter Geach occupies an interesting and unusual niche. Locating that niche and evaluating his achievements and shortcomings throws extremely revealing light on the methodological situation in contemporary *Sprachlogik* and logical semantics at large. In this note, I can only touch on some of the features of Geach's thought and its methodological backdrop. I begin with a few general observations.

Geach professes allegiance to the methods of modern symbolic logic and repeatedly expresses his admiration of the initiator of this logic, Gottlob Frege. At one point, he condemns those who refuse to use the strategy of illuminating ordinary language by translating it into quantificational forms by claiming that they 'seem to me to have put themselves outside the pale of rational discussion.'[1] There are nevertheless subtle differences between Geach's wonted procedures in this field and those of Frege and the typical latter-day Fregeans.

Why is it supposed to be useful to translate ordinary language into logical notation for the purpose of logical, philosophical, or linguistic clarification? The operative but often unexpressed reason seems to me to be that the semantical properties of logical languages are much better understood than those of natural ones. This is in the last analysis the reason why even some of those logicians and philosophers of language use this overall approach who do not themselves think that we can systematically discuss semantical matters or develop an explicit semantical theory. Cases in point include, I have argued or suggested elsewhere, Frege, early Wittgenstein, Quine, and Davidson.[2] Here I can give only a negative reason for emphasizing the gain in semantical rather than syntactical clarity as the real reason for resorting to logical symbolism in the study of the logic of natural languages. Even though there are a few ingenious linguists, for instance Emmon Bach, who think that logical symbolism can help us to understand the syntax of natural languages, the overwhelming majority of linguists deny that the syntax of sharply defined logical languages can help us in studying the syntax of any natural language like English. Those who believe in the

Harry A. Lewis (ed.), Peter Geach: Philosophical Encounters, 137–149.

translational strategy therefore must have something else in mind, unless they are simply wrong. Even though the semantical motivation of such philosophers of language as Frege, early Wittgenstein, and Davidson is far from obvious, I strongly believe that it nevertheless is very real indeed. In Frege's case, this is among other things shown by the way he explains in practice (i.e., in the *Grundgesetze*) his notation in terms of the references (e.g., truth-conditions) of the expressions of his logical notation. Elsewhere I have called attention to the remarkable connections between the so-called picture theory of Wittgenstein's *Tractatus* and the logical semantics of Tarski and Carnap.[3] And Davidson's tacit semantical motivation is seen clearly from his purported reliance on Tarski-type truth-definitions.

These philosophers of language offer interesting objects of comparison in examining Geach. Their approach suffers from two main weakness. One is what Merrill Provence Hintikka has aptly called the problem of miraculous translation. Recent linguistics (among other things) has amply shown how extra-ordinarily difficult a translation from ordinary language to logical notation really is. It is not so much that there are some particular sentences or other expressions which we do not know how to translate. Rather, even when we know perfectly well what the correct rendering of some particular natural-language expression in logical notation is, we may perfectly well fail to have the vaguest idea as to what the rules are that we are following in giving that translation. An example which I have discussed at length elsewhere and whose role in recent logico-philosophical discussion is due to Geach is exemplified by the following.[4]

(1) If Bill owns a donkey, he beats it.

We know perfectly well that the right translation of (1) is the general implication

(2) $(x)\ ((x \text{ is a donkey } \& \text{ Bill owns } x) \supset (\text{Bill beats } x))$.

Yet there is no end of questions one can ask of (1) and of the transition from (1) to (2) even after having agreed to this translation. Why does the indefinite article, which usually operates like an existential quantifier, act as a universal quantifier in (1)? How does the pronoun 'it' function in (1)? (It is not a pronoun of laziness, nor does it pick out an earlier reference to an individual, for its head 'a donkey' does not refer to one donkey more than to another.) Why is (1) grammatical while the following is not?

(3) *If Bill owns every donkey, he beats it.

Why does the mere inversion of left-to-right order change the preferred reading of (1), as witnessed by the following?

(4) Bill beats a donkey if he owns it.

A somewhat clearer pair of examples might be the following.

(5) Bill will beat a donkey if he manages to find it.
(6) If Bill manages to find a donkey, he will beat it.

Here it is clear that the indefinite article is preferably understood as an existential quantifier in (5) but as a universal one in (6). Why? Such questions can be multiplied further. They show that our translations from (say) English to logicians' canonical notation often are miraculous indeed. A mere *ad hoc* translation does not illuminate the workings of our natural languages except perhaps by making us aware of some of the problems concerning it.

The other main weakness in the position of Frege and the philosophers and logicians I may perhaps call neo-Fregeans is that they overestimate the unproblematic character of the semantics of logical languages. Indeed, for a variety of reasons none of the people I mentioned above has developed anything like an explicit semantical theory.

How is this approach to be viewed? How is Geach's work related to it? At first sight, he might seem to be a typical representative of the translational strategy just adumbrated. Not only does he emphasize the virtues of modern logic and praise its originator, Gottlob Frege. Not only does he follow the example of the translationist and leave most of the actual semantics of logical notation unexamined. (Like Quine, Geach has never done any work in logical model theory.) One at least one occasion, he explains the reasons for the translational approach precisely along the lines I have indicated: '[A] bright light is often thrown upon truth-conditions, and thereby upon the sense, of ordinary-language sentences by the endeavour to translate them into quantificational forms.'[5] Apart from style and spelling, this could have been said by Donald Davidson.

In some respects, it is in fact not very misleading to bracket Geach with the philosophers just mentioned. For instance, it seems to me that their failure to come to grips with the semantics and pragmatics of ordinary extensional logic is shared by Geach. Among other things, he misses along with the other translationalists the role of branching quantifiers in natural and in formal languages, even though he in fact came closer than perhaps anyone else to doing so (I shall argue later in this paper). Likewise, Geach is in the same boat with the other philosophers mentioned above in adhering much too closely (for my taste) to the Fregean paradigm of a logical notation (in the sense of a logical language).[6] For instance, he adheres as firmly as anybody to the Frege-Russell distinction between the allegedly different senses of 'is' in natural languages.[7]

I do not think that Geach's reliance on medieval models sets him aside in any deep way, either, even though it is one of the most obvious differences between his explicit writings and those of most comparable modern philosopher-logicians. Geach's relation to the schoolmen would require a separate discussion. Let me register here only my suspicion that many of the anticipations of modern views he finds in the medieval logicians are in fact rather shallow.

The real differences between Geach and the neo-Fregeans lie elsewhere. One of them is that Geach seems to have a much less sharp semantical vision underlying his actual work than they have. I have argued elsewhere[8] that Frege did not fail to discuss ordinary extensional semantics because he did not have a clear idea of what it is like. Indeed, it seems to me that he had a remarkably powerful and sweeping vision of just that. The real reason for his silence was that he did not believe that we can meaningfully discuss semantical matters.[9] (Precisely the same seems to me to be the case with Quine.) Geach seems to agree with Frege's conclusion much more than with Frege's underlying vision. He is somewhat more interested in the actual syntax of natural language and in its role as articulating semantical matters than Frege and neo-Fregeans typically are. Moreover, and more importantly, Geach's semantical ideas are much more eclectic and much more *ad hoc* than those of Frege or Quine. Geach often speaks for instance of truth and of an expression's being applicable to a thing. These are archetypically semantical ideas (in the sense I am using the term here). But in using such locutions, Geach is guided much less firmly by an overriding general semantical view than Frege or Quine. He is in semantical matters much more of an opportunist than the neo-Fregeans. I take this observation to be a commendation, not a criticism. If I have a criticism to offer here, it is rather that Geach could have pushed his opportunism further than he has in fact done.

This relative absence of a unified semantical vision may seem a disadvantage. In reality, it has nevertheless freed Geach to study certain problems which have no connection with the underlying Frege-type semantics.[10] Very roughly, these are the problems concerning natural languages which are due to the differences between their semantics and the semantics of Fregean formal languages. In so doing, it seems to me, Geach has done a tremendously important service to all philosophical logicians and philosophical semanticists. He has in effect called attention to semantical discrepancies between natural languages and those particular formal languages which Frege introduced and Russell popularized, even if he has formulated his point in the form of problems ('sophismata,' as his predecessors would have called them) which present difficulties to any view rather than in the form of objections to Frege and Fregeans. Typically, Geach's lingering loyalty to Fregean paradigms has prevented him from developing a large-scale systematic theory of the phenomena he has studied, and frequently it has led him astray concerning the detailed explanation of this or that phenomenon.[11] But his relative freedom from Frege's and Quine's semantical paradigm has enabled him to raise questions which can be pushed further so as to break the framework within which Frege and Quine and other neo-Fregeans are in effect moving. All his failures thus do not eclipse the tremendous contribution Geach has made in this area. I do not think that I am atypical in finding that I have to resort to Geach's examples and to discuss time and again problems which Geach has introduced into recent discussion.

It follows from what I have said that in order to document my evaluation of Geach's work I would have to do two things. First, I would have to show how Geach's ostensibly Fregean solutions to his puzzles fail to have a genuine semantical backing. Secondly, I would have to offer a better diagnosis of the same problems and to show how this better solution involves a violation of the Fregean paradigm. This is a larger task than I can undertake in one note. It is also complicated by the relative freedom from Fregean ideas Geach manifests in his actual practice of *Sprachlogik*. Often, he is willing to borrow from other conceptual frameworks for the purposes of his analysis, not only from the medieval writers but also from the concepts and terminology of traditional grammar. Geach is fully aware of the provisional character of such borrowings. He does not realize, however, to what extent the real reason for his resorting to such admittedly *ad hoc* categories is a symptom of the failure of his official Fregean methodology. Geach's commendable eclecticism thus makes it hard to discuss within a reasonably brief compass his actual analysis of the crucial problems (crucial in the sense of a crucial experiment) I have in mind.

In order not to leave my general remarks completely up in the air, let me nevertheless illustrate them by discussing both the weaknesses and the potential but largely untapped strength of one particularly important concept Geach has used. (For the latter purpose, I have to refer to recent developments in logical semantics which I unfortunately do not have to to expound here, much less to defend.) This concept is that of predicable. It is defined by Geach as follows: 'A *predicable* is an expression that gives us a proposition about something if we attach it to another expression for what we are forming the proposition about.'[12]

This characterization is in terms of concepts which look semantical, such as a proposition being *about* something. It does not refer to the expressions of our language only, but also to their relations to the entities this language is used to speak of. At first sight Geach's notion of predicable looks therefore rather like some of the basic concepts of Frege's, such as the notion of an expression which expresses a concept (in Frege's strangely wide sense of the term 'concept' or '*Begriff*'). Indeed, Geach's 'predicable' may be thought of as a mere variant of this Fregean notion. But when we look and see how Geach uses the concept we can see that Geach's notion belongs to an entirely different sphere of ideas from Frege's. Frege's notion is part and parcel of a sharply defined range of ideas. It is applicable only on the presupposition that this tacit semantical model of Frege's is applicable. For instance, the only sense that Frege presumably could have associated with the notion of a proposition's being 'about' some one thing is this particular thing's being the reference of some 'name' in the sentence expressing this proposition (or 'thought', as Frege would have said).

Geach uses the notion of predicable much more freely than a Fregean would ever be prepared to do. In much of his most interesting and subtle discussion, he allows predicables to be attached not only to names and other expressions that refer to a particular individual, but also to what he calls referring phrases and what I shall call quantifier phrases, such as 'any *X*', 'a *Y*', 'every *Z*', 'most *W*'s', 'precisely three *U*'s', etc. Frege would of course have balked at such applications of *his* expressions for concepts. Geach observes correctly that he is departing from Frege here, and equally correctly says that even if Frege's treatment is correct, it need not exclude other accounts of the same subject matter.[13] All this is correct, and indeed commendable, because it is just by departing from Frege that Geach has been able to develop ways of going (at least potentially) beyond Frege in certain important respects.

However, Geach's usage marks a much more radical departure from Frege's underlying ways of thinking than he points out and probably is even aware of. By stretching the use of the notion of a proposition's being *about* something beyond the cases Frege would have countenanced Geach is leaving behind all clear-cut semantical ideas of the kind I have claimed motivated Frege. There is no sense in the strict Fregean framework to saying that a proposition (thought) is about some man, all animals, any boy, or most women. By leaving the firm but narrow ground of Fregean semantical ideas behind him Geach has entered into an interesting but ill-defined area where one's semantical concepts are likely to be heuristic and pre-theoretical rather than such sharp notions that can be built into a satisfactory explicit semantical theory without a great deal of explication and unpacking.

Geach's actual uses of the notion of predicable illustrate these points. First, he lends his concept of predicable what can only be called a syntactical role. For instance, he uses it for some of the same purposes other analysts would use the explicitly syntactical concept of *scope*. For instance, instead of saying that in the sentence

(1) Each boy admires most girls

the quantifier phrase 'each boy' has a wider scope than 'most girls', Geach expresses the same point by saying that the predicable 'each boy admires—' does not genuinely occur in (1).[14] Rather, according to Geach we must assume that (1) is formed by attaching the predicable '—admires most girls' to the referring phrase 'each boy'.

What are we to say of such uses of the notion of predicable? Several observations are pertinent here.

(i) Geach's solution of the scope problem in (1) is not a result of any systematic theory in which the concept of predicable plays a role. The concept is used to codify our semantical intuitions, not to account for them or to put them in a wider theoretical framework.

(ii) Geach says himself that his use of the notion of predicable in sentences like (1) is equivalent to the usual treatment of scopes in the usual quantificational notation of formal logic. This shows that his treatment suffers from the same shortcomings as any purely translational account of the scopes of quantifier phrases in natural language. In the translational account, the whole theoretical weight falls on the translation rules, which so far are understood only very imperfectly. This reinforces the impression one had in (i) that what Geach is giving us is not an explanation of what happens in ordinary language, but rather an alternative notation in which we can say the same things as we can express in English. However, the theoretical (e.g., semantical) motivation of the alternative notation is not explained by Geach.

(iii) This is partly due to the fact that Geach never really explains in down-to-earth semantical terms what it means for a predicable to be attached to a quantifier phrase. However, this weakness could have been turned by Geach into a strength.

The notion of scope is not the only nor even the best treatment of the semantical distinction between the two combinatorially possible readings of (1). Indeed, the truth seems to me to be that what really matters is the order in which the different ingredients of a sentence like (1) are tackled when we process it semantically. The familiar formal logicians' notion of scope is only one way of marking this semantical order syntactically.

(iv) Indeed, the notion of scope is not very happy when it is applied directly to natural language. Not only is it the case that there are in natural languages no precise scope indicators. There is a sense in which the scopes of at least some quantifier phrases are indefinite (indefinitely large).[15]

Even though Geach seems to deny it, we can for instance begin a sentence, paragraph, or a book by saying something about 'an old man' and keep on referring to the same individual as 'he' or 'the old man'.[16] This means that the 'scope' of at least some quantifiers is in principle unlimited.

The right conclusion to draw here is that the very concept of scope is virtually useless in connection with the quantifier phrases of natural languages. What matters in natural languages is not scope as a syntactically marked chunk of sentence or discourse, associated for instance with a quantifier phrase, but the order in which different quantifier phrases are dealt with in our semantical information processing. And this could in principle be handled by means of Geach's 'predicable' notion better than by means of conventional formal logicians' notion of scope.

Geach is thus unnecessarily modest when he lets us understand that what his treatment can do can also be done by means of the quantificational notation of the usual Frege-Russell sort.

Similar things can be said of other uses Geach makes of his concept of predicable. Sentences closely related to (1) are dealt with by Geach in terms of what he calls *double* and *multiple* predicables.[17] They are supposed to apply to more than one name or quantifier phrase. This is still compatible with an attempted assimilation of predicables to linguistic expressions for what Frege called concepts, for a Fregean concept may have more than one empty slot. But on the Fregean scheme it does not make any sense to do what Geach does and what is crucial to some of his most interesting observations: it does not make any sense to claim that the order matters in which a multiple predicable is applied to two or more referring phrases.[18] (For instance, if we think of (1) as involving the double predicable '—admires—', then we must think of the second slot as being filled first and the first slot as being filled last. More generally, Geach has to assume that the order of application makes a difference in handling scope problems of the kind we noted above.) Geach's procedure means, on the one hand, another indication that his notion of predicable is not grounded on any solid semantical theory. (In the usual logical formalism there is not one two-place or multiple-place 'predicable', simple or complex, for which it makes a difference in what order its blanks are filled.)

On the other hand, Geach does not quite seem to have the courage of his logic to follow up the suggestions of his own notation. If it makes sense, Frege notwithstanding, to speak of the order in which the blanks of a multiple predicable are filled, it surely must make sense to speak of their being filled simultaneously. Indeed, there are perfectly good examples of what this must mean in ordinary discourse. Geach considers[19] two readings of sentences like

(2) Most electors voted for most candidates.

These two readings can be paraphrased as follows:

(3) It is true of most electors that they voted for a majority of candidates.

(4) It is true of most candidates that they were voted for by a majority of electors.

According to Geach, these can be thought of as being obtained from the double predicable '—voted for—' by filling its slots one by one in the two different alternative orders.

As linguists have pointed out, there nevertheless is a third reading which can be paraphrased as follows:

(5) It is true of a group of electors and of a group of candidates that the former voted for the latter, and the two groups consisted of most electors and most candidates, respectively.[20]

(If the point is not clear, consider the parallel example

(6) Three hunters shot three tigers.

This can surely mean that a party of three hunters shot between them three tigers.)

By means of examples it is easy to see that (5) can be false even though (3) and (4) are both true.[21]

Geach could apply most naturally his idea of multiple predicable to examples like the reading (5) of (2). All he would have needed to do is to say that in the third reading two referring phrases are inserted *at the same time* into the blanks of a two-place predicable, not one after the other.

What would have happened if Geach had done this? He would have reached a treatment of a special case of what are often referred to as branching quantifiers.[22] This would have been a most interesting novelty. The possibility of doing this within Geach's framework shows how perceptive he is and how independent his framework of ideas really is of the range of familiar ideas of conventional formal logic. For branching quantifiers go way beyond ordinary quantification theory (first-order logic). Indeed, examples like (2) have (mistakenly) been offered as evidence for the failure of formal logic to elucidate the semantics of ordinary language quantifiers.[23] In reality formal logic does not fail here, even though 'Fregean Logic', that is, linear first-order logic, does fail.

The Fregean framework of ideas fails radically in other respects, too, if branching quantifiers (or, in Geach's framework, simultaneous filling of the blanks of a predicable) is allowed. Among other things, the Fregean idea of a formal system of logic fails, for the logic of branching quantifiers is not axiomatizable.[24] Likewise, the important principle of compositionality, sometimes known as the Frege principle ('the meaning of a complex expression is a function of the meanings of its component expressions') is likewise vitiated by the presence of branching quantifiers.[25] Hence this venture beyond familiar ideas is a radical one indeed.

Hence Geach could not have assimilated his treatment to the usual Frege-type treatments, either, if he had countenanced the idea of filling the blanks of a multiple predicable simultaneously (as it were, independently of each other). And this would have been both a triumph and a disaster for him. Triumph because it would have showed that Geach's treatment is superior in certain ways to the Fregean one. Disaster, because it would have left Geach without a satisfactory semantical backing for his syntactical discussion. As long as an analyst of ordinary language like Geach stays within the sphere of Fregean ideas, he has the security of the semantical backing that Fregean logic possesses, even when this semantics is not spelled out. But when one ventures beyond this safe ground, one faces the task of giving a semantical account of what one is doing. In Geach's case, he could not have taken for granted the idea of a predicable's applying to or being true of things. In the familiar old applications, these notions have the tacit and frequently somewhat confused backing of familiar first-order semantics. Not so in those new applications which Geach could have undertaken and which would have taken him beyond the familiar range of Fregean ideas.

These observations concerning Geach's notion of predicable serve to illustrate my general diagnosis of Geach's approach. They illustrate on the one hand the independence of Geach's approach of the semantical foundation which is tacitly assumed in most uses of the familiar apparatus of modern logic. This is a weakness, for a merely translational strategy will not provide us with a genuine theory. On the other hand, the same independence of Geach's approach enables us to use it beyond his own applications and well beyond the scope of Fregean ideas.

Department of Philosophy,
Boston University, Boston

NOTES

[1] Peter Geach, 'Back-Reference', *Philosophia* **5** (1975), pp. 193–206. (See p. 206.)
[2] See my survey article, 'Semantics: A Revolt Against Frege', in G. Flöistad and G. H. Wright (eds.), *Contemporary Philosophy: A New Survey*, Vol. 1, Nijhoff, The Hague, 1981, pp. 57–82, with further references to the literature. Cf. also my paper, 'Frege's Hidden Semantics', *Revue Internationale de Philosophie* **33** (1979), pp. 716–722.
[3] See 'Quantification and the Picture Theory of Language', in *Logic, Language-Games, and Information*, Clarendon Press, Oxford, 1973, pp. 26–52, and 'Language-Games', in Jaakko Hintikka et al. (eds.), *Essays on Wittgenstein in Honour of G. H. von Wright*, North-Holland, Amsterdam, 1976, pp. 105–125.
[4] See my joint paper with Lauri Carlson, 'Conditionals, Generic Quantifiers, and Other Applications of Subgames', in Esa Saarinen (ed.), *Game-Theoretical Semantics*, D. Reidel, Dordrecht, 1979, pp. 179–214.
[5] *Op. cit.*, p. 206.
[6] Cf. G. E. M. Anscombe and P. T. Geach, *Three Philosophers*, Basil Blackwell, Oxford, 1961, especially pp. 131–151.
[7] Cf. my criticism of the distinction in '"Is", Semantical Games, and Semantical Relativity', *Journal of Philosophical Logic* **8** (1979), pp. 433–468.
[8] In 'Frege's Hidden Semantics' (note 2 above).
[9] Cf. Jean van Heijenoort, 'Logic as Calculus and Logic as Language', *Synthese* **17** (1967), pp. 324–330.
[10] Much of *Reference and Generality*, Cornell University Press, Ithaca, N.Y., 1962, provides examples of this.
[11] An example is Geach's insistence that 'the one indisputable class of cases where a pronoun does pick up the reference of its antecedent is supplied by pronouns of laziness,' that is, pronouns which merely stand for their grammatical antecedents. Examples supplied by Karttunen and others show that this claim simply cannot be defended. Lately, Geach has tried to avoid the problem by allowing a pronoun of laziness to be replaceable also 'by a repetitious phrase somehow reconstructable out of the antecedent' ('Back-Reference', note 1 above, p. 194). This turns the notion of a pronoun of laziness into so vague an idea as to be useless in the actual analysis of pronouns.
[12] *Reference and Generality* (note 10 above), p. 25.
[13] *Reference and Generality* (note 10 above), pp. 57–58.
[14] *Reference and Generality* (note 10 above), pp. 89–90.
[15] Cf. my paper 'Quantifiers in Natural Languages', in Esa Saarinen (ed.), *Game-Theoretical Semantics* (note 4 above), pp. 81–117. (See pp. 81–83.)
[16] See *Reference and Generality* (note 10 above), pp. 124–128; 'Back-Reference' (note 1 above), pp. 198–201. Of course Geach is not denying the possibility of using pronouns (and the-phrases in their anaphoric function) in this way. What he is denying is the appropriateness of describing their mode of operation in terms of references to one and the same individual. Even though it seems to me that Geach is wrong here, this does not affect the matter now at issue. Indeed, my present point is further reinforced if Geach is right.
[17] *Reference and Generality* (note 10 above), pp. 94–107.

[18] *Reference and Generality* (note 10 above), pp. 99–101.
[19] *Reference and Generality* (note 10 above), pp. 98–100.
[20] Whether this is taken to mean that each member of the group of electors voted for each member of the group of candidates (as is perhaps the most natural assumption to make here) or that the former group somehow voted as a body for the second one is immaterial to my argument at this point.
[21] Let the electors be Betty, Mary, and Jane, and the candidates Thomas, Richard, and Harold. Betty voted for Richard, Mary for Thomas and Harold, and Jane for Richard and Harold. Then (3) and (4) are true but (5) false, as you can easily ascertain. The converse implications are of course valid: (5) implies (3) and (4).
[22] See my paper, 'Quantifiers vs. Quantification Theory', *Linguistic Inquiry* **5** (1974), pp. 153–177; reprinted in Esa Saarinen (ed.), *Game-Theoretical Semantics* (note 4 above), pp. 49–79.
[23] See R. S. Jackendoff, *Semantical Interpretation in Generative Grammar*, MIT Press, Cambridge, Mass,. 1972, pp. 307–308.
[24] See my paper, 'Quantifiers vs. Quantification Theory' (note 20 above).
[25] See my paper, 'Theories of Truth and Learnable Languages', in Jaakko Hintikka, *The Game of Language*, D. Reidel, Dordrecht, 1983, pp. 259–292.

JAMES D. McCAWLEY

NATURAL DEDUCTION AND ORDINARY LANGUAGE DISCOURSE STRUCTURE

'Natural deduction' is natural in virtue of the way that it allows formal proofs to closely mirror the structure of familiar ordinary language arguments.[1] For example, the sentences of the ordinary language argument (1) correspond one-to-one with the lines of the proof (2):

(1) Whoever committed the murder left by the window. Anyone who left by the window would have had mud on his shoes. Suppose that the butler committed the murder. Then he left by the window. In that case, he has mud on his shoes. So if the butler committed the murder, he has mud on his shoes.

(2)

1	(∀*x*: *x* committed the murder) (*x* left by the window)	
2	(∀*x*: *x* left by the window) (*x* has mud on *x*'s shoes)	
3	│ *b* committed the murder	supposition
4	│ *b* left by the window	1, 3, ∀-exploitation
5	│ *b* has mud on *b*'s shoes	2, 4, ∀-exploitation
6	⊃ (*b* committed the murder, *b* has mud on *b*'s shoes)	3-5, ⊃-introduction

The division of (2) into a main proof and a subordinate proof (set off by the vertical line) and the division of both main proof and subordinate proof into suppositions or premises and inferred propositions corresponds to structure that is at least implicit in the natural language text, in that a person uttering that sequence of sentences performs different speech acts in uttering each sentence, depending on where the given sentence fits into the hierarchical structure indicated by the horizontal and vertical lines. Natural deduction is in fact of special interest to linguists for being the most explicit, detailed, and successful attempt so far at describing something that linguists often refer to but generally say exasperatingly little about, namely discourse structure. In the remainder of this note, I will be concerned with natural deduction as a framework for linguistic analysis of discourse more than as a system of logic, though the logician may find something of interest in my use of topological notions in discussing the structure of proofs.

Harry A. Lewis (ed.), Peter Geach: Philosophical Encounters, 151–160.

In McCawley (1981), I illustrate the rule of v-exploitation (3) by the natural language argument (4):

(3) $v(A_1, A_2, \ldots, A_n)$

$$
\begin{array}{l}
\quad \left| \begin{array}{l} \underline{A_1} \\ \ldots \\ B \end{array} \right. \\
\quad \left| \begin{array}{l} \underline{A_2} \\ \ldots \\ B \end{array} \right. \\
\quad \ldots \\
\quad \left| \begin{array}{l} \underline{A_n} \\ \ldots \\ B \end{array} \right. \\
B
\end{array}
$$

(4) Creepy Calabresi got off the plane in either Chicago, Kansas City, or Las Vegas. Suppose he got off in Chicago. Then we would have called his brother. But his brother wants to get rid of Creepy and he would have tipped off the feds. Suppose Creepy got off in Kansas City. Then he would have called his girl-friend. But his girl-friend is working for the IRS now, and she would have tipped off the feds. Suppose Creepy got off the plane at Las Vegas. Then he would have called the Fettucini Kid. But the Fettucini Kid has been arrested and the fuzz would have a stoolie taking the phone calls, and he would have tipped off the feds. So someone has tipped off the feds.

Let us consider how (4) might be recast as a derivation in Fitch's system. At least one discrepancy between (4) and its formalization is forced on us by the fact that *Someone tipped off the feds* appears only once in (4), whereas the corresponding symbol B appears $n+1$ times in (3) (once at the end of each of the n subproofs and once in the main proof), and hence the formalization of (4) will have to include four lines corresponding to *Someone tipped off the feds*. In addition, each of the segments of text that corresponds to a subproof in (4) contains at least one sentence that is presented as a fact rather than as an inference from the supposition of that subproof, e.g. *His brother wants to get rid of Creepy* is not an inference from the supposition that Creepy got off in Chicago but is something that the speaker is asserting categorically. In particular, the proposition that Creepy's brother wants to get rid of him (likewise the proposition that Creepy's girlfriend is working for the IRS and the proposition that the Fettucini Kid has been arrested) can serve as a premise for later inferences in the main proof, whereas of course the suppositions of the subordinate proofs and the subsequent lines of the subordinate proofs that have been inferred from those suppositions (such as the proposition that Creepy's brother has tipped off the feds) cannot.

Leaving aside the question of how the remaining inferences are made (e.g. how one infers that Creepy's brother tipped off the feds from the premises that Creepy called his brother and that Creepy's brother wants to get rid of him), we can offer as the closest counterpart to (4) in Fitch's system a derivation (5) in which material that in the ordinary language text appears in the middle of subordinate segments of text appears instead at the beginning among the premises, with the rule of *Reiteration* importing those propositions into the subproofs in which they are used, and the proposition that someone tipped off the feds appearing at the end of each of the subproofs:

(5)

1	v(p, q, r)	
2	p' (= Creepy's brother wants to get rid of Creepy)	
3	q' (= Creepy's girlfriend is working for the IRS)	
4	r' (= The Fettucini kid has been arrested	
5		p	
6		p''	
7		p'	2, reiteration
8		a tipped off the feds	
9		(∃x) (x tipped off the feds)	8, ∃-intro
10		q	
11		q''	
12		q'	3, reiteration
13		b tipped off the feds	
14		(∃x) (x tipped off the feds)	13, ∃-intro
15		r	
16		r''	
17		r'	4, reiteration
18		r'''	
19		c tipped off the feds	
20		(∃x) (x tipped off the feds)	19, ∃-intro
21	(∃x) (x tipped off the feds)	1, 5–9, 10–14, 15–20, v-exploitation

There is a trivial way in which the structure of proofs in the Fitch system could be altered so as to remove these discrepancies between the formal proof and the natural language text. In the Fitch system in its pristine form, proofs are *trees* in the graph-theoretic sense. More specifically, they are ordered labeled continuous trees[2] in which the terminal nodes are sentences and the non-terminal nodes dominate (sub-)proofs, with the sentences of any (sub-)proof distinguished as to whether they are suppositions or not. Thus, using the labels 'Proof' and 'S', marking suppositions and premises by a + and using numerals to identify the various propositions that appear in (5), the structure of (5) can be alternatively be represented as (6):

(6)

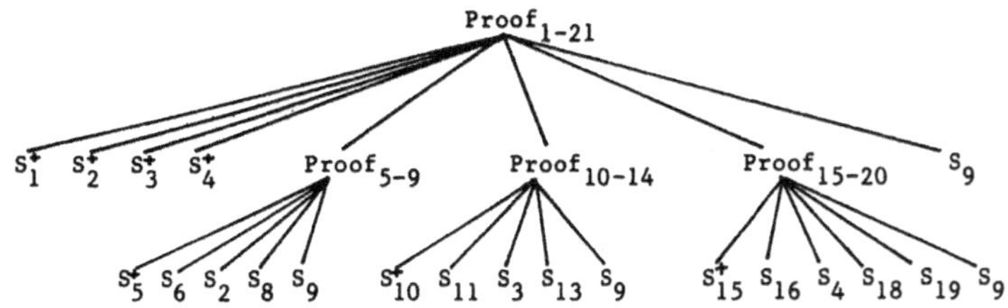

A graph with ordered terminal nodes is **continuous** if and only if whenever a node dominates two terminal nodes, it also dominates all nodes that are between those nodes. Let us consider a modified version of (6) that will fail to be continuous and will also fail to be a tree, in that certain nodes will have more than one immediate ancestor (i.e. the graph will 'branch upwards' as well as branching downwards). Specifically, let us replace by a single node all occurrences of any given proposition in (6), while retaining the domination relations of (6) and the linear ordering of the propositions given by the natural language text (4):

(7)

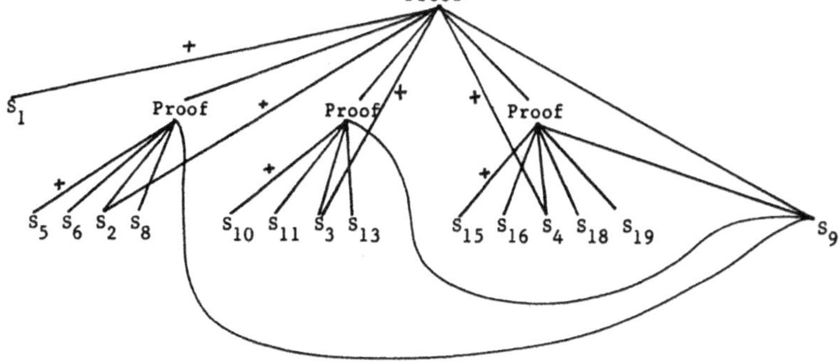

Rules of inference in the Fitch system can be regarded as conditions under which a proposition of a given form may occur in a given position in a proof. The entire proof is well-formed according to the given rules of inference if and only if (i) every line of the proof either is a supposition or conforms to one or other of the rules of inference, and (ii) the last line of the proof is directly dominated by the topmost Proof node, i.e. the proof must terminate in a line that is a conclusion of the main proof and not just a conclusion of one of the subproofs. Subject to three minor revisions, the same can be said even if the Fitch system is altered so that proofs are not trees but the more general sort of graph illustrated by (7). The minor revisions are: first, that now the rules of inference must be taken as giving conditions on the logical acceptability not of nodes of the graph (= lines of the proof) but of arcs connecting the nodes, i.e. if a given proposition plays several roles in the same proof, it must play all of those roles correctly if the whole proof is to be well-formed, and the different arcs terminating at a given node correspond to the different roles of the given proposition; second, that the status of a proposition as a supposition must accordingly be taken as a feature of an arc rather than of a node, since a proposition can be a supposition of only one proof at a time, e.g. in (7), S_3 is a constituent of both the main proof and the second subproof but is a supposition only of the former; and third, that whereas rules of inference in the pristine Fitch system are templates specifying what a given proposition must be preceded by if it is to be an acceptable inference, the propositions forming the justification for a given conclusion must now be allowed to either precede or coincide with it, e.g. the rule of v-exploitation (3) now is taken as allowing proofs like (7), in which the various occurrences of *B* in the subordinate proofs coincide with the conclusion *B* of the superordinate proof.

The justifications of the various steps (= arcs) in (7) are the same as the justifications of the corresponding steps in (5). In particular, it is the rule of Reiteration that allows S_2, S_3 and S_4 to belong simultaneously to the main proof and to the subordinate proof. It is not clear, though, that those propositions need be regarded as belonging to the subordinate proofs at all. They are used in the subordinate proofs, and the possibility of uttering them in the middles of the stretches of text corresponding to those subordinate proofs is effective in making clear that they are used in those subproofs, but since propositions of the main proof remain 'operative' in the subproofs, they need not actually belong to the subproof in order to be used there. Some discussion of the role of the rule of Reiteration is in order here. What use need be made of that rule depends on an important detail of the interpretation of the other rules of inference. I have tacitly assumed a policy that renders unnecessary many invocations of Reiteration that would otherwise be needed; specifically, I have assumed that, as in steps 4 and 5 of (2), the lines of proof that give the justification of a given step may be in a superordinate proof: lines 4 and 5 are derived on the basis of premises of the main proof that 'remain operative' in the subproof. The alternative policy would be to require the propositions from which a given proposition is inferred to be immediate constituents of the same (sub-)proof as it belongs to and to import additional propositions from superordinate proofs, as needed, by Reiteration. These two policies are equivalent with regard to what conclusions they allow one to derive from given premises. I have chosen the former policy because of the closer correspondence that it allows between formal proofs and natural language argumentation, in which superordinate steps are repeated (often in such a form as a *since*-clause) only for clarity's sake, much in the way that for clarity's sake one may use a proper name in places where a pronoun would otherwise suffice. Under this policy, there is indeed only a highly restricted class of proofs in the Fitch system in which Reiteration need be used (or simulated by combinations of other rules of inference),[3] namely those involving ∼-introduction (*reductio ad absurdum* proofs) in which one of the two contradictory propositions *B* and ∼*B* that figure in the subproof is imported from a superordinate proof:

(8) ∼-introduction

| *A*
|———
| . . .
| *B*
| ∼*B*

∼A

As the rule is formulated in (8), *B* and ∼*B* are not premises of the whole proof but must occur within the subproof. Moreover, it will not do to replace (8) with a rule (8′) in which one of *B* and ∼*B* is a premise, since that rule is weaker than (8), e.g. it does not allow a proof of the law of non-contradiction:

(8′)
$$\begin{array}{l} B \\ \begin{array}{|l} A \\ \hline \ldots \\ \sim B \end{array} \\ \sim A \end{array}$$

Thus, one could dispense with Reiteration (and simulations of it, cf. note 3) only by generalizing (8) so that it covered both inferences that conform to the template (8) and inferences that conform to something on the order of (8′), in which one of *B* and $\sim B$ is in the superordinate proof. As what can be regarded for the moment as a brute-force measure to stretch present notation so as to cover this generalized form of $\sim$-introduction, I will adopt the device of putting square brackets around symbols that can be matched indiscriminately to lines of the proof in which they appear or to lines of a superordinate proof:

(8″)
$$\begin{array}{l} \begin{array}{|l} A \\ \hline \ldots \\ [B] \\ \sim B \end{array} \\ \sim A \end{array}$$

[*B*] in (8″) means in effect '*B* has already been established', in that what is 'established' in a given proof remains 'established' in discourse that is subordinate to that proof. My talk of propositions counting as 'already established' at a given point in a proof is reminiscent of Lauri Karttunen's treatment (1974) of pragmatic presupposition. For Karttunen, a pragmatic presupposition of a sentence is a demand that that sentence makes on its *context*, where 'context' is understood in the technical sense of 'set of propositions that are taken as 'already established' for the purposes of the given discourse, at the given point of the discourse'. The context consists of 'background knowledge' that the participants in the discourse can take as shared by all of them, plus whatever propositions have been asserted by one or other of them and not challenged or withdrawn, i.e. the assertion of a proposition causes it to be added to the context as long as the participants at least tacitly consent to its being added to the context. To say that the sentence *John regrets that he never married* pragmatically presupposes that John never married is to say that it is normal to utter that sentence only relative to a context that entails the proposition that John never married. In McCawley (1979), I extend Karttunen's treatment to sentences such as *Mary is sure that John never married and she thinks that he regrets that he never married* by proposing that at any given point of a discourse there may be several 'contexts', one for each 'world' to which reference is made, and that the pragmatic presuppositions are demands on the context associated with the world that serves as reference point for the linguistic item that carries the pragmatic presuppositions, e.g. the pragmatic presupposition carried here by *regret* demands that the context for 'Mary's belief world' entail that John never married, and the speaker has caused the proposition that John never married to be added to the belief-world context by asserting that Mary is sure that John never married, thus insuring that the demand made by *regret* will be met.

Karttunen's notion of context is applicable to texts in which proofs are presented. At any point in any (sub-)proof, the 'context' corresponding to that subproof consists of those propositions that have been supposed or inferred earlier in that (sub-)proof or in a superordinate proof. Each step in the proof causes the proposition derived in that step to be added to the context for the (sub-)proof in which it occurs, though not to the contexts for superordinate proofs. At the beginning of each subproof, the context for that subproof consists of the context for the immediately superordinate proof, e.g. in (2), the subproof begins with a context consisting of the propositions given as lines 1 and 2; the context for the subproof is then supplemented in turn by lines 3, 4, and 5, while the context for the main proof continues to consist just of lines 1 and 2. The requirement that the lines of a proof that provide the justification for a particular step be in either the same (sub-)proof or a superordinate proof can then be restated as a requirement that those steps be in the context (in Karttunen's sense) of the given step.

This restatement becomes more than a mere change of terminology if one allows the context at the outset of the proof to be non-null, i.e. if one allows inferential steps to depend not just on earlier steps of the proof but also on propositions that are taken as 'background knowledge'. A large proportion of textbook examples of the various rules of inference in fact involve exactly that. Consider, for example, everyone's favorite example of *reductio ad absurdum* proof, namely the classical demonstration that $\sqrt{2}$ is an irrational number. Several propositions of number theory figure in the derivation of a contradiction from the supposition that $\sqrt{2}$ is rational: the proposition that every rational number is a ratio of two integers that have no common factor, the proposition that $(2c)^2 = 4c^2$, the proposition that if $x = y$, then $x/2 = y/2$; etc. As much of 'elementary' number theory as one feels like employing is taken as 'already established' for the purposes of this proof.

In this article, I have pointed out a number of ways in which a well-known system of natural deduction can be altered so that, while leaving unchanged the conclusions that can be derived from given premises, we are able to keep to a minimum the descrepancies between natural language argumentative texts and formal proofs that purport to reconstruct the inferential structure of those texts. I regard the elaboration of such variants of natural deduction systems as important for the contribution that they can potentially make to a better understanding of the relationship between formal logic and the psychology of reasoning. The closer that the formalization of a system of logic can be made to agree with the structure of natural language texts the fewer irrelevant distractions there will be for the scholar who seeks to identify the principles and structures that figure in psychological processes of reasoning.

Department of Linguistics,
University of Chicago

NOTES

[1] In this article I take as reference point the system of rules of inference presented in Fitch (1952), except that, as in my (1981), I assume a system of restricted rather than unrestricted quantifiers, and I treat $\wedge$ and $\vee$ as combining with arbitrarily many propositions at a time rather than with just two and accordingly, for convenience' sake, write all propositional connectives, even $\supset$, before rather than between propositions that they are combined with. In the names of rules of inference, I have replaced Fitch's 'elimination' by the term 'exploitation', on the grounds that the logical operator may continue to play a role after the application of the rule of inference and is thus not really 'eliminated', e.g. if you infer p from $\wedge pq$ by '$\wedge$-elimination', the $\wedge$ of the $\wedge pq$ is still available for later exploitation: you can use the same premise to infer q 6 lines later if you so choose.

[2] On the role of trees of this specific type in linguistics, see McCawley (1968) and Wall (1972).

[3] I allude here to the possibility of importing a superordinate line B into a subproof by conjoining it with any earlier line C of the subproof or of a superordinate proof, thus

introducing $\wedge BC$ into the subproof by $\wedge$-introduction, and then deriving B from $\wedge BC$ by $\wedge$-exploitation.

[4] In Section 11.2 of McCawley (1981), I use the notion of multiple contexts in discussion of the problem of 'intentional identity' raised by Geach (1967). I speculate there that an adequate solution to Geach's problem will necessarily involve both logic and (quite literally) 'sociology of knowledge' and note that the very problem raised by Geach is also presented by perfectly ordinary descriptions of scientific research.

REFERENCES

Fitch, F. B.: 1952, *Symbolic Logic*, Ronald, New York.

Geach, P. T.: 1967, 'Intentional Identity', *Journal of Philosophy* **74**, 627–632.

Karttunen, L.: 1974, 'Presupposition and Linguistic Context', *Theoretical Linguistics* **1**, 182–194.

McCawley, J. D.: 1968, 'Concerning the Base Component of a Transformational Grammar', *Foundations of Language* **4**, 243–269.

McCawley, J. D.: 1979, 'Presupposition and Discourse Structure', David Dinneen and C. K. Oh (eds.), *Syntax and Semantics 11: Presupposition*, Academic Press, New York, pp. 335–337.

McCawley, J. D.: 1981, *Everything That Linguists Have Always Wanted to Know about Logic (but Were Ashamed to Ask)*, University of Chicago Press, Chicago.

Wall, R.: 1972, *Introduction to Mathematical Linguistics*, Prentice Hall, Englewood Cliffs.

MICHAEL DUMMETT

DOES QUANTIFICATION INVOLVE IDENTITY?

How are 'every', 'some' and their like related to 'the same as'? Is there any ground for saying that an understanding of the former—of expressions of generality—presupposes an understanding of the latter—of expressions of identity? Let us begin with formalised languages, those framed by use of the apparatus of predicate logic. In a second-order language, identity is definable, so no problem arises. If, in a first-order language, identity is taken as primitive, it is at least unproblematic that an interpretation of the language must involve *a* relation of identity over the domain. But suppose that a language is framed within first-order logic *without* identity: is there then any need to assume that an interpretation of the language will involve or invoke a relation of identity over the domain?

The idea that there is such a need is forcefully expressed in the following quotation from Quine's review of *Reference and Generality*[1]:

> This doctrine [Geach's denial of the existence of an absolute relation of identity] is antithetical to the very notion of quantification, the mainspring of modern logic. Quantification depends upon there being values of variables, same or different absolutely; grant quantification and there remains no choice about identity, not for variables. For a language with quantification in it there is but one legitimate version of '$x = y$'.

Now why should this be so? Why, if the language contains no identity-predicate, should the mere assignment of a domain to the variables involve assuming a relation of identity between its elements? The response that suggests itself it that, given a domain, the *possibility* of defining a relation of identity over it—the minimum reflexive relation—is always open. A more detailed form of argument to this effect might run as follows. If we are engaged in giving a verbal *statement* of the interpretation of the object-language, we have first to specify the domain of the variables. To give the interpretation of the non-logical constants, we have to be able to refer in the metalanguage to elements of that domain, or to pick out subsets of it. In order to know whether a given interpretation is admissible, that is, intelligible, we must know when two terms of the metalanguage pick out the *same* element of the domain, since the requirement that one and the same element, con-

Harry A. Lewis (ed.), Peter Geach: Philosophical Encounters, 161–184.

sidered as picked out by each of two distinct singular terms of the metalanguage, should behave differently in respect of the satisfaction of some predicate of the object-language will render the interpretation contradictory and so inadmissible. Hence, to give an interpretation relative to a domain presupposes a relation of identity defined over it. And, according to the present argument, the same applies to the case in which we have only a mental apprehension, rather than a verbal statement, of the interpretation. In such a case, we must grasp the domain, and have some means of picking out — mentally alluding to — its elements; and, in apprehending the interpretation of any predicate relative to this domain, we must in the same way know under what conditions two such allusions are to the same or to distinct elements.

In any case, as Quine has pointed out, for any first-order language L we can *define* a relation $=_L$ which will satisfy the axioms for identity. This relation is to be defined as follows:

$$x =_L y \leftrightarrow [\bigwedge_{j=1}^{k} \bigwedge_{i=1}^{n_j} \forall z_1 \ldots \forall z_{i-1} \forall z_{i+1} \ldots \forall z_{n_j} (P^{(j)} z_1 \ldots z_{i-1}\, x\, z_{i+1} \ldots z_{n_j} \leftrightarrow P^{(j)} z_1 \ldots z_{i-1}\, y\, z_{i+1} \ldots z_{n_j})].$$

Here $P^{(1)}, \ldots, P^{(k)}$ is an enumeration of the finitely many primitive predicates of L, where each predicate $P^{(j)}$ $(1 \leqslant j \leqslant k)$ is of degree n_j; given the actual values of $k, n_1, \ldots, n_k$, we can of course write out the definition in full.[2] The definition captures the notion of being indiscernible relative to the language L. Given that $=_L$ is an identity relation in the sense of one satisfying the standard axioms for identity, what problem remains? We may try enunciating two possible theses:

(i) The defined symbol $=_L$ *must* always be interpreted as genuine identity.

(ii) The defined symbol $=_L$ *can* always be interpreted as genuine identity.

Of these, (i) is obviously false. Given any model M of a theory T in L, if M has an infinite domain, we can always find another model M′ of T by adding a new element b to the domain which behaves with respect to the interpretations of the predicates of L just like some fixed element a of the original domain; $=_L$ will then come out as holding between a and b. On the other hand, (ii) is equally obviously true. Given any model M of a theory T in L, $=_L$ will be interpreted as an equivalence relation over its domain, indeed a congruence relation with respect to the predicates of L. We can therefore find a new model M^* of T whose domain consists of the equivalence classes into which the domain of M was partitioned by the relation which interprets $=_L$; in M^*, $=_L$ will come out as genuine identity. So far, then, we have found no problem.

The quotation from Quine occurred as part of an attack on Geach's doctrine of relative identity, and I turn to Geach's article 'Ontological Relativity and Relative Identity' for a formulation of that view.[3] Geach claims that for no language L can $=_L$ denote absolute identity. Since he does not believe that there is any such thing as absolute identity, he has a little difficulty in explaining what it is that he is saying that $=_L$ cannot be interpreted to mean. One attempt he makes is to say that, while $=_L$ expresses indiscernibility within L, absolute identity would consist of indiscernibility within any possible language; he goes on to criticise quantification over all possible languages as suspect in view of the semantic paradoxes. I will not spend time discussing how grave this suspicion is: we do not need to invoke the totality of all possible languages. In giving the familiar proofs of the falsity of thesis (i) and the truth of thesis (ii), I did not appeal to any such totality: the 'genuine identity' referred to in those theses was simply the identity of the metalanguage. A relativist may seize on this, and remark, 'You cannot escape relativisation to a language: you must speak always of identity as expressed in some language, not as independent of any particular language'. This remark, taken in one way, is patently true, indeed platitudinous; taken in another way, it is patently false. It is platitudinous in the sense that, if I am speaking English, and use the word 'identical', I mean it to be understood in the sense that 'identical' bears in English, just as with any other word that I use; or, again, if we, speaking English, contemplate a speaker of a language L' asking, in L', whether the relation denoted by $=_L$ for some third language L is really identity, we shall perceive that he is asking whether the interpretation of $=_L$, as stated in L', coincides with the relation expressed by the identity-predicate of L'. The remark is patently false in the following sense: that if we ask in English whether $=_L$ does or does not denote the relation of identity, there is nothing wrong with our question; it is not ambiguous or incomplete. We do not need to qualify the word 'identity' as 'identity in English', any more than we need to do when we say that the rules of baseball are identical with those of rounders.

Geach says, 'We can never so specify what we are quantifying over that we are secure against an expansion of our vocabulary enabling us to discriminate what formerly we could not'. What, then, is his thesis? We might state it as follows:

(iii) For any theory T in a language L, there exists an extension T' of T in an expansion L' of L such that the truth of T' demands that the interpretation of $=_{L'}$ does not coincide with that of $=_L$.

Geach uses 'theory' to mean 'interpreted theory' and 'language' to mean 'interpreted language'. We begin, therefore, with a model M of T, and advance to a model M' of T' of which M is a submodel. Now, clearly, if the domain of M is infinite, we shall be able to find a model M' of T itself whose domain properly contains that of M, such that $=_L$ does not denote identity over the new elements. But that is not what Geach had in mind: he meant that $=_{L'}$ will not be interpreted as coinciding with $=_L$ over their common domain. Whether this can happen or not depends not only upon how the domain has been specified, but also on the interpretation of the predicates of L. If, in M, they were so interpreted that $=_L$ does not come out as the identity relation over the domain D_M, then the situation envisaged by Geach is possible. If, however, the predicates of L were so interpreted that $=_L$ came out as the identity relation over D_M, then no such possibility exists. The general thesis (iii) is therefore plainly false.

Nevertheless, Geach has a formal argument to support his view. He attributes to a believer in absolute identity the thesis:

(iv) For any language L, we can so construe L that, if $=_L$ holds between x and y, then x and y are absolutely identical.

If 'absolutely identical' is understood to mean 'genuinely identical', i.e. 'identical' (as the word 'identical' is used in the metalanguage), then this is just another formulation of that thesis (ii) which we have seen to be evidently true. Geach has, however, a refutation of it. Suppose that the theory T, expressed in the language L, is ontologically committed to the existence of men. Let L_0 be a proper sublanguage of L, and suppose that $=_{L_0}$ holds between men if and only if they have the same surname. Let T_0 be the restriction of T to L_0. Then, Geach says, on thesis (iv) T_0 is ontologically committed to the existence of creatures for whom possession of the same surname is a criterion of absolute identity. But, in that case, T is also committed to the existence of such creatures, 'since a theory picks up the ontological commitments of each subtheory'. But this, he triumphantly concludes, is absurd; hence thesis (iv) must be false.

This argument depends upon an oscillation between taking a language, or a theory, to be interpreted and taking it to be uninterpreted. Given a model M of T (formulated in L), by restriction we obtain a model M_0 of T_0 (formulated in L_0) having the same domain as M. In the case imagined, $=_{L_0}$ will not, in M_0, denote identity. But that does not disprove thesis (iv), which merely stated that there would be *some* model of T_0 in which $=_{L_0}$ denoted identity. If we consider such a model, say M_1, it need not be a submodel of M: but it is only for *interpreted* theories that a subtheory can be claimed to carry over its ontological commitments to the main theory. Indeed, this claim needs very careful statement if it is to have any cogency at all: M_1 might in fact happen to be a submodel of M, namely by containing, within its domain, just one man for each surname; and then there would be no reason at all why the theory T, with the model M, should 'pick up' the requirement that $=_{L_0}$ should express identity over the domain of M.

The fact is that, even though Geach himself engages in it, discussion of the matter along the lines we have so far pursued misses the point of Geach's real concern. Theses (i) and (ii) are formulated on the assumption that, in the metalanguage, we can intelligibly speak of *the* identity-relation over the domain of an interpretation. Given this assumption, it requires no elaborate argument to refute thesis (i), and it is vain to attempt to refute thesis (ii); so Geach has no need of any arguments which appear to be aimed at refuting either thesis on its own terms. What he really wishes to do is to challenge the terms in which they are formulated, that is, to challenge the assumption that we may reasonably speak of *the* identity-relation over a given domain.

One of Geach's principal contentions is that it can *always* happen that, for some expansion L' of a language L, $=_{L'}$ comes out as a finer equivalence relation than $=_L$; as we continue to expand the language, we might get finer and finer equivalence relations ad infinitum. How can this be, we may ask, if we happened to start with a language L for which $=_L$ comes out as the identity-relation over the domain, or if we happened at some stage to arrive at a language for which this was so? The answer is that we are begging the question in supposing this to be an intelligible possibility: we are presuming that, given a domain, there must be some one determinate relation over it which constitutes genuine identity, *the* identity-relation associated with that domain. In specifying the domain, we do not, according to Geach, provide for any such relation; if we did, then there would be a limit to the process by which, for L taken as ranging over a sequence of more and more inclusive languages, $=_L$ may come to express a finer and finer equivalence relation: but, since we do not, there is no such limit.

Geach's central concern is thus with the way in which we specify the domain of an interpretation. Logicians usually say merely that, in giving an interpretation, we need to specify a non-empty domain, without bothering themselves with the means by which this is to be done. (Sometimes they add that the domain must be taken to be a *set*, and, in this, a severe restriction may be implicit: but, to make it explicit, we should have to go very deeply into the notion of a set.) How, then, do we specify a domain? Suppose that our metalanguage is English, or some other natural language. Then one natural way to specify a domain would be by the use of a count noun, either unrestricted or restricted by some predicate embedded in a relative clause: thus we might say that the domain was to consist of dogs, or of all dogs now alive in California. For present purposes, however, this seems an unpromising approach. It is a cardinal doctrine of Geach, and one that we should hardly want to deny, that, for any count noun '*A*', the expression 'the same *A* as' has a determinate sense: so, for a domain specified by means of the count noun '*A*', the relation expressed by 'is the same *A* as' will surely constitute an identity-relation which is the limit of all possible interpretations of $=_L$, for variable L, over that domain. If, e.g., '*A*' is 'dog', we can hardly introduce an interpretation under which *a* is the same dog as *b*, but *a* satisfies some predicate which *b* fails to satisfy.

It is for this reason that, instead of considering a specification of a domain by means of a count noun, Geach treats instead of a specification by means of a list, finite or infinite; I shall make no cavil about the idea of an infinite list. He insists that a list, used for this purpose, may contain repetitions. In order that we may use it to specify a domain, we do not need to know whether it contains repetitions or not. Indeed, it does not make sense to ask whether any element is listed twice—we can only ask, for this or that count noun '*A*', whether two entries mention the same *A*; the answer may be different, according to the count noun we select; and there is no *one* count noun which is the right one to use for this purpose.

These remarks hang oddly together with the thesis to which Geach subscribes, as do I, that to every proper name is associated some determinate criterion of identity, given that he acknowledges that a list is a string of proper names; indeed, without affecting the issue, we could allow the list to contain singular terms other than proper names. It is not that Geach has gone back on this thesis concerning proper names, since he reiterates it in the very article, 'Ontological Relativity and Relative Identity', under discussion. If to two proper names occurring in the list there are associated conflicting criteria of identity, it would not appear to make sense to say that they denoted the same object; and, when the same criterion is associated with two names, that criterion determines the sense of asking whether a repetition is involved, without the need for any arbitrary choice. Given an assignment, to each entry on the list, of an appropriate criterion of identity, it seems that an identity-relation over the elements of the domain specified by the list had thereby been determined, one that will, as before, set bounds to the fineness of the interpretation of any $=_L$.

In any case, we must enquire whether, in specifying the domain by means of a list, we thereby impose any constraints upon the interpretations that may be given relative to this domain, or whether we are allowed to make any entry behave differently, with respect to the interpretation of some predicate, from any one other entry. If the proper names making up the list are thought of as having senses associating with each of them a criterion of identity, we should surely want to impose the constraint that any two names which, according to the relevant criterion of identity, denoted the same object must behave alike in respect of the interpretation. Even if the names are not said to have such senses, there must be some means employed for fixing their reference, since otherwise there would be no content to calling them 'names'; and, at least in the case in which, for some two entries, the *means* of fixing their reference completely coincided, we should expect a ban on allowing them to behave differently with respect to any interpretation. In any case, whether or not any such constraints are imposed, it would seem that the *list* itself, taken together with those constraints, if any, sets a bound to the fineness of any equivalence relation expressed by some $=_L$. For, even if there are no constraints, we surely cannot get any finer equivalence relation than that which holds between the object corresponding to each entry and itself, but not between any such object and the object corresponding to any other entry. After all, even if we do not mind a list's being repetitive, we shall want it to be exhaustive; we want to be sure that, for each item we wish to include, there is at least one entry to which it uniquely corresponds. It seems, therefore, that we must employ some criterion for what merits separate inclusion in the list; this criterion must be such that, for any count noun '*A*', if we want to admit interpretations such that, if *x* is not the same *A* as *y*, then *x* may satisfy some predicate which *y* fails to satisfy, then there is no entry on the list to which *x* and *y* both correspond. For, of course, the picture aroused by the conception of specifying a domain by means of a list is that the interpretations of the predicates over this domain are given by citing the entries, or *n*-tuples of entries, on the list which satisfy those predicates.

It is, however, exactly this which Geach denies. The insistence on the possibility that the list may contain repetitions was really beside the point: the real point is that it may contain compressions. 'If', he says, 'we list the things we are quantifying over by their names, one of these names may turn out to be not a proper name but a shared name, of objects that we now can discriminate but previously could not'. How does it come about that we make this new, finer, discrimination? Not, surely, by an expansion of the formal object-language. If we discover that some name, '*a*', occurring on the list is really a shared name, we shall, presumably, replace it as quickly as possible by two or more new names, say '*b*' and '*c*'. Until we have made this distinction between *b* and *c*, we have no means of introducing into the expanded language a predicate which is true of *b* but not of *c*, since we have no means of formulating the intended interpretation of such a predicate; the discrimination cannot, therefore, be a consequence of the fact that $=_{L'}$ does not hold between *b* and *c*, although $=_L$ did, since we cannot even make the stipulations that have this effect until we have first distinguished between *b* and *c*. How, then, does this discovery that '*a*' was really a shared name differ from a *change* in the domain of quantification? In what sense does the domain remain the same?

What, indeed, does such a discovery consist in? We may, admittedly, have made a mistake in listing '*a*' in the first place: we may have been using the name '*a*' indifferently for each of two twins, whom we wrongly took to be the same person, and now we discover our mistake. It must be acknowledged to be obvious that no specification of a domain is immune to a mistake of this sort: but, while the domain we specify when we replace '*a*' by '*b*' and '*c*' is that which we were all along aiming to specify, the original specification was simply wrong, and did not fully succeed in its aim. Evidently, Geach did not have cases of this sort in mind; so let us change the example. Suppose, now, that we live in a society familiar with the phenomenon of identical twins, but in which the word 'person' is so used that a pair of twins is construed as a case of the bilocation of a single person: '*a*' names such a bilocated individual. Later, under foreign influence, this society adapts its use of the word 'person' so that twins are counted as distinct people, and this particular pair are now named '*b*' and '*c*'. This time, no mistake is involved, at least not a *mere* mistake: there is conceptual change. Again, we must concede that no specification of a domain is proof against revision under the pressure of such conceptual change. But, this time, it is unclear in what sense the revised list specifies the *same* domain as the original one; in attempting to view it as doing so, we only assimilate this case to that in which a mistake is involved. In any case, our new capacity—better, propensity—to discriminate between the twins as distinct individuals is not due to any enrichment of the object-language. On any reasonable choice of the language L, as originally interpreted over the domain given by the list containing '*a*', $=_L$ will now be regarded as failing to hold between the twins; e.g. if 'Fx' means 'x is in England' and 'Gx' means 'x is in Spain', it may be that 'Fx' is true of one twin and false of the other, and 'Gx' true of the other and false of the one, although, originally, we rated 'Fx & Gx' true of the object denoted by '*a*'.

So far, then, it appears that Quine was right: the interpretation of a first-order language L demands that a domain be specified for the variables; in specifying such a domain, we cannot avoid determining an identity-relation over it which constitutes the finest possible interpretation of $=_L$, and determines the admissibility of proposed interpretations of the predicates of L relative to that domain. However, Geach's principal concern seems to be with natural, not with formalised, languages: and so we have to ask, for natural languages, the analogue of the question we have been considering for formalised languages. It could even be thought that Geach could quite consistently have conceded the point about formalised languages, say by allowing that the specification of a domain for the variables of such a language always involves the use of some count noun '*A*', and hence that 'is the same *A* as' will express the identity-relation associated with that domain, and yet have maintained his contention for natural languages. For reasons that I shall explain, I do not think that this is quite true: but that is far from obvious at first sight.

Natural languages differ from formalised languages in the manner by which they express generality. Their means of doing so differs in two respects from that of a quantificational language. (i) They lack the machinery of quantifiers and variables: the noun-phrase, containing the indefinite article or some word like 'some' or 'every', stands in the argument-place of the predicate; sometimes a pronoun serves as a variable bound by this noun-phrase, sometimes the effect of further occurrences of the bound variable is obtained by a range of devices including the use of connectives between words or phrases that fall short of being clauses. (ii) Usually the operative noun-phrase contains a common noun, viz. one admitting a plural, expressing more than a formal concept, specifically a count noun. (By a noun expressing a formal concept is meant one with no informative content, that is, one the applicability of which, when it is correctly used, cannot be meaningfully questioned. The word 'thing' is the prototype of such a word: if someone says, 'That was a terrible thing to happen', his hearer cannot appropriately reply, 'It was terrible, all right, but I shouldn't call it a thing'.) I shall consider only restricted generality, of the kind expressed by a phrase containing a count noun, and not that expressed by one containing a mass term, nor yet unrestricted generality, both of which raise different problems. For our purposes, feature (i) is unimportant: our problems would be unaffected by the introduction of bound variables, namely if, instead of saying, 'A dog rushed out and bit Henry', we said, 'For some dog *x*, *x* rushed out and *x* bit Henry'. The important point is not whether or not variables are used, but that there is not a fixed number of sorts of variable, given in advance, each with its associated domain; instead, the range of generality is indicated by the count noun occurring in the noun-phrase.

A count noun is naturally taken to be a noun '*A*' that can occur in contexts of the form 'There are just *n* *A*'s' or 'There are just *n* *A*'s which are *F*', when the latter means 'There are just *n* *A*'s each of which is *F*'. A count noun must therefore admit both a singular and a plural, and be capable of occurring as a predicate 'is an *A*'; and, necessarily, the phrase 'is the same *A* as' must have a determinate sense.

What *is* the analogue, for natural languages, of the question we asked about formalised languages? It emphatically is *not* the question: Is there an absolute relation of identity? One version of a negative answer to *that* question would be to say that the phrase 'is the same as' has a clear sense only when the context determines a count noun '*A*' such that we can replace the phrase by 'is the same *A* as'. Far from wishing to controvert *this* thesis, I accept it as true. A stronger version is to say that it sometimes holds, for particular count nouns '*A*' and '*B*', that we can truly say that something is the same *A*, but not the same *B*, as something else. I do not wish to controvert even this stronger thesis: under some possible uses of certain count nouns, it is indisputably true.

For any count noun '*A*', let us say that 'is the same *A* as' denotes an identity-relation. The analogue of our original question is, now: Is there, for any expression of restricted generality of the kind with which we are concerned, a definite identity-relation associated with the range (or domain) of generality? It may seem quite trivial to answer this question. An expression of restricted generality of the kind with which we are concerned is effected by means of such a phrase as 'Every *A*', 'Some *A*', 'Every *A* which is *F*', 'Some *A* which is *F*', etc., where '*A*' is a count noun. It may, then, seem obvious that, for any such phrase, the identity-relation associated with the range of generality will be that expressed by 'is the same *A* as'. However, we shall see that this cannot be so certainly concluded in every case: although I shall return an affirmative answer to the question, it will not be quite so easily arrived at.

Geach characterizes a word '*A*' as a count noun if it is subject to the following pattern of explanation. First we introduce the expression 'is the same *A* as', as standing for some equivalence relation. Secondly, we use the predicate 'is an *A*', which is definable as meaning 'is the same *A* as something': this, he says, is a case of derelativisation, comparable to the definition of 'is a father' as meaning 'is the father of someone'. In neither case could we proceed the other way round: we could not first introduce the predicate 'is a father', and then explain what it was to be someone's father; likewise, 'is the same *A* as' is not paraphrasable as 'is an *A* and is (absolutely) the same as'. Next, we explain sentences of the form 'There are just *n* *A*'s' as holding good just in case we can assign the natural numbers from 1 to *n* 'to such objects in a domain as are *A*'s'—that is, to those objects of which 'is an *A*' is true—according to the principle that the same number is to be assigned to an object *x* and to an object *y* if and only if *x* is the same *A* as *y*.

The expression 'is the same *A* as' may or may not be explicable by some verbal definition: Geach considers an artificial example in which it is. We introduce the word 'surman' by laying down that 'is the same surman as' is to hold between two men if and only if they have the same surname. The expressions 'is a surman' and 'There are *n* surmen' are then explained in accordance with the general stipulations.

I wish to single out two particular classes of what are ordinarily reckoned as count nouns that are not subject to Geach's pattern of explanation. Consider, first, the word 'father'. This certainly functions as a count noun in the ordinary sense: it admits a plural, and we can sensibly ask, 'How many fathers are present tonight?' Geach is precluded from saying that it accords with his pattern of explanation, since he cannot claim that 'is a father' is to be explained as meaning 'is the same father as someone'; on the contrary, he is committed to the more plausible view that it should be explained as meaning 'is the father of someone'. Indeed he expressly acknowledges this in his paper 'Names and Identity',[4] saying, '"is a brother" is a derelativization not of "is the same brother as" but of "is brother of"'. There is no reason to hold that 'is the same father as' is logically prior to 'is a father'; on the contrary, 'is the same father as' is most naturally explained as meaning 'is a father and is the same man as'. Geach actually concludes that 'is the same brother as' is 'an ill-formed expression generated by false analogy'.[5] He says that 'It could be legitimate only if "is a brother" were its derelativization'. The suggestion appears, therefore, to be that a noun that does not fit the preferred pattern of explanation by reference to which Geach characterizes count nouns cannot meaningfully appear in the gap within 'is the same... as'. The suggestion is implausible in itself; if it were correct, it would follow that there is not the connection which Geach claims in 'Ontological Relativity and Relative Identity'[6] between what Frege called statements of number and what Geach admits as count nouns; more specifically, it would follow that not every noun which can immediately follow such an expression as 'There are three...' can appear in the gap of 'Is the same . . . as'. The point does not apply only to count nouns '*A*' such that 'is an *A*' is to be explained as 'is an *A* of something', but, equally, to such a noun as 'baker'. There is an intuitive absurdity to saying that 'is a baker' is to be explained as meaning 'is the same baker as someone', which is borne in on us when we ask how we might explain 'is the same baker as': the obvious way to explain the latter phrase is as meaning 'is a baker and is the same person as', while 'is a baker' has a straightforward explanation as meaning 'is a person who earns his living by baking bread'. The latter explanation, if taken as primary, yields just the right meaning for 'is the same baker as', showing that there is no reason to take the latter as logically prior to the former. I shall call count nouns '*A*' such that 'is the same *A* as' is to be explained as meaning 'is an *A* and is the same *B* as', for some other count noun '*B*', *derivative count nouns*: they form one important class not subject to Geach's pattern of explanation.

Consider, now, the noun 'nationality'. There is some uncertainty whether we should allow 'is the same nationality as' to occur only in contexts such as 'Persian is the same nationality as Iranian', or also in contexts like 'Henri is the same nationality as Pierre'. Some would say that sentences of the latter form exhibit an ungrammaticality of the same sort as that which they would detect in such a sentence as 'She was wearing a kind of kimono' (instead of 'She was wearing a kimono of a kind'), and which is seen, grossly, in 'He kind of slumped in his chair', and that we ought only to say, 'Henri is of the same nationality as Pierre', or 'Henri has the same nationality as Pierre'; certainly the suggestion is in line with French and German, although both admit the 'a kind of . . .' formation. I do not wish to legislate on the grammatical point: we can consider both alternatives.

'Nationality' is certainly a count noun: it admits a plural, and we can sensibly ask, 'How many nationalities are represented at this conference?'. If we rule out sentences like 'Henri is the same nationality as Pierre', then 'is the same nationality as' can express only an equivalence relation defined over nationalities. In this case, 'is a nationality' is indeed equivalent to 'is the same nationality as something': nevertheless, the noun 'nationality' is not subject to Geach's pattern of explanation, for two reasons. First, 'is the same nationality as' does seem to be paraphrasable as 'is a nationality and is the same as'; indeed, just because it is over *nationalities* that the equivalence relation is defined, we can hardly think of 'is the same nationality as' as being introduced before 'is a nationality' is understood. Secondly, 'is a nationality' is not *only* explicable as 'is the same nationality as something'; it is better explained as meaning 'is the nationality of someone'. Frege long ago described a pattern according to which a count noun like 'nationality' may be thought of as being introduced. We first introduce 'has the same nationality as' or 'is of the same nationality as' as expressing as equivalence relation between human beings, in this particular case one definable as 'is a citizen of the same country as'. Next, we form the operator 'the nationality of...', so used that 'the nationality of *x* is the same as that of *y*' is equivalent to '*x* has the same nationality as *y*'. Finally, we explain 'is a nationality' as meaning 'is the nationality of someone'. (The only feature of Frege's account which I have omitted, and the only controversial one, is the detailed means of taking the second step, of introducing the operator 'the nationality of...'.) I shall call count nouns subject to Frege's pattern of explanation *abstract count nouns*: other examples are 'weight', 'status' and 'species' (in the last case, the operator takes the form 'the species to which... belongs').

If we adopt the second alternative, and admit 'Henri is the same nationality as Pierre' as grammatical, thus regarding 'is the same nationality as' as expressing an equivalence relation between human beings, then indeed the phrase, so understood, must be viewed as having priority in the order of explanation over other phrases involving the word 'nationality'; it will not be equivalent to 'is a nationality and is the same as'. But the noun will still not exhibit Geach's pattern of explanation, because 'is a nationality' will not be equivalent to, let alone be definable as, 'is the same nationality as someone': Henri is the same nationality as Pierre, but neither is a nationality. In fact, the word 'nationality', so used, is subject to precisely the same Fregean pattern of explanation as before, save that the equivalence relation is expressed by 'is the same nationality as' in place of 'has the same nationality as'.

For any count noun '*A*', whether of the type of 'surman', of 'baker', of 'nationality', or of none of these, I shall call the relation expressed by 'is the same *A* as' a *relative identity-relation*. It will be noted that, for different replacements of '*A*', a phrase of the form 'is the same *A* as' may function in different ways, according as (a) it does or does not stand for a relation between things of which 'is an *A*' is true, and (b) it permits suppression of the count noun '*A*' without ambiguity, wherever it is properly used.

We have looked at two important classes of count nouns not subject to Geach's pattern of explanation; their existence shows that not all words that would ordinarily be classed as count nouns are subject to that pattern. It is, however, comparatively unimportant that some words naturally thought of as count nouns do not conform to that pattern of explanation; the important claim made by Geach is that some do so. But we need to ask whether, if a count noun '*A*' has been introduced in accordance with Geach's pattern of explanation, that manner of introduction completely determines its use in the language. For instance, have Geach's stipulations concerning his artificial word 'surman' completely specified the use of this noun? Geach thinks so: he insists that, in sentences involving this word, 'we need not use' other 'words in some artificial sense that I have negligently failed to specify'. At first sight, however, there are many possible contexts that Geach has not provided for. We expect that any count noun '*A*' will figure, not only in the three contexts Geach explicitly considers, viz. 'is the same *A* as', 'is an *A*' and 'There are just *n* *A*'s', but also in contexts of the following kinds: 'is an *A* which is *F*'; 'There are *n* *A*'s which are *F*'; 'some *A*'; 'every *A*'; 'an *A*' in subject position (including grammatical direct and indirect object position, after a preposition, etc.); 'some *A* which is *F*'; 'every *A* which is *F*'; and 'an *A* which is *F*' in subject position. It does not appear that these contexts have been automatically provided for as soon as the three contexts which Geach deals with have been stipulated for. And, on reflection, it seems that all these other contexts will have been wholly or largely provided for if we lay down which predicates may be significantly applied to *A*'s, and under what conditions each such predicate is true of some one given *A*. With derivative count nouns, this presents no problem: we shall apply to bakers just the same predicates that we apply to men, subject to the same satisfaction-conditions. It is different with 'nationality' or 'species'. The things that can be said of nationalities are not at all the same as those that can be said of people, and an understanding of the word 'nationality' depends upon knowing what things can be said of a nationality and under which conditions it is true, for instance that someone can be said to acquire or to be deprived of a nationality, and the like. Similarly with 'species': things can be said of a species that cannot meaningfully be said of any of its members, for instance that it is extinct, and things that can be said of a species and can also be said of its members, e.g. that it is degenerate or that it is 10,000,000 years old, are said of the species in a different sense. I do not mean to place a strong construction on 'in a different sense' here; I mean only that the condition for a species to be degenerate is not to be explained in terms of the condition for any of its members to be degenerate.

Once we have laid down which predicates can be applied to *A*'s, and under what conditions they are true of a given *A*, we are in a position, not merely to understand the use of '*A*' in any of the contexts I listed, but also to introduce proper names of *A*'s. In this connection, the expression 'proper name of an *A*' must be treated with caution. By Geach's stipulations, Jimmy Carter is a surman, and hence, in one sense, 'Jimmy Carter' is the proper name of a surman; but this is not the sense in which 'Jimmy Carter' is the name of a man. This can be seen from the fact that 'Jimmy Carter is now the President of the United States' does *not* follow from 'Jimmy Carter used to grow peanuts and the same surman is now the President of the United States'. In a special sense, 'Jimmy Carter' is the name of a man but is not the name of a surman; and I shall express this sense by saying that 'Jimmy Carter' *designates* a man and does not *designate* a surman. The sense, namely, is this: that '*y* is *F* and Jimmy Carter is the same man as *y*' entails 'Jimmy Carter is *F*', whereas the same entailment fails when we replace 'man' by 'surman'. This supplies the sense in which Geach holds, and I entirely agree, that with every proper name is associated some criterion of identity, usually, perhaps always, expressible by 'is the same *A* as' for some count noun '*A*'; and, in this sense, the language does not as yet contain any proper names that designate surmen. If we wished to introduce them, we should have to lay down the conditions under which a given surman satisfied a predicate significantly applicable to it; and it is that stipulation which Geach appears to have omitted to make.

It is apparent from Geach's practice that he wishes to apply to surmen the same predicates as those we apply to men; so we may enquire after their satisfaction-conditions. Two possibilities occur immediately: (i) a predicate is true of a surman just in case it is true of every man who has the corresponding surname; or (ii) a predicate is true of a surman just in case it is true of some man who has the corresponding surname. Since Geach speaks of surmen living in Leeds, (i) is unlikely to accord with his intentions, since, under (i), it is improbable that any surman lives in Leeds. But, now, if we adopt (ii), are we to say that 'Some surman is both alive and dead' is true or false? If we apply the condition directly to the complex predicate 'is both alive and dead', it comes out false; yet there plainly exist x and y such that x is alive and y is dead and x is the same surman as y. What we have run into here is scope trouble: is condition (ii) to be applied to the predicates 'is alive' and 'is dead' taken separately, and the sentence then construed in the usual way as arrived at by conjunction and existential generalisation, or is it to be applied directly to the complex predicate? One possible choice would be to lay down that the scope is always to be taken as narrowly as possible, that is, in the example, in the first of the two ways just canvassed; in this case, the sentence 'Some surman is both alive and dead' will come out true.

Now this would be a perfectly coherent way of speaking, even if it would have some rather paradoxical-sounding consequences; but, on reflection, it seems unlikely to be a correct interpretation of Geach's intentions. For one thing, he says that 'if x is a surman in Leeds, then x has a heart in his breast, guts in his belly, and so on'. When scope difficulties are involved, the interpretation of a sentence is very sensitive to how we construe its internal structure; but if we take 'x has guts in his belly' to mean something like 'For some y and for just one z, y is a gut and z is a belly, and y is in z and z is part of x', this predicate will not, under the proposed convention, be true of any surman; no surman will have just one belly, but will have as many of them as there are men bearing the corresponding surname. Much more important, however, is the fact that Geach is convinced that he has no need to *specify* the conditions under which such a predicate is to be true of a given surman: he continues, 'These predicables will be true of x in their ordinary everyday sense; to make these predications true, we need not use words in some artificial sense that I have negligently failed to specify'. How has it come about that I have (I hope) at least made it appear that some such specification is required, whereas it appears obvious to Geach that none is needed?

The reason is that I have been assuming that, when a predicate is applied to a surman, the identity-relation associated with such a predication will be that given by 'is the same surman as'. What is meant here by speaking of 'the associated identity-relation'? The fact is that, just as the phrase 'proper name of an A' has to be treated with caution, so has the phrase 'predicating something of an A'. In *one* sense, when I say, 'Jimmy Carter comes from Georgia', I am predicating something of a surman; but, in *another* sense, I am not, because I have attached the predicate to a name which does not, in our special sense, designate a surman, but, rather, designates a man. If *x* is a surman, and *x* lives in Leeds, and *x* is the same surman as *y*, does it follow that *y* lives in Leeds? Evidently not, when '*x*' and '*y*' are taken as ranging over men; but how is it when they are taken to range over surmen, that is, when we construe '*x* is a surman', not as a predicate applying to certain elements of the domain, but as indicating that '*x*' is to range over surmen?

The answer is that it depends how we construe the expression 'ranging over surmen'. The critical question is what is the identity-relation associated with the given expressions of generality. The phrase '*x* is the same *A* as *y*' expresses such an identity-relation just in case 'For some *x* and *y*, *x* is *F* and *x* is the same *A* as *y* and *y* is *G*' is taken to entail 'For some *x*, *x* is both *F* and *G*'; e.g. 'is the same surman as' expresses the identity-relation associated with the expression 'some surman' just in case it is a sufficient condition for the truth of 'Some surman is both alive and dead' that, for some *x*, *x* is alive, and, for some *y*, *x* is the same surman as *y*, and *y* is dead. Geach's conviction that there is no stipulation that he has failed to make arises from the fact that he does *not* want to make the relation expressed by 'is the same surman as' play this role in connection with expressions like 'some surman'; and we may express this by saying that he does not wish to introduce into the language any expressions of generality for which the relevant range consists of surmen—he does not want to quantify *over* surmen at all. That is why he can say that all predicates applied to surmen are to be taken 'in their ordinary everyday sense': when an expression like 'some surman' is used, the associated identity-relation will be given, not by 'is the same surman as', but by 'is the same man as', and so 'Some surman is both alive and dead' will come out as false, since we cannot find a case in which *x* is a surman, and *x* is alive, and, for some *y*, *y* is dead and *y* is the same *man* as *x*.

Geach does not wish to quantify over surmen in any other sense than that of restricting the domain to those men who satisfy the predicate 'is a surman'. Just as there is a broader and a narrower sense of 'proper name of an *A*', and a broader and a narrower sense of 'predicating something of an *A*', so there is a broader and a narrower sense of 'quantifying over *A*'s'. In the broader sense, to quantify over *A*'s is to quantify over a domain each element of which satisfies the predicate 'is an *A*'; but, in the narrower sense, it involves using an expression of generality with which is associated the identity-relation expressed by 'is the same *A* as'. Because Geach wishes to quantify over surmen only in the broader, not the narrower, sense, the word 'surman', as he intends it to be used, will function precisely like a derivative count noun such as 'baker' save in the two special contexts 'is the same surman as' and 'There are *n* surmen'. These two perturbations of the familiar pattern will produce some paradoxical-sounding results; for instance, 'There is just one surman in Llanellyn' and 'Some surman in Llanellyn is over 30' may both be true, while 'Every surman in Llanellyn is over 30' is false. This is not to suggest that there is any incoherence in the use of 'surman', as proposed by Geach; reflection should, however, make it evident that 'surman' will not behave in the same way as any existing word in the language.

It is thus apparent that, for every expression of restricted generality of the kind with which we are concerned, there must be *some* identity-relation associated with it. It will not, however, hold in all cases that with an expression like 'some *A*' is associated the relation expressed by 'is the same *A* as', which is why the short way with our question about expressions of generality in natural language is not in fact a correct response. To understand an expression of the form 'some *A*', it is not enough that one should have mastered the expression 'is the same *A* as': knowing how to use 'is the same surman as' does not tell us whether 'Some surman is both alive and dead' is to be reckoned true or false. What we have to know is which is the right identity-relation to appeal to in judging statements involving 'some *A*': whether that expressed by 'is the same *A* as', or some other. Geach is anxious to insist that, when *x* is the same *A* as *y*, there may always be some count noun '*B*' such that *x* is not the same *B* as *y*. The correct answer is, I think, that it depends whether 'is the same *A* as' expresses the identity-relation associated with the domain of the variables '*x*' and '*y*': if it does, then there can be no such count noun '*B*'. But suppose that, for the sake of argument, we concede this claim of Geach's. It remains that, when the phrase 'some *A*' is used, we must know whether or not 'is the same *A* as' expresses the associated identity-relation: otherwise we cannot determine the truth of sentences of the form 'Some *A* is both *F* and *G*'. On the way of construing sentences involving the word 'surman' that I have attributed to Geach, it counted as a relevant objection to the truth of 'Some surman is both alive and dead' that, although *x* is alive, and *y* is dead, and *x* is the same surman as *y*, still *x* is not the same *man* as *y*. But, now, suppose that I observe, of some man now taking tea with me, that he is the same man as the one who presided over last night's meeting: can I infer that some man both presided over that meeting and is now having tea with me? I cannot make this inference unless I know that the fact that it is the same *man* is conclusive here. Suppose it said that, although it is the same man, it is not the same whatnot, where 'whatnot' is some suitable count noun: how am I to know that this is not a valid objection to the inference? I cannot know this, unless I know that, for the relevant domain—that invoked in using the expression 'some man'—and for these predicates, the relation expressed by 'is the same man as' plays a quite special role, a role which other identity-relations which may be defined over that domain do not play; that it is *the* identity-relation associated with that domain.

Geach gives an actual example of a range of words which he claims are count nouns subject to his pattern of explanation. These are specifications of the ambiguous word 'word': e.g. 'token word', 'dictionary-entry word', 'orthographic word', 'phonetic word', etc. Now I think it is plain that such expressions do *not* behave like 'surman' on the way of construing it sketched above—nor, indeed, given Geach's explicit stipulations, on any other—since the predicates we apply differ in different cases. A dictionary-entry word may be said to be invariable, but not an orthographic word; an orthographic or phonetic word may be ambiguous, but not a dictionary-entry word; a token word, but not a phonetic word, may be almost illegible. The most important fact to note about this example, however, is that there are two distinct ways of understanding the account which Geach gives of these expressions. One is to take 'orthographic word', 'dictionary-entry word' and 'phonetic word' as related to 'token word' as 'surman' is related to 'man': 'x is the same orthographic word as y' would then be defined to mean 'x is a token word and y is a token word and x has the same spelling as y', while 'is an orthographic word' would be defined to mean 'is the same orthographic word as something'. It would then follow, as Geach says, that every element of a domain specified as consisting of words was, simultaneously, a token word *and* an orthographic word *and* a dictionary-entry word, etc. On this interpretation, we start with a domain of token words, and define, over it, a number of equivalence relations, expressed as relative identity-relations, such as that expressed by 'is the same orthographic word as'. Here 'token word' plays in relation to 'orthographic word' exactly the role that 'man' is supposed to play in relation to 'surman'. But, then, what about the nouns 'token word' and 'man' themselves? Are they to be taken as subject to Geach's pattern of explanation for count nouns? If so, over what domain is the relation expressed by 'is the same token word as' or that expressed by 'is the same man as' defined? If Geach were to deny that 'token word' and 'man' exemplify his pattern of explanation, we should have another, and most important, class of count nouns not subject to that pattern.

It does not seem, however, that he would take that view; for he speaks of 'token word' as a count noun alongside 'dictionary-entry word' and the rest, and, for him, it is a criterion for '*A*' to be count noun that 'is an *A*' should have to be explained in terms of 'is the same *A* as' by derelativisation. This conflicts with the suggested interpretation of Geach's account of nouns like 'orthographic word', for, under that interpretation, 'token word' was taken as already understood and as being used to fix the domain over which the various relative identity-relations were to be defined; we should then have no need to *introduce* 'token word' by specifying the sense of 'is the same token word as' as standing for an equivalence relation over that domain. This suggests a different interpretation of Geach's account: one under which 'token word' plays no distinguished rôle in fixing the domain over which the various relative identity-relations are to be defined, and under which no other count noun plays such a rôle; the domain is given *first*, antecedently to our defining over it *any* identity-relation. This is in harmony with the thesis I have already rejected, the thesis which it is the main burden of this essay to confute: that a domain may be specified without associating any particular identity-relation with it.

Much of what Geach says supports this second interpretation. He writes,

The word 'word'... is ambiguous. It may mean 'token word', or 'dictionary-entry word', or various other things. Despite this, I may specify as the universe of discourse the words in a given volume in my room at Leeds; for I could give each word in the volume a proper name and get a finite list of them. The ambiguity I have just mentioned is an ambiguity over what shall count as *the same* word; but since a list specifying a domain... need not be nonrepetitive, this need not worry us. The count of token words, of orthographic words,... and of dictionary-entry words, may be different in each case; all the same, each thing in the universe is the same token word as itself *and* the same orthographic word as itself *and* the same dictionary-entry word as itself, and thus *both* is a token word *and* is an orthographic word *and* is a dictionary-entry word. I dismiss the protest that this result is incoherent because the entity in question must be of only one of these three kinds; there is no 'must' about it. We have in view an entity that belongs to the field of these different equivalence relations, and therefore comes under three different counts using different count nouns; *each* of the count nouns applies—that is how count nouns are used. It is on the contrary the question 'But which is it *really*?' that is incoherent and unintelligible.[7]

Now it is true enough, on Geach's account, that of each element in the domain we can truly predicate 'is a token word', 'is an orthographic word' and 'is a dictionary-entry word'; but, if we adopt the first interpretation of that account, under which the domain is originally fixed as consisting of token words (say because the proper names occurring in Geach's list have associated with them the criterion of identity expressed by 'is the same token word as'), it does not in the least follow that the question, 'Which is it *really*?', is devoid of sense. On the contrary, to the question, 'What is it?', asked of an element of the domain, the answer, 'It is a token word', will have a distinguished place, because it alone will indicate the identity-relation associated with the domain. This suggests, therefore, that, of the two interpretations, it is the second that corresponds with Geach's intentions: it suggests that we are supposed to select a domain for our variables in advance of fixing *any* identity-relation for it, and that 'is the same token word as' denotes one of the equivalence relations which we only subsequently define over this domain.

It is just here, therefore, that Geach's views concerning natural language link up with those concerning formalised languages with which I began: for the domain of definition of the relations expressed by phrases of the form 'is the same *A* as', Geach requires the same amorphous sort of domain, given by a list whose entries may later be discerned to be shared, rather than proper, names, as that which he wanted to invoke in speaking of an interpretation of a formalised language. Of the equivalence relations mentioned by Geach in the example, that expressed by 'is the same token word as' happens to be the finest; but we may later see fit to introduce another, defined over the same domain, say one expressed by 'is the same runcible word as', which is yet finer. If, originally, we listed the elements of the domain by giving a separate entry to each distinct token word in the volume, it will now prove that some of our entries each correspond to distinct runcible words, and hence are not, after all, proper names, but only shared names. This is why the plausible suggestion made earlier, that Geach could have conceded the point for formalised languages, but still have maintained his position for natural language, is not in fact correct: if his thesis about formalised languages crumbles, so does his thesis about natural language.

As I see it, the truth of the matter is this. The different criteria of identity associated with count nouns and with proper names are not criteria for the obtaining of different equivalence relations between determinate objects. Reference to an object requires that we have some criterion of identity for the object referred to; hence the criterion of identity associated with a count noun must be understood before reference is possible to any object for whose identity it is the appropriate criterion. These criteria of identity may be thought of, at least in the first place, as criteria for the truth of statements of identity between things referred to ostensively. The question, 'Is this the same as that?', where the demonstratives are supplemented by pointing gestures, is indeed indeterminate: we have to insert a count noun after 'the same' to make it a determinate question. This is enough to show that demonstrative pronouns do not have to be understood as tacitly associated with any particular criterion of identity on each occasion of their use, and that, for this very reason, they are not, in general, to be taken as picking out determinate *objects*. Below the level at which we employ genuine predicates of objects, predicates such as 'is loyal to his friends' or 'has visited Nigeria', which can be understood only as defined over a domain of objects, there is a level at which we employ quasi-predicates which need not be so understood, but can be attached to demonstratives unqualified by any count noun: and it is in respect of this more basic level of discourse that we can truly say that no particular relation of identity is presupposed. But the question, 'Is *a* the same as *b*?', where '*a*' and '*b*' are proper names, *requires* no supplementation by a count noun, though it will always *admit* one, since a criterion of identity is already associated with the proper names. What is more, as was asserted by Quine in the quotation with which we began, though it is by no means so obvious, the same holds good of the determination, for particular values of the variables, of a formula '$x = y$' of a formalised language; for the specification of the domain of the variables likewise requires the invocation of an associated identity-relation.

New College,
University of Oxford

POSTSCRIPT

This essay was written in November 1976, and given in that month as a lecture at the University of California, Berkeley, at Stanford University and at the University of Mexico. It overlaps extensively with an earlier unpublished essay, 'Geach and Quine on Identity', and is printed here in almost exactly the form in which it was originally written. Since writing it, I have had the opportunity of reading H. W. Noonan's *Objects and Identity*, submitted in 1977 as a Cambridge Ph.D. thesis, and Nicholas Griffin's *Relative Identity* (Oxford, 1977). I have not, however, thought it advisable to complicate and lengthen this essay by relating the views expressed in it to those of either writer. With Griffin's book, the views advanced here do not, I think, have a great deal in common. It is otherwise with Noonan's thesis. His approach to the topic and mine are radically different; but, in reading his thesis, I found many of the points I had made here made also by Noonan, of course quite independently and sometimes in a very different manner. To give just two instances, he gives great prominence to the distinction between the narrower and the broader sense in which a name may be said to be the name of an A (the former being expressed by him by saying that it is the name OF an A); he also uses the very same example, of the count noun 'nationality', that I used. Naturally, there is also a great deal in Noonan's thesis on matters not discussed here; it is much to be hoped that it will appear in a form that will give it a wider readership. I hope it is clear that in this essay I have attempted no comprehensive treatment of the topic of relative identity; my aim has been solely to answer the question embodied in the title, and the corresponding question for natural language. (*Added on receipt of proof, July 1990*) Noonan's thesis was published as a book, with the same title, by Nijhoff (The Hague) in 1980.

NOTES

[1] W. V. Quine, review of P. T. Geach, *Reference and Generality, Philosophical Review*, 1964, 100–104.

[2] The definition as stated is adapted for the case when the only non-logical constants of L are predicates. It will be obvious how it is to be modified for the general case in which L contains also function symbols.

[3] P. T. Geach, 'Ontological Relativity and Relative Identity', in M. K. Munitz (ed.), *Logic and Ontology*, New York University Press, New York, 1973, pp. 287–302.

[4] P. T. Geach, 'Names and Identity', in S. Guttenplan (ed.), *Mind and Language*, Clarendon Press, Oxford, p. 157.

[5] *Op. cit.* pp. 156–157.

[6] 'Ontological Relativity and Relative Identity' (see note 3 above), p. 290.

[7] *Op. cit.* p. 294.

ANSELM W. MÜLLER

CONCEPTUAL SURROUNDINGS OF ABSOLUTE IDENTITY

When for the first time I read a paper to a philosophical audience—I had come to England only a short while ago, and was still getting used to both a language and a style of argumentation—, it was Peter Geach who encouraged me to develop further the thoughts contained in that paper. Today, after a dozen of years, I wish to take up his encouragement and connect some of those thoughts, which concerned the idea of necessary properties, with the question whether identity is absolute or relative. If my considerations turn out to conflict with Peter's answer to this question, I am nevertheless sure that he will recognize in their contribution to this volume a sign of deep respect, of friendship, and of gratitude for what I have learnt from him.

I

Peter's thesis that identity is relative amounts to this: 'When one says '*x* is identical with *y*', this . . . is an incomplete expression; it is short for '*x* is the same *A* as *y*', where '*A*' represents some count noun understood from the context of utterance . . .' (1972, p. 238).

In *Reference and Generality*, this thesis is supported by the claim that it provides a solution to the following puzzle: Heraclitus bathed in some river yesterday, and bathed in the same river today; whatever is a river is water; but Heraclitus did not bathe, today, in the same water that he bathed in yesterday. On the thesis that identity is absolute, this would involve a contradiction. (Cf. 1962, p. 150f.)

I wish to suggest a different solution: When we say, of some *x*, that *x* is a river, we do not just ascribe a property to *x*. In saying what kind of thing *x* is, we imply that certain properties are essential to it in the sense that the existence of *x* without these properties is impossible. Some such properties (e.g., in the case of a river, this or that position on the earth) are not themselves constitutive of being that *kind* of thing. Others are. Being water, i.e. being made up of water, is thus constitutive of being a river. Being made up of the same water on different days is, of course, not; nor is, therefore, the permanent

Harry A. Lewis (ed.), Peter Geach: Philosophical Encounters, 185–193.

presence of any (portion of) water essential to a particular river. Hence the following is a possible situation:

(1) For some *x*,
x is (essentially) a river and
x is (essentially) water and
x is (composed of) this water today and
x was not (composed of) this water yesterday and
Heraclitus bathed in *x* yesterday and
Heraclitus is bathing in *x* today.

How does it come about, on my view, that in this situation Heraclitus can be said to have bathed in some river yesterday and to be bathing in the same river today, but not to have bathed in some water yesterday and to be bathing in the same water today? The answer is made difficult by the fact that we have no clear conception of a portion of water that is not determined by boundaries like the walls of a vessel. On the other hand, expressions like 'the same water' or 'this water' seem, in our context, to invoke some such notion. And *if* it is admitted, we can formulate what we do *not* wish to claim, as follows:

(2) For some *x*,
x is (essentially) a portion of water and
Heraclitus bathed in *x* yesterday and
Heraclitus is bathing in *x* today.

Rather we wish to accept Heraclitus' insight into the fleeting nature of things in the following form:

(3) For no *x*,
x is (essentially) a portion of water, and
Heraclitus bathed in *x* yesterday and
Heraclitus is bathing in *x* today.

How does (3) compare with (1)?

Let me first observe that, unless they are to contradict each other, 'being (essentially) a portion of water' has to be understood as claiming more than 'being (essentially) water'. However the former expression is to be explained, in being applied to anything *x*, it classifies *x* as a certain kind of thing and *eo ipso* specifies the criterion of *x*'s identity (which is different from the one specified, e.g., by 'being (essentially) a river'); whereas 'being (essentially) water', while implied by 'being (essentially) a portion of water'—as well as by 'being (essentially) a river'—has no such force.

A second observation: Unless (1) and (3) contradict each other, no river is ever identical with any portion of water. Since a river can, at any given time, be viewed as coextensive with a certain portion or water, this means that more than one thing can be in the same place at the same time.

II

Is not this result by itself enough to refute my solution? I do not think so. Locke was cautious enough to say that two things *of the same kind* cannot be in the same place at the same time. Let us consider the claim that some body x is, at time t, in the same place as y, but $x \neq y$. What could the excuse that x and y differ in kind amount to? Need they not, at least at t, share *all* their properties, and hence be identical, at least at t? But, if at t, then, surely, always and *tout court*!

I do not wish to examine the steps of this argument. My reply is a distinction concerning the use of 'sharing properties'. In a sense, x and y will share all their properties, if found in one and the same place at t; in particular, all we can (by perception, experiment etc.) observe there and then will be attributable to y as well as to x. In another sense of 'property', being some kind of thing (like being a river) may be counted a property; if, as I hold, 'being a river' attaches essentially to any object it attaches to at all, and implies the essential attachment of certain properties (in the first sense) to that object, then x may differ from y in not being the same kind of thing as y and in not being connected essentially with those properties with which y is connected essentially—even if x is not, at t (or, maybe, ever), distinguishable from y by observation.

Does not our linguistic practice—e.g. the identification of things by pointing—presuppose that no two things, even of different kinds, are coextensive? What, I think, it presupposes, on the whole, is this: Any given place is occupied, at any time, by at most one thing that we possess a word for or are prone to form a concept of. This presupposition is, on the whole, fulfilled because our ways of articulating reality are largely determined by the limits of both our interests and our capacities.

We cannot identify, or pick out, a body by an expression like: 'that which now occupies this area of space'. The point is not new to anyone who is acquainted with Wittgenstein's observations on ostensive definition. But what precisely are their implications for an adequate conception of the functioning of names (proper and common), and of identity? An explanation like 'the same as this', conjoined to a pointing gesture, is not, indeed, sufficient to give meaning to a word, even if we make the assumption that 'this' is understood as relating to a body (rather than, say, a colour) and that no misunderstanding occurs concerning the area intended. (What I have said in the preceding paragraph seems to guarantee the sufficiency of such an explanation. But then, it is *not* just the demarcation of a definite area of space that guarantees it: I pointed to the more or less reliable fulfilment of a presupposition. Also, cases where an *explanation* is *needed* will *not* be ones covered by the guarantee!)

III

How will the explanation be supplemented? By the further use that is made of the name that is being explained. Or by a classification ('this piece of jewellery'—so it is a ring, not a parcel of gold, that I am pointing to). But the first of these does not mean that (as, e.g., Strawson has held) an individual can be traced under different sortals (unless, perhaps, the sortals are related like 'cat' and 'animal', or 'ring' and 'piece of jewellery'). And the second does not mean that (as Peter would say) the classification must be understood as relativizing the identity ('the same *piece of jewellery* as this').

Rather, as Peter himself says, a name is connected with a criterion of identity. A definite use of a given name for something x and a capacity of distinguishing a body that is identical with x from one that is different from x, stand and fall together. But then, once we have grasped what the name is a name of, we need not *also* specify an A such that the question whether something y is identical with x must, in order to admit of an answer, be turned into the question whether y is the same A as x. As long as the explanation of a name does not contain, or somehow get across, that classification which allows us to decide which further applications of the name are correct and which are incorrect, the explanation is incomplete: no individual (or type of thing) has been specified and named. It follows from this that an individual cannot 'be traced under different sortals' if this is to mean that an answer to the question 'Where is it now?', or 'Does it still exist?', may turn out different according as we have chosen the sortal under which to trace it.

Suppose someone was told: 'What you see Heraclitus bathing in is the Cayster'; the next day he is surprised to hear that Heraclitus is again bathing in the Cayster: he thinks the Cayster has by now been dispersed in the Aegean. It would, no doubt, be tactful, but wrong, to say: 'The Cayster was (or is?) indeed the same *portion of water* as something that is now in the Aegean, but it is *the same river* as what you see here.' No, 'Cayster' is the name of something that cannot be traced under the sortal 'portion of water' and cannot be the same anything as anything but a river.

IV

It seems to me that, if identity is relative, so is existence. Suppose we accept a formulation like 'Joe is the same man as will be executed tomorrow, and the same collection of cells as will be buried in the prison cemetery the day after'. Since the collection of cells will, unlike the man, exist after tomorrow, it seems to follow that Joe will cease to exist as a man (or: cease to be a man, or that man) tomorrow but cease to exist as a collection of cells (or: cease to be a, or that, collection of cells) much later.

But our use of proper names (and thus our concept of an individual) is such that the question when Joe ceases to exist is not given an answer *relative to* this or that concept which Joe is being considered as exemplifying. One might try to express this 'absoluteness' of existence by saying: It is when he ceases to *be a man* that he ceases to *be*. But it sounds dubious to say of a man that he ceases to be a man. (Not that we could, on my view, say it, properly speaking, of any other kind of thing!)

Let me consider an example of Peter's (1972, pp. 242–245): a universe of discourse consisting of 'the words in a given volume on my shelves'; but imagine them written with pencil. Suppose now I erase an occurrence of 'fuzzy' from the top line of page 44 (the word occurs once more in the volume). To the question 'Has the word 'fuzzy' ceased to exist?', I should have to answer: 'An individual (a token-word) of that type has ceased to exist. The type-word 'fuzzy' has not.' (We may leave open the question what sort of equivalence between token-words is to determine a type.) Peter, on the other hand, if I understand him right, would not allow himself *this* distinction, since his universe of discourse is to contain only words standing in variously relativized relations of identity and diversity. So, I suppose, he would have to say that 'fuzzy' ceases to exist *as* a token-word but not *as* a type-word.

If we stick to concepts of 'absolute' existence and non-existence (and at least the use of names of bodies seems to imply such concepts for the case of bodies), we shall have to admit a correspondence between our use of predicables and an acceptance of what there is (and thus, if you like, some sort of reflexion of our 'ideology' in our 'ontology', which Peter intends to avoid by relativizing identity: 1972, p. 244f). It may be possible to explain (away) the existence of a type-word in terms of occurrences of token-words which stand in a certain equivalence relation to each other. Nothing comparable suggests itself for the case of bodies: If we have the concept of a piece of gold—whether or not we also discriminate between pieces of gold according to their shapes—, our ontology recognizes, as long as there is any gold at all, things that are (essentially) pieces of gold and whose existence conditions are different from those of the golden rings which some of the pieces of gold go to make up. It would seem entirely arbitrary to reduce the existence of a piece of gold to (e.g.) the existence of someone's wedding ring and the existence of the crown into which his dentist turned it after his divorce. (Why not reduce the other way round?)

V

Have we not fallen into some sort of conventionalism or anti-realism in making what there is dependent on the substantival predicables *we* happen to apply? And is not Ockham's razor called for to prevent Heraclitus from bathing in many things at once (e.g. in the Cayster as well as in an indefinite number of portions of water)?

It should be clear that, on the view I put forward, the existence of parcels of gold, as well as that of golden rings, does not depend on our having those concepts. Rather, in applying this or that notion of a kind of thing, e.g. by using names proper or common in a specific manner, we 'delineate' entities, much like someone delineating an area on a plane. The entities are not created by the formation of the notions in question, but neither do they present themselves as units independently of the application of these notions. And this view is, as far as I can see, on all fours with Peter's anti-abstractionist position. If *it* is called anti-realism, or conventionalism, I do not mind being an anti-realist, or a conventionalist. However, I should have thought that conventionality is present rather in our practice of singling out established objects of discourse as objects *tout court*. (E.g.: If 'Cayster' is explained by a pointing gesture, no-one will in fact consider whether a portion of water might be meant. Of course, I do not think that the conventionality that shows up here is any kind of arbitrariness; it may have deep roots in basic human needs and interests, and in limitations of our perceptual and other capacities.)

If this is a correct view of objects, or at least of what ought to be treated as real in the realm of bodies, we shall be at a loss for a criterion by which to decide *where* Ockham's razor should be applied —apart from the question *why* it should be applied. Admittedly, we wish to beware of being committed to the existence of objects of which contradictory predicates can be proved true. But as a general preventive device Ockham's razor seems to me to be too blunt an instrument.

VI

Some of the passages in Peter's work that relate to the topic of identity suggest to me that I may have misunderstood him. For we could perhaps construe phrases like '—is the same type-word as—', or '—is the same dictionary-entry word as —', on the model of '—is (= has) the same colour as—', or '—is (= has) the same weight as —'. (Cf., e.g., 1972, p. 245, where he says about '—is the same surman as—': 'this means, simply, '—is a man with the same surname as—, who is also a man'.) On this interpretation, a statement of relative identity about x and y would be the indication of a kind of property such that x and y shared a property of that kind; i.e.: a generic specification of a predicate which applied to x and to y.

To my mind this interpretation of 'relative identity' would make the notion acceptable. But I am uncertain whether this notion would do the work required by a thesis of relative identity that captured Peter's intentions. Two points, in particular, should be noted about the notion of relative identity as just explained.

(1) Absolute identity will still be with us, irreducible, presumably, to

this relative identity. And, if possible, a reduction would make absolute identity a *very* special case of relative identity. Suppose we tried this explanation: x is the same thing as (or absolutely identical with) y iff, for no kind of property K, x is/has a different K from y, i.e.: iff *everything* true of x is true of y. Peter would object to this reduction because its quantification over predicates, not restricted to a circumscribed vocabulary, will land us in paradoxes. (I have not been able to discuss the relativity argument he draws from this kind of consideration in 1972, pp. 238–241.) From my point of view, there might be an additional problem: It is not clear that x's *being (essentially)* (an) F can be rendered as 'being (essentially) (an) F' *being true* of x. (Cf. the second paragraph of section II above.) Whether reducible or not, absolute identity between x and y is characterized by implying, logically, that the conditions of existence are the same for x and y. So there is absolute identity if there is 'absolute' existence. (Cf. section IV above.) (Since Peter is in favour of a fixed 'ontology' (1972, p. 245), he might, himself, find room for a notion of absolute identity be means of the notion of an I-predicable that does not cease to be one on any extension of the language (= of the 'ideology'!).)

(2) A connected point: Even on the present understanding of 'relative identity' we shall have to distinguish *entities* in cases where Peter's application of Ockham's razor would leave only a distinction of substantival terms. A few examples: Apart from a token-word x, which is of the same type as (some other token) y, we may recognize the type-word instanced by x, which is absolutely identical with the type-word instanced by y. But I need not insist on the existence of such 'abstract objects'. There are the two (and more) portions of water that are 'relatively identical' in constituting, at some time or other, one and the same river—*as well as* the river they constitute. Tom's wedding ring was of the same piece of gold as is the crown on one of his teeth. But the ring was not only an ('absolutely') different thing from the crown, but also from the piece of gold (with which, too, it was relatively identical, perhaps even necessarily so, sharing all its properties, in some ordinary sense—as long as it existed!). The only way I can see of avoiding such multiplication of (temporarily) coextensive bodies would be to consider the lot as 'constructed' on the basis of equivalence relations between four-dimensional time-slices; and this proposal could not seriously be held to reveal our notions of kinds of material things as a mere manner of speech.

VII

I am, of course, aware that the position sketched on the preceding pages is a kind of essentialism, and that many philosophers (with none of whom Peter is more than relatively identical) have developed not, indeed, *a priori* arguments but an *a priori* aversion against essentialism, which it is difficult to overcome. Sections II–IV provide arguments in favour of my position and, thus, against a relativistic solution to problems about identity. An assessment of the conflicting views may have to await closer examination of the content of, and the relations between, such concepts as: kind of stuff, kind of thing, count noun, substantival term, essential property, criterion of identity, and others that belong to the conceptual surroundings of identity. I should be glad to have contributed a little towards an exploration of their constellation.

Department of Philosophy,
University of Trier

REFERENCES

Geach, P. T.: 1972, 'Identity', in *Logic Matters*, Basil Blackwell, Oxford, pp. 238–247.

Geach, P. T.: 1962, *Reference and Generality*, Cornell University Press, Ithaca, New York.

C. J. F. WILLIAMS

ON SAMENESS AND SELFHOOD

I hope in this paper to establish that, in an important sense of 'empirical', the concept of personal identity is not wholly an empirical concept. It cannot therefore be equated, as some philosophers have maintained, with the concept of some relation or set of relations whose existence could only be discovered empirically. I also hold that the concept of identity is not the concept of a relation at all. If, therefore, it is implicit in thoughts that we have of ourselves as remembering, intending, fearing, etc., which are crucial to our thought of ourselves as persons, it is difficult to see how it could be replaced, as has also been envisaged, by some relational concept or concepts. I do not believe that the concept of personal identity can either be reduced to some other concept, or allowed to fall into disuse as a relic of outworn, metaphysical beliefs no longer important to us.

I

Derek Parfit wishes us to take certain important questions, notably 'the question of survival', and prize them apart from 'the question of identity'. This done, the question about identity — so he believes — has no further interest. It is the more curious that he gave the well-known article in which he makes this claim the title 'Personal Identity'.[1] Be that as it may, 'what matters in survival', to use another of the phrases which he has made familiar, is not, in his view, the identity relation. In particular it does not have, as the identity relation has, the property of transitivity.[2] It is to be described in terms of different relations, having different logical properties — relations which Parfit refers to as 'continuity' and 'connectedness'.

But is identity a relation in the first place? Wittgenstein denies this in the *Tractatus*. As Roger White has recently made clear,[3] this denial is the logical development of the views about identity which Frege expresses in 'Sense and Reference'. Frege showed that an identity statement can be both true and informative only if the expressions having the same reference but different sense. Examples of expressions

Harry A. Lewis (ed.), Peter Geach: Philosophical Encounters, 195–212.

occurring on either side of the identity sign and expressions which have this character are most easily to be found amongst definite descriptions. Indeed Russell claims that unless an identity statement contains at least one definite description it will be a trivial tautology: 'The author of *Waverley* is the same as the author of *Ivanhoe*' and 'Scott is the same as the author of *Waverley*' convey useful information, but if 'Scott' and 'Sir Walter' occur as genuine proper names in 'Scott is the same as Sir Walter', that proposition is as tautological as 'Scott is Scott'. This doctrine of Russell's has a curious consequence. The definite descriptions that occur in every identity statement that is informative, every identity statement, that is, for which we have a real need, give way, in Russell's analysis, to formulae which combine quantifiers, truth-functions and the identity relation. The only positions in these formulae where we have an identity sign are positions where the argument places governed by that sign are occupied by bound variables of quantification. As White has remarked, it is an odd relation whose sign can never combine informatively with proper names, but only with variables.

Wittgenstein's proposal rationalizes this. Identity of object, he implies, is not something which needs to be said: it shows itself in identity of sign. Thus

(1) John is my son-in-law and John is my dentist

shows by repetition of the name that my son-in-law is the same as my dentist. Similarly 'John is my son-in-law and James is my dentist' would show by change of name that my son-in-law and my dentist are different people. Where bound variables are concerned Wittgenstein uses the same method to show identity: a repeated variable shows that we have the same object, change of variable that we have a different one. Only here, I would maintain, what is shown is also said. What (1) shows, 'For some *x*, *x* married my daughter and *x* looks after my teeth' explicitly says, and this is equivalent to

(2) My son-in-law is the same as my dentist.

Whatever information of this kind we need to convey, and normally convey, by the use of 'is the same as' and its synonyms, can be conveyed in a perspicuous notation by quantification and its variables. The usual forms in the *Begriffsschrift* for propositions which convey information about identity and difference will be '$\Sigma x K\varphi x\psi x$' and

'Σ*x*Σ*y*Kφ*x*ψ*y*'. Identity is thus shown not to be a relation—not, that is, something best expressed by means of a two-place predicate. White's exposition of Wittgenstein's views on this is most helpful:

Now what are we to say about a sign whose only primitive use—the use to which all other uses may be reduced—is within the scope of a quantifier flanked by two variables? Here it naturally suggests itself that we should treat it as belonging with the apparatus of quantification, and that an appropriate representation of it within the *Begriffsschrift* would be one in which it was made part of the quantificational apparatus.[4]

White further points out that the Wittgensteinian doctrine that identity is not a relation is comparable with the Fregean doctrine that existence is not a first-level concept. The canonical form for expressing existence is quite different from that appropriate for first-level concepts, and the proposed canonical form for expressing identity is quite different from that appropriate for expressing relations. 'Σ*x*φ*x*' is not easily confused with 'φ*a*', nor 'Σ*x*Kψ*x*χ*x*' with 'ψ*ab*'.

II

Suppose some children are watching a ventriloquist and his dummy, and are taken in by the performance. The dummy, as they perceive it, is singing 'Land of Hope and Glory', and the ventriloquist is vigorously conducting an orchestral accompaniment. They, at first, do not realise that the person who is singing 'Land of Hope and Glory' is the same as the person who is conducting the orchestra. Little by little, however, they catch on to what is really happening. By noticing movements in the ventriloquist's throat and failures of correlation between the sounds produced and the movements of the dummy's mouth they come to see that the person who is singing is one and the same as the person who is conducting. In making this judgment of identity they use publicly observable criteria of one sort or another. But the ventriloquist does not use such criteria. He knows that he is both singing and conducting, and he knows accordingly that the singer and the conductor are one and the same person. By ascribing to himself, not on the basis of publicly observable criteria, nor indeed on the basis of any criteria at all, two simultaneous activities, he is aware automatically of the identity of the person singing with the person conducting. So what it means to say that singer and conductor are identical is different from what it means to say that certain observable facts obtain, despite these facts' being the criteria for the children's eventual judgment of identity. Certainly these criteria played no part in the ventriloquist's judgment of identity. His judgment was based on two self-ascriptions. He judged that he was singing and he judged that he was conducting. Moreover, he judged that both he was singing and he was conducting; and so, by existential generalization, he was able to judge that, for some person *x*, both *x* was singing and *x* was conducting. That, as we have seen, has the basic form for the expression of identity.

We can compare this case with a situation discussed by McTaggart.[5] Suppose a man judges that he is late for an appointment (my example, not McTaggart's). Then he is able, straight off, without any more information, to judge that someone is being judged to be late for an appointment and that someone is judging that someone is late for an appointment. Moreover he can proceed directly to judge that the person who is being judged to be late for an appointment is the same as the person who is judging that someone is late for an appointment. This, again, is tantamount to the judgment that, for some person *x*, both *x* is being judged to be late for an appointment and *x* is judging that someone is late for an appointment. And this judgment is possible for him because he is aware that both these descriptions apply to him himself.

Identity as such is clearly not an empirical concept. No observations are required to confirm or disconfirm the proposition that the number which is the third of nine is the same as the number which is the positive square root of nine. Since, in Wittgenstein's view, identity can be expressed by certain rules for the use of the variables of quantification, this is just what we should expect. Identity, like quantification, will be topic-neutral, and for this reason will not be an empirical concept. But identity statements of the form '*x* is the same as *y*' have plausibly been claimed to be elliptical for statements of the form '*x* is the same *A* as *y*',[6] and while *being the same number as* is obviously not an empirical concept, *being the same piece of string as* obviously is; and *being the same person as* looks more like the latter in this respect than the former. These distinctions carry over to the Wittgensteinian expression of identity. Corresponding to those fillings-in of the elliptical identity expression, 'is the same — as', there are fillings-in of the quantifier, 'For some — *x*'. Instead of saying 'The third of nine is the same number as the positive square-root of nine' we may say, in the spirit of the *Tractatus*, 'For some number *x*, both *x* is the third of nine and *x* is the positive square root of nine.' Answering to the doctrine that regards identity expressions as elliptical, we have the doctrine that disapproves of unrestricted quantification. On this view it is the concept *person*, carrying with it criteria of identity, about which we have to ask whether it is or is not empirical.

Like the concept of *pain* the concept of *personal identity* is undetachable from the observable criteria for its use in other-ascribing contexts. But its meaning is not exhausted by these criteria. As with the concept of pain, so with the concept of personal identity, a man could not be said to possess this concept if he were capable of applying it only to others on the basis of observable criteria, and not also capable of applying it to himself without the use of criteria. This does not mean that it is not in any way an empirical concept, any more than *pain* is not in any way an empirical concept. But it means that there is more to the concept than the empirical criteria for its use.

III

Maybe I have already said enough to establish that the concept of personal identity is not exhausted by the empirical criteria which we may on occasion employ in exercising it. But the concept of a person is the concept of something which endures through time; and if its use in reidentification were entirely dependent on empirical criteria, my claim would seem thin. Personal identity has often been held to be involved in, even to be constituted by, memory. We must examine the way in which memory provides a basis for the reidentification of persons over time, and see whether what has been said about simultaneous identification applies here too.

Our ventriloquist was able to judge straight off that he was the one who was singing and also the one who was conducting. He had no need to employ criteria of identity. This does not mean that his judgments could not be mistaken. It is not impossible to imagine hallucinatory conditions in which a man thinks that he is singing, or conducting an orchestra, when he is doing no such thing. A man who mistakenly, in a drugged state perhaps, believed that he was singing 'Land of Hope and Glory' at a time and place when someone else was in fact singing this song, might mistakenly believe that he was doing this as well as what he was in fact doing, namely, conducting its orchestral accompaniment. He might thus conclude, again mistakenly, that the person who was singing was the same as the person who was conducting.

The man, on the other hand, who judges that he is late for an appointment cannot be mistaken in judging that the person who is judging that someone is late is the same as the person who is being judged to be late. The only way he could be mistaken in thinking this would be by mistakenly thinking that he was judging that he was late for the appointment—and what sense can be made of this?

Reidentification over time differs from simultaneous identification in that it is never immune from error. The man who at time t_1 is judging that someone is late is the same as the man who at t_1 is being judged to be late. The man who performed some act at time t_2 may or may not be the same as the man who thinks at time t_3 that he remembers performing it. I can make first-person simultaneous identifications which are out of reach of mistake: no identifications over time, not even first-person ones, are infallible in this way.

First-person identifications over time, even if fallible, are nevertheless capable of being made without resort to criteria.[7] Suppose that what is in question is who locked the garage door before we went to church. I may claim that I was the one who locked the door, simply on the basis of remembering doing so. Remembering may be the basis of my claim, but it is hardly something I employ as a criterion. So when I judge that it was I who locked the garage door I employ no criterion in making this judgment. It would be wrong, however, to classify *this* judgment as a judgment of identity. I do not identify the person who locked the garage door by telling you that it was I who did it, although my telling you so may enable you to identify him as the person who has just said this.[8] But just as I am able, on the basis of memory and without employing criteria, to judge that I was the one who locked the garage door, so I can judge without resort to criteria that I am, say, now sitting down. I am thus able, without use of criteria, to make a reidentifying judgment that someone who is now sitting down is someone who before church locked the garage door, although I may well be mistaken in doing so. This is a genuine identification of a person over time. But we shall see in section VI that identity is connected with memory in a more intimate way than this.

IV

Locke maintains that if he 'had the same consciousness' that he witnessed Noah's flood as he has that he is writing 'this now', he could not doubt that he who is writing is 'the same self' as one who witnessed the flood.[9] What does he mean by 'having the consciousness that'? Three possibilities suggest themselves: 'having the consciousness that I saw the ark and Noah's flood' may mean 'remembering seeing the ark and Noah's flood', or 'believing that I remember seeing the ark and Noah's flood', or 'seeming to myself to remember seeing the ark and Noah's flood'. Philosophers have held that we can properly say that I remember, e.g., seeing something, only if I actually did see it. (This is contested by Geach in his paper 'Reincarnation',[10] but let us assume for the moment that this analysis of 'remembers' is correct.) If, on the other hand, I merely seem to myself to remember seeing the ark and Noah's flood, let us say that I 'phenomenally remember' seeing them. Now I take it that I can phenomenally remember witnessing something if I merely seem to recall witnessing it, without actually believing that I witnessed it. As Aristotle pointed out,[11] there are some 'seemings' which involve belief and some which are compatible with doubt or positive disbelief. If I phenomenally remembered witnessing Noah's flood, I hope I should be certain that I had witnessed no such thing and regard the apparent memory as some pathological phenomenon. But between phenomenal and genuine memory there is an intervening possibility. I may actually believe that I remember seeing Noah's flood. My so believing would in no way guarantee that I do remember it (Locke is unclear about this), and I could phenomenally remember it without so believing. But if we take Locke's 'having the consciousness that he saw the ark and Noah's flood' as believing that he remembers seeing them, then indeed he is right to say that his having this consciousness precludes doubt that he is 'the same self' as a person who had this experience. I cannot at once believe that I remember doing *X* and doubt whether I am the one who did *X*. What I think when I think I remember doing *X* includes the thought of being the same person as did *X*. To the extent that Locke is concerned with what I think when I think I remember something, his account is perfectly acceptable: to the extent that he is concerned with whether what I think is correct, he clearly oversteps the mark.[12]

V

Locke draws our attention to the fact that a proposition like 'John believes that he remembers witnessing Noah's flood' entails the corresponding proposition 'John believes that he is the same person as witnessed Noah's flood'. This is shown by his making the point that belief in the one is incompatible with doubt of the other. But, in order to forestall objections by Parfit, I shall move the discussion from the area of belief to that of phenomenal memory. Not only does believing that I remember doing *X* include believing that I am the same person as did *X*, but phenomenally remembering doing *X* includes the thought of being the same person as did *X*, even if this thought is not assented to. It will be useful to set out the connection between 'phenomenal memory' and personal identity by means of the following derivation:

(3) It seems to John that he remembers witnessing Noah's flood
(4) It seems to John that he remembers that he witnessed Noah's flood
(5) It seems to John that, for some *x*, *x* remembers that *x* witnessed Noah's flood.

The formula 'It seems to John that he remembers . . .' is intended to capture the notion of 'phenomenally remembering': it can seem to John that *p* without John's believing that *p*. My claim is that each of propositions (4) and (5) is entailed by, though it does not entail, its predecessor in the series. I shall defend this claim in due course.

First, however, I need to establish that (5) attributes to John a thought involving identity. In (5), as I shall argue, the expression 'remembers that . . .' may be understood, in Quine's sense, transparently. So understood, an expression of the form '*x* remembers that *y* witnessed Noah's flood' can be used to assert a relation between values of the variables '*x*' and '*y*':'—remembers that... witnessed Noah's flood' is a two-place predicate of the form 'φ(—, . . .)', corresponding to the open sentence form 'φxy'. Where '*x*' and '*y*' take the same value, as in 'John remembers that John witnessed Noah's flood' the proposition is a reflexive one of the form 'φxx'. The existential quantification of this, '$\Sigma x \varphi xx$', gives the form of what is said in (5) to seem to John to be the case. The proposition which expresses this is thus the result of substituting a variable for the name in a reflexive proposition and binding the variable thus substituted by the existential quantifier. This prompts me to return to, and examine a little more closely, the Wittgensteinian analysis of identity. 'John married my daughter and James looks after my teeth' may be regarded as the result of filling the argument-places in the two-place predicate '— married my daughter and... looks after my teeth', by the names

'John' and 'James', respectively. It is thus of the same form as 'John hates James', i.e. 'φxy'. By the same token (1) is obtained by filling up each of the two places of the same two-place predicate by the name 'John'. It is thus of the same form as the reflexive proposition 'John hates John', i.e. 'ϕxx'. The proposition which, as I argued, not merely shows identity, but asserts it, namely (2), is the existential generalization of (1). That is to say, it is of the form '$\Sigma x K \psi x \chi x$'. But propositions like (2) which exemplify this form *ipso facto* exemplify the simpler form '$\Sigma x \varphi xx$', just as (1), as I have been arguing, exemplifies the form 'φxx'.

What is required for the expression of identity, according to this Wittgensteinian theory, is merely a repeated variable bound by a single quantifier, and the simplest form in which this apparatus can be used is in propositions of form '$\Sigma x \varphi xx$'. In the *Tractatus* this is taken for granted. The micro-structure of propositions like 'The author of *Waverley* is the author of *Ivanhoe*', which would need conjunction to make it explicit, is ignored, and Wittgenstein goes straight for the macro-structure: 'I do not write "$(\exists x,y).f(x,y).\ x=y$", but "$(\exists x).f(x,x)$"; and not "$(\exists x,y).f(x,y).\sim x=y$", but "$(\exists x,y).f(x,y)$"'.[13]

Both (2) and what is said in (5) to seem to John to be the case, namely, that, for some *x, x* remembers that *x* witnessed Noah's flood, are of the logical form '$\Sigma x \varphi xx$', the minimal form necessary for the expression of identity. (5) is entailed by (4), which is entailed by (3). None of these contain the words 'is the same as' or 'is identical with', but in virtue of the Wittgensteinian analysis of identity, each attributes to John, explicitly or implicitly, the thought of identity. The identity in question is, of course, personal identity. Locke may therefore be held to have been correct if his view can be interpreted as the view, simply, that the idea of personal identity is conveyed by the form of consciousness called memory. If I phenomenally remember doing something, I thereby have the thought that I am the same as the person who did it: the latter thought is in fact no more than an existential generalisation of the propositional version of the former.

A footnote to this claim—Locke's account of personal identity makes liberal use of the noun 'self', and philosophical talk about 'selves' has been ridiculed by philosophers who dislike too much divergence from ordinary language.[14] A point that is sometimes made is that, unlike English and German, French has no word which can serve even in philosophical writing both as a suffix for the formation of reflexive pronouns and as a common noun. '*Même*' cannot be detached from '*moi-même*', '*elle-même*', etc., to be used as a name for a philosophical topic. We have '*Das Selbst*' and 'The Self', but not '*Le Même*'. Leibniz's *Nouveaux Essais* represents Locke's use of 'self' by '*le soi*', '*le moi*' or '*soi-même*'. But '*même*', as well as serving as a suffix for the formation of reflexive pronouns, is the ordinary French word for 'same'. 'Same' and 'self' in English have completely different functions (although we do have 'self-same' as an alternative for 'same'). But in this matter at least the French language may perhaps be thought to show itself to be more 'logical' than English, to follow more closely the structure of thought, as Wittgenstein's French professor claimed it did. For that same repetition of name or variable which constitutes a reflexive proposition is also essential to showing or saying perspicuously that something is the same as something else—as the 'illogical' ordinary idiom expresses it.

VI

I must now defend the claim that each of propositions (3) and (4) entails its successor in the derivation (3)–(5). What is ascribed to John in (3) is a personal memory, but in (4) a factual memory. The factual memory could exist without the personal memory, but does the converse hold? Parfit maintains that the statement 'My apparent memory of *having* an experience is an apparent memory of *my* having an experience' rests on a mistake.[15] If this is a mistake, it is surely at least as much a mistake to hold that my apparent memory of having an experience involves the apparent memory *that I had* the experience. The background of Parfit's claim is Sidney Shoemaker's notion of 'quasi-memory'.[16] What is 'quasi-remembered' actually took place. Quasi-memory is veridical in this sense. But the person who had the original experience or performed the original action need not be the same as the person who now quasi-remembers it. If quasi-memory frequently occurred, and was known to occur, John would not believe that he remembered doing X, although he might believe that he quasi-remembered doing X, without reference to some further criteria. In Shoemaker's terms, he would have to have good reason for believing that the 'M-causal chain' which linked his present quasi-memory with the actual performance of X had undergone no branching in the intervening period. Only so could he rule out the possibility that though someone did in fact do *X*, and he quasi-remembered it, the person who did it was other than himself. And his reasons for so believing would involve the application of empirical criteria. So if instead of (3) I had 'John believes that he remembers witnessing Noah's flood', my derivation would not begin with a proposition which ascribes to John a thought involving a non-criterial application of the concept of identity. This was the reason for setting out the derivation (3)–(5) in terms of what seems to John to be the case, rather than in terms of what John believes to be the case. Moreover, Parfit's phrase 'apparent memory' suggests what happens when it seems to John that he remembers something, rather than what happens when John believes that he remembers something. And the arguments to which he appeals are appropriate to the interpretation of 'apparent' as 'phenomenal' memory.

Parfit appeals to what other writers, Wiggins and Palma as well as Shoemaker, have said on the subject.[17] Unfortunately, their arguments fail to establish the point. Wiggins wishes to defeat the claim that '*X* remembers doing *Y*' entails '*X* remembers himself doing *Y*'. He confuses the issue by equivocating between opaque and transparent senses of 'remembers'. His example depends on its being true that

(6) The first parachutist remembers jumping

although it is false that

(7) The first parachutist remembers the first parachutist jumping.[18]

But (7) is false only if it is taken opaquely. Only if it is taken transparently, however, is it entailed by

(8) The first parachutist remembers himself jumping,

and it is the entailment of (8) by (6) which Wiggins is out to disprove. We may grant that it is not part of the content of the first parachutist's memory that it was the first parachutist who jumped. It does not follow that it is not part of that content that it was he himself who jumped. Since the argument is about content, it is relevant to the question of what is entailed by (3), by a statement about what someone seems to remember. But it is a bad argument, and does nothing to defeat the claim that (3) entails (4).

The argument which Parfit borrows from Palma is also about content. As we have seen, Parfit holds that the view that 'my apparent memory of *having* an experience is an apparent memory of *my* having an experience' rests on a mistake. The mistake alleged is that of thinking that part of what I seem to remember about this experience is 'that I, the person who now seems to remember it, am the person who had this experience'. This last quotation is ambiguous. If the phrase in apposition to 'I', namely 'the person who now seems to remember it', is thought of as describing a part of what I say I seem to remember, we can agree with Parfit that it is a mistake. But it need not be taken this way, any more than Austin's 'Yesterday I saw a man born in Jerusalem' need imply that yesterday I was a witness at a birth. If I say that my memory of having an experience is a memory of *my* having it, I do not have to include in what I remember the fact that the person who now remembers it is the same person as the person who had it. That I am the same person is something I can infer from the fact that I seem to remember my having it, not part of what I seem to remember. Palma does indeed show that Flew and others[19] are wrong in holding that 'what is remembered is that the speaker is the same person as did, felt, saw whatever it was'.[20] But this does not imply, as he and Parfit think it does, that a wedge can be driven between remembering having an experience and remembering myself having an experience. Admittedly, identity is not part of the content of what I remember. As we saw in Section III a judgment made on the basis of memory, e.g., that it was I who locked the garage door, is not itself a judgment of identity. But since remembering locking the garage door is the same, *pace* Palma and Parfit, as remembering myself locking the garage door, judging that I did this can give rise to a judgment of identity when combined with another judgment about myself, as that I am now sitting down. However, we are now in a position to see that identity is more intimately connected with memory even than this. In order to obtain, at least implicitly, a judgment of identity, I do not need to predicate of myself anything other than what I seem to remember about myself, except the fact that I seem to remember it. Memory has only to become self-conscious in order to provide the materials for identity. An identity judgment is implied, not only by a conjunctive proposition like 'I am sitting down and I locked the garage door', but by a reflexive proposition like 'I remember that I witnessed Noah's flood'. This is because the form of an identity proposition is not '$\Sigma x K \varphi x \psi x$', but the more basic '$\Sigma x \varphi x x$'.

VII

A proposition of the form '$\Sigma x\varphi xx$' may be entailed by one of the form 'φaa', but what I have now to show is that (5) is entailed by (4). A possible defence might be constructed along these lines: What is said by (5) to seem to John to be the case is the existential generalization of what is said by (4) to seem to John to be the case. (5), it might be held, is of the form 'It seems to John that $\Sigma x\varphi xx$' and (4) of the form 'It seems to John that φaa'. There is not, of course, a general rule permitting the inference of propositions of the form 'It seems to x that p' from corresponding propositions of the form 'It seems to x that q' where 'p' is entailed by 'q'. However, when the substitutes for 'p' and 'q' are an existential generalization and its instantiation, it is plausible to suggest that such an inference is legitimate. It cannot seem to me that Jane is coming down the road (or singing in the kitchen) without its seeming to me that someone is coming down the road (or singing in the kitchen). If (4) and (5) are related in this way the entailment seems justified.

However, not every proposition of the form 'It seems to x that q' entails a corresponding proposition of the form 'It seems to x that p' where p is an existential generalization of 'q'. One way of appreciating this is to look at an example which, like (3) and (4), contains the indirect reflexive pronoun 'he', much discussed by H-N. Castañeda.[21] Let us recall a story from the Second Book of Kings.[22] King David, after Nathan the prophet had told him the story of the rich man who stole the single ewe lamb belonging to a poor man in the same city, judged that the man who had done this thing deserved to die. But by the poor man Nathan intended Uriah the Hittite, by the ewe lamb his wife Bathsheba, and by the rich man David himself. So the man of whom David judged that he deserved to die was in fact David. David did not at this stage judge that he himself deserved to die. That did not come, if at all, until after Nathan had concluded his prophecy with the shattering 'Thou art the man'. It does not matter what the expression was that David used, before his eyes were opened, to pick out the subject of the predicate 'deserves to die'. However he designated the man, his thought that the man so designated deserved to die did not involve the thought that he himself deserved to die, while he remained ignorant of the fact that he himself was the man.

Using the context 'judged that—' transparently, we may say in these circumstances: 'David judged that David deserved to die'. The existential generalization of this is 'For some *x*, *x* judged that *x* deserved to die'. Again, if David had had time to reflect on his own judgment, it might have been true to say, still using transparent constructions, 'David was aware that David judged that David deserved to die'. But before Nathan had said 'Thou art the man', it could not have been said that David was aware that, for some *x*, *x* judged that *x* deserved to die. David was not *then* aware that one and the same man was judging that someone deserved to die and was being judged to be deserving of death. It is this implication of the fact that two occurrences of a variable are here bound by a single existential quantifier that makes it necessary to qualify the type of existential generalization which legitimizes entailment of psychological propositions in the way suggested. Where the existential quantifier in '*p*' binds more than one occurrence of a variable, we cannot infer propositions of the form 'It seems to *x* that *p*' from corresponding propositions of the form 'It seems to *x* that *p*', even where '*q*' is an existential generalization of '*p*'.[23]

Yet another objection can be made against the suggested defence of the inference of (5) from (4). This inference cannot possibly be of the form 'It seems to *x* that *q*; therefore it seems to *x* that *p*', whatever the relation between '*p*' and '*q*'. For (4) is not of the form 'It seems to *x* that *q*'. Substitutions for '*q*' in this schema have to be complete propositions; but 'he remembers that he witnessed Noah's flood' is not a complete proposition. It is only a fragment of a proposition, containing as it does reflexive pronouns which cannot do their work until given an antecedent to refer back to. The inference from (4) to (5) cannot be of this type.

Nonetheless the inference of (5) from (4) is valid. Castañeda has warned us repeatedly against inferring propositions like 'David was aware that he judged that he himself deserved to die' from propositions like 'David was aware that David judged that David deserved to die'—and rightly so. But he is wrong to forbid the converse inference.[24] His doing so is the result of considering such propositions only in the sense which comes from reading '*x* was aware that...', '*x* judged that...', etc., as opaque contexts. It is, however, quite possible to give these contexts a transparent interpretation. Thus interpreted 'David was aware that David judged that David deserve to die' is true, provided David had these thoughts about the man who in fact is

David, however that man was designated in David's thoughts. If what David had said in his heart had been 'I judge that I deserved to die', it would still have been true that David was aware that David judged that David deserved to die. And if so, it would surely have been true that David was aware that someone judged that someone deserved to die. An existential generalisation which does not bind more than one occurrence of a variable can still be used to legitimize this latter inference. We can safely infer propositions of the form 'It seems to *x* that $\Sigma y \Sigma z \varphi yz$' from corresponding propositions of the form 'It seems to x that φab'.

From (4), then, we could safely infer that it seems to John that someone remembers that someone witnessed Noah's flood. But in (4) the 'someone' who does the apparent remembering is designated by the reflexive pronoun 'he', and so is the 'someone' who is apparently remembered as witnessing the flood. What seems to John to be the case could be expressed by him as 'I remember that I witnessed Noah's flood'. This double use of the pronoun 'he' or 'I' does justify, as double use of 'John' in purely referential position would not, our conflating the two 'someones' into a single 'some'. If David had been aware that *he* judged that *he* deserved to die, not only would he have been aware that someone judged that someone deserved to die, but he would also have been aware that, for some *x*, *x* judged that *x* deserved to die. Just so, the fact that it seems to John that *he* remembers that *he* witnessed Noah's flood entails, not only that it seems to John that someone remembers someone witnessing this, but that it seems to John that, for some *x*, *x* remembers *x* witnessing this. In other words, (4) does indeed entail (5). Accordingly, the derivation of (5) from (3) goes through unscathed.

VIII

What this shows is that the concept of memory involves the concept of identity. And since, in ascribing to himself the apparent memory of doing or undergoing something, John employs no empirical criteria of personal identity, he does not do so either in ascribing to himself the thought ascribed to him in (5). So the concept of personal identity can be used in a judgment without empirical criteria being employed. The empirical criteria which are employed in judgments of personal identity cannot therefore constitute the whole meaning of 'is the same person as'.

How does all this bear on Parfit's wish to prize the question of survival apart from the question of identity? At times Parfit seems willing to allow that the relations of continuity and connectedness are different from the 'relation' of identity, and merely to claim that it is the first two that are important, that 'matter in survival'. This claim that what is important in survival is continuity and connectedness, not identity, looks like an evaluative judgment rather than a judgment about meaning. But in other places he insists that identity is not 'a further fact' over and above the facts of continuity and connectedness, and this is a view about the meaning of 'same person'—a mistaken view. Continuity and connectedness are relations whose existence can be established only empirically. Identity is not. 'Is the same person as' cannot, therefore, mean the same as 'is continuous and connected with', and facts involving the one cannot be the same as facts involving the other. Nor can the concept of personal identity be dismissed as unimportant or replaceable in its psychological and social role, given certain 'advances' in medical technology, by those of continuity and connectedness. I cannot ascribe to *myself* the apparent memory of doing or undergoing something without being in a position to make a non-empirical application of the concept of personal identity. The same goes for ascriptions to *myself* of fear and hope and intention. Could a person who made no such ascriptions have the concept of *himself* as a person at all? The idea of *persons* may not be altogether happily identified with the idea of *selves*; but the logic of identity enables us to see that Locke's generous use of the word 'self' in explicating the idea of *person* shows a true feeling for what is indeed important in this area of philosophy.[25]

University of Bristol

NOTES

[1] *Philosophical Review* **80**, 1971, 3–27 (reprinted in J. Glover (ed.), *The Philosophy of Mind*, Oxford Readings in Philosophy, Oxford University Press, Oxford, 1976, pp. 142–162). The remarks quoted are on p. 147 of the reprint, to which page references given below also refer.

[2] On p. 152 of his essay 'Names and Identity' (in S. Guttenplan (ed.), *Mind and Language*, Clarendon Press, Oxford, 1975, pp. 139–158) Geach remarks: 'I have heard there are those who would regard the identity expressed by personal proper names as being conceivably non-transitive'. It is important to note that this is not Parfit's view: the

necessary transitivity of identity is one of his reasons for dispensing with it as 'what matters in survival'.

[3] R. M. White, 'Wittgenstein on Identity', *Proceedings of the Aristotelian Society* **78**, 1977, 157–174. In 'Is Identity a Relation?', *ibid.* **80**, 1979, 81–100, I have argued that Wittgenstein's view that identity is not a relation is correct, but not his belief that we need no sign for identity. I have developed these themes more extensively in *What Is Identity?*, Clarendon Press, Oxford, 1989.

[4] *Op, cit.*, p. 168.

[5] Cf. McTaggart's article s.v. 'Personality' in James Hastings (ed.), *Encyclopedia of Religion and Ethics*, T. and T. Clark, Edinburgh, 1915. (It was, of course, Peter Geach who drew my attention to this.) My ideas in this and the following section owe much to two papers by Richard Swinburne: 'Personal Identity', *Proceedings of the Aristotelian Society* **74**, 1973, 231–248, and 'Persons and Personal Identity', in H. D. Lewis (ed.), *Contemporary British Philosophy* (Fourth Series), George Allen and Unwin, London, 1976, pp. 221–238.

[6] See P. T. Geach, *Reference and Generality*, Cornell University Press, Ithaca, New York, 1962, § 31.

[7] This was argued convincingly by Sidney Shoemaker in Chapter IV of *Self-knowledge and Self-Identity*, Cornell University Press, Ithaca, New York, 1963.

[8] *Cf.* G. E. M. Anscombe 'The First Person', in S. Guttenplan (ed.), *Mind and Language*, Clarendon Press, Oxford, 1975, pp. 45–65.

[9] John Locke, *Essay concerning Human Understanding*, Book II, ch. xxvii, § 11.

[10] *God and the Soul*, Routledge and Kegan Paul, London, 1969, p. 11.

[11] *De Anima* III 428^b 2–4.

[12] That he does in fact so overstep the mark is clear from the following passage, *inter alia*: 'Wherever a man finds what he calls himself, there, I think, *another* may say is the same person' (*ibid.* § 26, my italics).

[13] 5.532. The Pears-McGuinness translation has '$(\exists x.y)$' instead of '$(\exists x,y)$' at its third occurrence in the English version of this proposition.

[14] *Cf.* A. G. N. Flew, 'Selves', *Mind* **58**, 1949, 355–358.

[15] *Op. cit.*, p. 152.

[16] *Cf.* S. Shoemaker, 'Persons and Their Pasts', *American Philosophical Quarterly* **7**, 1970, 269–285.

[17] *Ibid.*, nn. 22, 25, 21.

[18] D. Wiggins, 'Locke, Butler and the Stream of Consciousness: and Men as a Natural Kind', in Amelie Rorty (ed.), *The Identities of Persons*, University of California Press, Berkeley and London, 1976, pp. 139–173; see p. 142. (This appeared in *Philosophy* **51**, 1976, 131–158. It is referred to as unpublished in a footnote to Parfit's 1971 *Philosophical Review* article; though this footnote is amended in the reprint in Glover's collection. For references, see note 1 above.)

[19] Who should not be thought to include Reid, whom Parfit in the same note quotes disapprovingly as saying 'My memory testifies not only that this was done, but that it was done by me who now remember it'. Reid's phrase 'who now remember it' suffers from the same scope-ambiguity as Parfit's phrase 'the person who now seems to remember it'. Palma's article 'Memory and Personal Identity' is in the *Australasian Journal of Philosophy* **42**, 1964, 53–68.

[20] Antony Flew, 'Locke and the Problem of Personal Identity', *Philosophy* **26**, 1951, p. 55.

[21] '"He" The Logic of Self-Consciousness', *Ratio* **8**, 1966, 130–168; (*cf.* also Castañeda's 'On the Phenomeno-Logic of the I' in *Proceedings of the XIVth International Congress of Philosophy* Volume **III**, Herder, Vienna, 1969, pp. 260–266, with the bibliography appended). Miss Anscombe has pointed out (*op. cit.*, p. 46) that this use of 'he', etc., which Castañeda distinguishes by placing an asterisk after the pronoun, corresponds to the Greek word 'ἑ, οὑ, οἱ'.
[22] II Kings (= II Samuel), ch. 12.
[23] I have argued this point at greater length in 'Is Identity a Relation?' (see note 3 above). I call there for a restricted interpretation of 'existential generalization' which makes the suggested rule of inference acceptable. See also *What Is Identity?*, Chapters 3 and 4.
[24] *Cf.* 'On the Logic of the Attribution of Self-Knowledge to Others', *Journal of Philosophy* **65**, 1968, p. 442.
[25] I am grateful to those who took part in a class I gave in the University of Notre Dame and to those who commented on a paper I read in the University of Bristol—both in 1978—for assisting me to develop my thoughts on this subject.

PETER GEACH

PHILOSOPHY OF RELIGION

As I have already stated (p. 3), I began thinking about the philosophy of religion under the influence of McTaggart's *Some Dogmas of Religion*. I have ever since believed that the holding of some dogmas as true is essential to any religion's being worth serious consideration: dogma is essential to religion as a shell to an egg or a skeleton to a human body, without it we have only a shapeless jelly. Much as Wittgenstein has influenced my thought, he had no direct influence in this matter; we never talked on such topics, and so far as can be determined he seems not to have held anything like this attitude towards dogmas. Undogmatic Christianity is a plain absurdity; as McTaggart pointed out, the recorded teaching of Christ includes dogmas; and as Hobbes pointed out, the earliest Christian writings contain a credal formula just as dogmatic and propositional as any later creeds: 'Jesus is the Christ, the Son of God'.

Very soon after my conversion to the Catholic Faith I took an important decision of principle: I would not of my own initiative write about religious matters, but would take an invitation to speak or write on such topics as a command from my Master. I feared that at best my work would be marred by my own vanity and combativeness; but I hoped that if I waited for orders before writing in the defence of the Faith, He who had laid the task upon me would give me grace towards its fulfilment. The invitation to give the Stanton Lectures in Cambridge came quite unsought; onerous as the task appeared in prospect, I dared not decline it.

I need not repeat here what I have said about philosophy of religion in my replies to Norman Malcolm and Hugo Meynell. I add only this: Truth is what matters. A religion that makes no claim to offer truth and to oppose the contrary errors deserves no serious consideration. It appears that some Catholics of our time equate the condition of a Pilgrim Church on Earth with the condition of always learning and never coming to knowledge of the Truth; a condition that is treated in the New Testament as undesirable (2 Timothy iii.7).

Harry A. Lewis (ed.), Peter Geach: Philosophical Encounters, 213–214.

Nothing in the end matters except Truth; for God himself is Truth; God is that which all true speaking and thinking points towards and from which all false speaking and thinking points away. (I have written about this in my Joint Session paper 'Truth and God', Aristotelian Society Supplementary Volume **56**). To God all falsehood is hateful, especially if it is supposed to help religion; God has no need of our lies.

The University of Leeds

NORMAN MALCOLM

PHILOSOPHICAL CONFUSION AND SIN

1

In *Providence and Evil* Peter Geach says: 'What we cannot hope to understand in this life is sin. Sin is the surd or absurd element in the universe' (1977, p. 63). Geach goes on to present a most interesting analogy between philosophical confusion and sin.

> Though we cannot understand sin, we can understand its not being understandable; at least not to us who are sinners. A parable will, I hope, convey my meaning. In his professional life a teacher of philosophy often has to present to his pupils thoughts of other philosophers that appear to him not merely false but radically confused; so that no clear, well-formed sentence of the philosopher's own language can convey the thing that was intended by the subjects of his criticism. If he clarifies their thought in his exposition, he is falsifying it by giving it a clarity it did not possess; but if he presents their thought it its native confusion, how is he to get across to his pupils what the confusion is?
>
> . . .
>
> So the philosophy teacher has the inescapable task of trying to describe confusions clearly, and convey to his pupils just what the confusions were. This is difficult enough in practice, in concrete cases; what the theory of the matter should be—a theory that could describe what confused thoughts are without itself lapsing into confusion—I have not the faintest idea. And the fact that *A* himself is liable to fall into confusion does not make *A* better qualified to describe other people's confusions, but rather worse. It may be that feeling temptations to a confusion is humanly necessary to a proper understanding of it; but it is not necessary to succumb in order to understand. The man of whom perfect understanding of philosophical confusions might be expected is the paragon who, although tempted at all points as we are, is yet without sin.
>
> . . .
>
> Looking at our inability to understand or adequately describe philosophical confusion does, I suggest, enable us in a measure to comprehend the incomprehensibility of sin, and moreover to see why it is that being a sinner yourself does not help you to understand sin.
>
> We do not, I conclude, know what it would be like to understand sin; to enter into the sinner's confusion and self-deception without misperceiving it or getting involved in confusion and self-deception oneself. (1977, pp. 63–65)

I wish to focus first on Geach's remarks about philosophical confusion. There is some ambiguity in what he says. Is understanding a philosophical confusion very hard to achieve; or is it impossible to achieve?

Harry A. Lewis (ed.), Peter Geach: Philosophical Encounters, 215–227.

Geach says both things. On the one hand, describing a confusion clearly 'is difficult enough in practice, in concrete cases' (p. 64); on the other hand, Geach speaks of 'our inability to understand or adequately describe philosophical confusion' (p. 65). I think that 'is difficult enough in practice' was a slip, although perhaps a significant one. The analogy with sin would seem to require that we are absolutely unable to understand philosophical confusion, just as we are supposed to be absolutely unable to understand sin. Furthermore, Geach puts the problem of understanding philosophical confusion in a way that suggests the presence of a logical paradox: for if the teacher clarifies the confused thinking in his exposition, then 'he is falsifying it by giving it a clarity it did not possess'; but if he presents the thinking 'in its native confusion', he is himself speaking in a confused way and not clarifying anything.

2

Let us look at some confusions other than philosophical ones. A man gets lost in London. Afterwards he explains it this way: 'I thought Trafalgar Square was Piccadilly Circus'. Isn't there a logical paradox here? For at the time of his confusion he could not have described his confusion by saying to himself: 'I think Trafalgar Square is Piccadilly Circus'. He certain wouldn't have had that thought; for he knew perfectly well that Trafalgar Square and Piccadilly Circus are different places. If someone had said to him at the time he was lost, 'You think that Trafalgar Square is Piccadilly Circus, don't you?', he would have denied it. Nevertheless, in telling us afterwards that he thought Trafalgar was Piccadilly, or that he took Trafalgar to be Piccadilly, he may be giving us an adequate description of how he went wrong and what his confusion was. Do we want to say that he gave us a false account of his confusion? I don't believe so. After a little reflection, what we should suppose is that if he had been asked, at a certain moment in his confusion, 'Are you thinking that *this place* is Piccadilly?', he would have said, 'Yes; isn't it?'. In fact the place referred to was Trafalgar. It was this mistaken belief that was at the center of his confusion, not the belief or thought, 'Trafalgar is Piccadilly'. Thus, a decription of his confusion can be given that has no appearance of paradox. The account, 'I (He) thought that Trafalgar Square was Piccadilly Circus', is superficially misleading because it suggests that the man himself had the thought, 'Trafalgar is Piccadilly'—and this would be a false account of his confusion. But one doesn't have to be misled by the verbal form of that account; in ordinary conversation few would be. An accurate, non-falsifying, unconfused interpretation of the man's confusion is available.

The kinds of confusion are many. A drunken man clings to a lamp-post, caressing and addressing it as if it were his horse. Does he *mistake* it for his horse? How could he? Certainly it is unlike the case of a sober man who sees a shape dimly in a fog and takes it to be his horse. Nor would we want to say that the drunken man is *pretending* that the lamp-post is his horse. Here indeed it may be impossible to give a clear account of the drunken man's mental confusion, for to do so would be to give it a clarity it doesn't possess.

Let us now consider some kinds of mental confusion that are manifestations of mental illness. For example, an upper middle class woman is beautiful and wealthy. Her husband adores her. She has two lovely young children. She is successful in a scientific profession. She says of herself that she has everything a woman could desire. She is, however, deeply depressed. She cries a great deal of the time. She says she feels 'empty'. But she does not know what is missing, what is wrong with her. At first neither she nor her psycho-analyst can give a clear account of her mental state. But it is possible that in the course of an extended analysis some insight will be achieved. A detailed account of desires, frustrations, emotions, may emerge—an account that the patient will acknowledge as accurate, and in terms of which her depression begins to be intelligible to her as previously it was not.

Then there is the psychotic patient who says to his analyst: 'I am in this room. I am also in Victoria Station. Perhaps I am two people. I don't know.' Taken at their face value these remarks are incoherent. But even here prolonged work with the patient may gradually bring to light deep emotional conflicts and phantasies in relation to which the patient's words take on unexpected meaning.

In these examples of mental illness the patient's confusion was at first opaque; but with skill, persistence, and luck, the analyst might help the patient to obtain some understanding of the nature of the mental disturbance.

I believe, with Wittgenstein, that there is some similarity between the treatment of a philosophy confusion and the treatment, in psycho-analysis, of mental illness (see 1958, 255). (I do not mean by this, any more than Wittgenstein did, that philosophical confusion is a manifestation of either neurosis or psychosis.) One point of similarity is the way in which an initial state of confusion can gradually be clarified. The nature of the confusion is of course different in the two cases. Philosophical confusion is primarily intellectual; in mental illness the confusion is primarily emotional.

3

Let us look at philosophical confusion. Once when I was talking to an undergraduate class about our perception and knowledge of material objects, I made a casual reference to the brown desk on the rostrum in the front of the room. To my surprise, one student declared with passion: 'That desk is not *brown*. It is *wood*!' Apparently this young man was assuming that a sentence of the form, '*A* is *B*', always means that *A* is made of, or composed of, *B*. How could he think that? Well, probably he had before his mind an instance or two where 'is' really is equivalent to 'is composed of'; and it had not occurred to him that 'is' might be used in some other way. This is an example, in miniature, of philosophical confusion. It did not arise from mature reflection. But it does display the propensity, typical in philosophy, of leaping to a generalization about the use of a word, instead of making a decent survey of its actual use.

I do not see any reason for believing that it is impossible to understand the confusion from which emerged the student's comically absurd exclamation. Might not the observation that he was assuming that 'is' always means the same as 'is composed of', be an adequate description of what went wrong in his thinking? Perhaps he had not explicitly formulated that assumption to himself. But may it not still be true that he *made* it? Suppose that, on reflection, the student acknowledged that he had made this assumption although perhaps without full awareness of what he was doing, and also admitted it to be a mistake. Wouldn't this be a case in which philosophical clarification was achieved, without falsification? Isn't the pointing out of false or unwarranted assumptions a type of criticism that Peter Geach has frequently and effectively produced in his own writings and lectures?

It occurs to me that Geach might want to say that the human propensity to leap to an unwarranted generalization is something we are unable to understand—that it is as unintelligible as is the human inclination to tell a lie in order to get out of a tight spot. Here there does seem to be a similarity between a propensity that contributes to bad philosophy and a propensity to sin. But it isn't clear what could be meant by the assertion that we are unable to understand these propensities. For they are familiar to us. We recognize them as natural human tendencies. In what sense do we *fail* to understand them? After all, explanations come to an end. In referring something to a natural human tendency haven't we reached rock bottom?

In any case, there seems to be no support here for the position from which Geach apparently started; namely, that there is a paradox in the very notion of clarifying a philosophical confusion. In my example of immature philosophy, an adequate account of the student's confusion of thought was achieved by calling attention to his wildly false assumption about the use of 'is'. A confusion of thinking was diagnosed, without the diagnosis being itself confused.

I wish to turn now to a problem that has serious perplexed many sophisticated philosophers. It is the problem of how it is possible to believe what is false, or to think what is not the case. Plato refers to the problem in the *Theaetetus* (188c, 189b). I am using Cornford's translation in his book *Plato's Theory of Knowledge* (1935, pp. 114–115); but I will condense the remarks that Plato attributes to Socrates. Socrates says the following:

> It is impossible that a man should see something, and yet what he sees should be nothing. Yet surely if what he sees is something, it must be a thing that is. And if he hears a thing, he hears something and hears a thing that is. And if he touches a thing, he touches something, and if something, then a thing that is. And if he thinks, he thinks something, doesn't he? And when he thinks something, he thinks a thing that is? So to think what is not is to think nothing. But surely to think nothing is the same as not to think at all. If so, it is impossible to think what is not.

Moore was worried about this problem. In *Some Main Problems of Philosophy* (1953, p. 289), he speaks of the following difficulty:

> It seems as if purely imaginary things, even though they be absolutely self-contradictory like a round square, must still have some kind of *being*—must still *be* in a sense—simply because we can think and talk about them. It seems quite clear that, in a sense, there *is* not and cannot be such a thing as a round square: but, if there is not, how can I possibly think and talk about it? And I certainly can think and talk about it. I am doing so now. And not only can I make and believe propositions about it: I can make *true* propositions about it. I know that a round square, if there were such a thing, would be both round and not round: it is a fact that this is so. And now in saying that there is no such thing as a round square, I seem to imply that there *is* such a thing. It seems as if

there must *be* such a thing, merely in order that it may have the property of *not* being. It seems, therefore, as if to say of anything whatever that we can mention that it absolutely *is not*, were to contradict ourselves: as if absolutely everything we can mention must *be*, must have some kind of being.

The paradox of false belief can be stated in general terms as follows: If you have a belief, what you believe is *that p* (where any proposition can be substituted for *p*). If you believe that *p*, then *that p* is the *object* of your belief. Your belief *has* to have an *object*, since you believe *something*. If you believe *that p*, then *that p* must exist, or be the case; otherwise your belief would be without an *object*, and would not be a belief. Given any belief *that p*, the object of your belief is *that p itself*, and not any substitute, such as an image, thought, or sentence. Therefore, the object, *that p*, exists or is the case, when you believe *that p*. But to say '*that p* is the case', means nothing more nor less than that *p* is true. Thus, whenever you believe something your belief is true, and false belief is impossible.

As with many philosophical problems this one seems at first to be a joke. Nevertheless it presents itself to a philosopher as a serious conceptual problem. In the *Tractatus* Wittgenstein was concerned to deal with the question of how one can think what is not the case. In his later work he discusses it in a number of places and in a number of ways: for example, in the *Investigations* (1958; 442–449, 462–464, 518); in *Zettel* (1967; 69, 70); in *The Blue and Brown Books* (pp. 30–41).

A closely connected problem focuses on the meaning of words. It is natural to suppose that the word 'red' means the colour red; that the word 'rain' means the phenomenon of rain; that the name 'Thames' means the river Thames; that the name 'Jimmy Carter' means the man Jimmy Carter. When you use those names in sentences you mean those things. There is something, namely, the colour red, that is the meaning of the name 'red', i.e. there is an existing thing, red, that is the meaning of that name. So when you use the word 'red' in a meaningful sentence, red must exist; otherwise your sentence would be meaningless. Even if you assert of a book that it is *not* red, you *refer* to *red*. Therefore, red colour must exist in order for you to used the word 'red' meaningfully.

These presentations are quite compelling; and they *look* coherent. That there is some bad confusion of thought is evident at first only from the absurdity of the consequences—namely, that one cannot think what is false, and that one cannot employ a word '*X*' meaningfully unless the thing *X* exists. Distinguished philosophers have racked their brains trying to untangle the confusion. This example would seem, at first sight, to support Geach's claim that it is impossible to understand or to adequately describe a philosophical confusion—for to clarify the thinking in one's exposition of it would falsify it by giving it a clarity it did not possess, whereas simply to present the thought in its native confusion does nothing to reveal what the confusion is.

I think there is both something wrong and something right in Geach's claim. What is wrong is that he neglects to consider that there are different *stages* in addressing a philosophical confusion. The first stage consists in presenting the problem in its original tempting aspect. Here no suggestions are offered as to where the confusion lies. One cannot both exhibit a knot and also undo it, all at one go. Geach speaks as if everything had to be done at once. But there is a second stage, which consists in trying to locate the errors of thought. This will be arduous and prolonged. Misleading analogies and imagery may be noted, as well as illicit assumptions and the misuse of expressions. At this stage philosophers will differ greatly in their skill, fertility, patience, and depth.

Yet there is also something right in what Geach says. For it is sometimes impossible to fully understand what a philosopher means by an assertion that plays a crucial role in his argument. For example, what is a person trying to say when he asserts that if I think of something, then that something must *exist*; or that if I believe *that p*, then this object of my belief, *that p*, must be the case; or that if I use the word 'red' in a meaningful assertion such as the assertion that this table is not red, then red colour must exist? I myself have the feeling that I half do and half don't understand these assertions. A necessary part of the study of this problem would consist in drawing attention to the obscurity of these assertions. It is an important insight in philosophy to realize that one doesn't know what one meant, an insight that loosens the hold of what previously seemed to be an indisputable truth.

It would be foolish for me to attempt in this essay a thorough study of the problem of believing or thinking what is not the case. I will limit myself to a couple of points. The argument of the *Theaetetus* asserts that if you see, hear, or touch something, then that something must exist. Now there is *a* use of 'see', 'hear', and 'touch', that conforms with that rule. But there are also uses of those verbs that do not—for example, their use in describing an hallucination or a dream. If in telling a dream I say, 'I saw a tiger leap into the room, heard him growl, and felt his teeth pierce my leg', it does not follow from this true report of my dream that there was actually a tiger in the room. So the move, without any qualification, that Socrates makes from 'If you see, hear, or touch something, that thing exists', to 'If you think something, that which you think exists', has a premise that is false unless suitably qualified. But if it were suitably qualified, the premise would be tautological.

A more important point is that the reasoning seems to obtain much of its force from regarding thinking, believing, or meaning something, as requiring some sort of *contact* with the something. If we focus on the expression, 'some sort of contact', the obscurity of the thought becomes evident. *What* sort of 'contact'? Surely not actual physical contact. Is it 'mental contact'? And how are we to understand *that*? It is true that I cannot actually take hold of the door key without making contact with it; but false that I cannot *search* for it unless I am in contact with it. In the *Investigations* Wittgenstein mentions a similar disanalogy: I can look for a man when he isn't there, but I can't hang him when he isn't there (1958, 462)!

This imagery of *contact* was strong in the *Tractatus*. In speaking of the correlations between the elements of a picture and things in the world, the *Tractatus* says: 'These correlations are, as it were, the feelers of the picture's elements, with which the picture touches reality' (2.1515). I have no doubt that this imagery contributed importantly to the doctrine of the *Tractatus* that the thing a name means is *the meaning* of the name (3.203): for if the thing were not there for the name 'to reach out to' (2.1511), the name would not *touch*, would not make contact with, anything in the world. The name would not hook on to anything in reality, and so would be meaningless.

In the *Investigations* Wittgenstein describes this as the error of supposing that a word has no meaning if nothing *corresponds* to it (1958, 40). To the reply that of course this does not hold for words that stand for *complex* things, but does hold for those words that stand for 'the simple elements of reality', Wittgenstein responds with the criticism that nothing whatever is simple or complex in an absolute sense, but that for different purposes and in different contexts we may designate one and the same thing either as simple or as complex (1958, 46–59).

The criticism does two things. It makes one see that one didn't know what one meant in speaking of 'the simple elements of reality', and therefore didn't know what one was thinking in supposing that language has to contain words that stand for such elements. But it also makes one realize that the distinction between 'simple' and 'complex', rather than being absolute, is relative to human decisions.

The criticism has the double effect of revealing the original thinking as confused, but at the same time helping one to a better understanding of the distinction of language about which one was confused. It destroys an air-castle, but also it clears up the ground of language from which the nebulous thought arose (1958, 118).

Undoubtedly this territory is very familiar to Peter Geach. My purpose in referring to it is to suggest that if we distinguish between the initial presentation of a philosophical problem and the subsequent diagnosis of misleading imagery and illicit assumptions, then Geach's contention that it is impossible to understand a philosophical confusion is undermined. What is true in his contention is that the original confused thoughts are not, in the second stage, transformed into clear thoughts. But what can and does happen is that specific points of confusion are identified; and this is itself an advance in understanding. In addition to, or as a part of, this uncovering of pieces of 'plain nonsense' (1958, 119), there may occur an improved insight into the actual working of those terms of language about which we tend to have false impressions when we first approach them in a philosophical frame of mind. All of this adds up to a considerable understanding of the original confusion. If this is so, then Geach has failed to provide a 'parable' that helps us 'to comprehend the incomprehensibility of sin'.

4

Let us turn to this latter theme. What are some of the various forms of sin? In his Letter to the Galatians Saint Paul gives a list:

> Immorality, impurity, licentiousness, idolatry, sorcery, enmity, strife, jealousy, anger, selfishness, dissension, party spirit, envy, drunkenness, carousing, and the like. (*Galatians* 5: 19–21)

In contrast:

> The fruit of the Spirit is love, joy, peace, patience, kindness, goodness, faithfulness, gentleness, self-control . . . . (*Galatians* 5: 22–23)

In First Corinthians some forms of sin are compared with love:

> Love is patient and kind; love is not jealous or boastful; it is not arrogant or rude. Love does not insist on its own way; it is not irritable or resentful; it does not rejoice at wrong, but rejoices in the right. (*1 Corinthians* 13: 4–6)

Is it true that we cannot understand jealousy, envy, hatred, selfishness, arrogance, resentment? 'What is the reason for *A*'s hatred of *B*?' 'Oh, didn't you know that *B* raped *A*'s daughter?' Here the motivation is so strong that *A*'s hatred is immediately intelligible. We would not be surprised if *A* sought revenge. We would be amazed if instead *A* was gently forgiving. There is a human point of view according to which such a manifestation of forgiving love would be far more incomprehensible than savage hatred.

I think there is an error in Geach's claim that 'What we cannot hope to understand in this life is sin,' and that 'Sin is the surd or absurd element in the universe'. It seems to me that human behaviour is intelligible or unintelligible only in relation to some framework of thought and feeling. Many people find vandalism unintelligible: 'How can the youngsters who live in public housing destroy the windows, doors, plumbing, in the very buildings in which they live? This is incomprehensible; completely irrational'. But if we consider the matter from this angle: that those youngsters are unemployed, bored, without recreational facilities, having no outlet for their energies other than this destructiveness, which also offers the thrill of danger—then their vandalism becomes intelligible.

Many of us regard political terrorism—the bombings, the taking and killling of innocent hostages—as monstrous, incredible. 'How can they do this to people who have not harmed them; whom they don't even know? Those terrorists must be insane'. But if we think of the terrorists as people who believe that the present structure of society is inherently unjust, that it guarantees the oppression of the greater part of mankind for the benefit of a privileged minority, that legal reform is impossible because the legal institutions are controlled by the wealthy minority, that a just society can be created only if disorder is produced in such magnitude that the present system will collapse—then we will no longer regard the terrorists as monsters of viciousness or as simply insane, even if we do not accept their assumptions, or their faith that a new age of justice will emerge from the ruins of the present political and social structures.

Geach's conviction that human sinfulness is an incomprehensible element in the world derives from a certain viewpoint, namely, that God has created ourselves and all things; that 'God is to be loved and admired above all things because he is all truth and all beauty' (1977, p. 79); that since 'we love truth and beauty by our nature... it is only vice and folly that stop us from loving God above all things' (*ibid*). It is inconceivable that we should not take delight in the almighty, loving God, and put our trust in him and give him our complete obedience. Since he has commanded us to love other human beings as much as we love ourselves, it is incomprehensible that we should be self-centred, arrogant, envious, revengeful. We should, in the words of Psalm 37, 'Refrain from anger, and forsake wrath'.

I heard a bishop say, in a sermon, something like the following: When I reflect on God's great gifts to me, especially his gift of love, *I cannot imagine* that I should fail to commit myself entirely to him: yet I know that I do fail!

It is in terms of this stark contrast between, on the one hand, one's immeasurable gratitude to God and, on the other, the vivid awareness of one's failure to obey the divine commandments, that the latter takes on the aspect of an incomprehensible folly. If, however, one does not think and feel within this religious framework, then the bearing of a grudge or the desire to get even for an insult or an injustice, present themselves as both natural *and* intelligible. The thoughts and emotions of a great part of mankind, including those whose religious inclinations are only half-hearted, are not centered in a religious framework. From a non-religious viewpoint, the love which, in Paul's words, 'endures all things', is itself incomprehensible.

It strikes me that Geach's declaration that sin is incomprehensible, has in part the character of what Wittgenstein calls an '*Äusserung*'. Human utterances and actions sometimes function as *expressions* of joy, grief, surprise. When Geach reflects on the greatness and beauty of God and on how mankind owes God complete gratitude, devotion, and perfect obedience; and when Geach *compares* this with the actual self-centeredness of human beings—then he feels *amazement*. How can this be! His declaration that sin is incomprehensible is an outcry of astonishment. In the terms of *this comparison*, Geach's astonishment comes across as a completely believable reaction.

But if one's viewpoint is thoroughly secular then human pride, self-esteem, ambition, will be what is *given*, and the responses of envy, anxiety, bitterness, rage, to defeats and frustrations, to injustice and betrayal, will appear as entirely intelligible. My objection to Geach's contention that these attitudes are unintelligible does not consist in a rejection of his Christian standpoint. When he asserts that sin is *the* absurd element in the universe I interpret him, perhaps mistakenly, as thinking that the unintelligibility of sin has the status of an absolute metaphysical truth. Whereas it is only within the terms of a particular contrast that sin is seen as incomprehensible. From another viewpoint nothing is more understandable than thoughts, feelings, acts of jealousy, hatred, revenge.

Geach might wish to hold that the Christian viewpoint is correct and that the other one is wrong. Here I am reminded of Wittgenstein's example, in *On Certainty*, of the people who are guided by the predictions of oracles rather than by the propositions of physics, and of his question: 'If we call this 'wrong' aren't we using our language-game as a base from which to *combat* theirs?' (1969, 609). I do not see that the conflict between the world-pictures of religion and of non-religion is something that admits of resolution by rational demonstration that one of them is more true, more correct, more reasonable, than the other.

To summarize: Peter Geach proposes an analogy between philosophical confusion and sin, saying that our inability to understand or adequately describe philosophical confusion provides a parable for our inability to understand sin. There are kinds of mental confusion, as for example some states of drunkeness or of delirium caused by fever, which are so jumbled and unstable that no coherent description of the confused mental states is possible. There are also mental confusions which are manifestations of mental illness, and which in their original appearance are unintelligible; but through intensive analytical work the nature of the confusion may become intelligible both to analyst and patient. Here there is a point of resemblance with philosophical confusion. Geach speaks of the latter as if there would be a logical paradox involved in both presenting the confusion in its original form and at the same time producing a clarifying account of what the confusion is. My reply is that the appearance of paradox disappears once we distinguish between the stage of initial presentation of the confused argument and the subsequent stage of diagnosing the errors of thought. Thus, although the clarification of a sophisticated and deep-seated philosophical confusion is extremely difficult, requiring skill, persistence, and imagination, there is nothing in the nature of philosophical confusion as such that makes clarification impossible. Finally, in regard to Geach's main theme, I argue that human sin (e.g. resentment, selfishness, envy, revenge) is not incomprehensible *per se*, but presents itself as incomprehensible only when looked at from a particular standpoint, which in the case of Geach is the world-picture of Christian thought and feeling.

King's College, London

REFERENCES

Cornford, F. M.: 1935, *Plato's Theory of Knowledge*, Routledge and Kegan Paul, London.

Geach, P. T.: 1977, *Providence and Evil*, Cambridge University Press, Cambridge.

Moore, G. E.: 1953, *Some Main Problems of Philosophy*, Allen and Unwin, London.

Wittgenstein, L.: 1958, *Philosophical Investigations* (Translated by G. E. M. Anscombe), Basil Blackwell, Oxford.

Wittgenstein, L.: 1960, *The Blue and Brown Books*, Basil Blackwell, Oxford.

Wittgenstein, L.: 1967, *Zettel* (Edited by G. E. M. Anscombe and G. H. von Wright; translated by G. E. M. Anscombe), Basil Blackwell, Oxford.

Wittgenstein, L.: 1969, *On Certainty* (Edited by G. E. M. Anscombe and G. H. von Wright; translated by Denis Paul and G. E. M. Anscombe), Basil Blackwell, Oxford.

HUGO MEYNELL

ON IMPROVING CHRISTIANITY

The distinction made by P. F. Strawson between 'revisionary' and 'descriptive' metaphysicians could with equal appropriateness be applied to theologians and philosophers of religion.[1] The 'revisionary' theologian or philosopher of religion would be one who proposes a new conception of God, and of man and the world as related to him; while his 'descriptive' colleague tries to clarify or justify those which already exist. The latter would not be concerned with propounding a new set of doctrines, or even a radically new way of understanding old ones; his attention would rather be to get clear about the nature and implications of what is already believed.

Given such a distinction, it is obvious enough that Peter Geach's writings on the philosophy of religion belong to the descriptive and not to the revisionary type.[2] He forcefully expounds the implications of Christianity as a coherent system of beliefs, perhaps especially those implications which have been evaded or obscured owing to their inconvenient or distasteful character, or to their actual or alleged incompatibility with current opinion in philosophy or the sciences.[3] His work is a powerful corrective to the tendency, so common among contemporary writers on religion, to confuse what is revisionary with what is descriptive, perhaps concealing the confusion with what has been called 'persuasive definition'. Thus someone might say or imply that 'true' or 'real' Christianity did not include belief in the Trinity, or in hell; when in fact what he was up to was *commending* a *selection* of traditional Christian beliefs which did not include these two, while the form of words that he was using misleadingly suggested that they did not in fact form a part of the content of traditional Christian belief, or were not organically connected with other aspects of it. Antony Flew's work on the philosophy of religion, though it does differ from Peter Geach's in a number of respects, has in common with it the exposure of this kind of subterfuge.[4] And it is not surprising that the believer and the unbeliever in any traditional form of religion are liable to share an interest in getting clear what its content and implications are, against those who have some interest, whether commendable or not, in

Harry A. Lewis (ed.), Peter Geach: Philosophical Encounters, 229–246.

obscuring the issue. And while it is disreputable to obscure the issue, not all the motives for doing so are equally corrupt; one might well hold in good faith, as Matthew Arnold did, that while it is of great importance for the future of civilisation that some aspects of traditional Christian belief should be maintained, others would have to be rejected as intellectually or morally intolerable by informed modern men.

The authors of the recent collection of essays, *The Myth of God incarnate*,[5] may be said to share Arnold's view of this matter. They wish to propound a selection from or modification of traditional Christian belief when does not include belief in the incarnation; as theologians, they are definitely and quite consciously revisionary. (One may occasionally detect in them, it is true, a tendency to lessen the shock of the effect of their departures from orthodoxy by too easy acceptance of arguments to the effect that there is no such thing as orthodoxy.[6] They admit that much that is of value in Christianity has as a matter of fact been bound up with belief in the incarnation; but they suggest that it could all the same be retained if the belief were abandoned, or, perhaps, if it were kept rather as an expression of Christians' attitude to Jesus, and of what he meant to them, than as a supposed literal truth about him. For the Old Testament Israelites, feelings about and understanding of God's redemptive power were specially related to and focussed on the person of Moses; and later the life and teaching of Hosea expressed for them God's self-sacrificing love in a vivid and compelling way. The story of Jesus, whatever mixture of fact and myth it may contain, can still do these things and more for Christians, even if the belief that he is somehow really and truly God incarnate, and not merely expressive for Christians in an incomparable way of God's nature and his purposes for mankind, is abandoned.[7]

But why should the doctrine of the incarnation in its traditional sense be abandoned? Roughly, one may say that three types of reasons are offered by the authors—historical, philosophical, and practical. The rest of this paragraph will be a summary of them. Historically speaking, it does not appear to be at all probable that Jesus thought of himself as God incarnate. It seems that he proclaimed the imminent reign of God, rather than the significance of himself; it was his followers who made him the object rather than the subject of proclamation, and one may trace step by step how the process of his deification continued through the New Testament period and afterwards.

The New Testamant authors cannot properly be said to have *any* one Christology; certainly they did not affirm the doctrine of the incarnation which later became accepted. The belief that there was one coherent set of ideas from the first, by contrast with which some later ideas may be regarded as deviant, will not bear historical examination. Each New Testament writer had a stock of conceptions and images through which he expressed the special significance of Jesus for himself and his community; ideas relating to incarnation form only a relatively small proportion of these, and one may set out, in principle if not in detail, how they were taken over from the surrounding paganism. And even if the doctrine of the incarnation were far more ancient and universal in the Church than in fact it is, there would be the philosophical objection that belief that God could be incarnate in a man, or that one and the same being could be at once human and divine, is incoherent. Jesus was a man of his time, with its characteristic mental set and limitations of knowledge; so much is implied by his true humanity. But to be God is to be unlimited in these respects. To speak of the *kenosis* which God underwent in becoming man, as some theologians have done in the attempt to reconcile belief in the incarnation with modern historical science, is to say what makes no sense in the last analysis; if a divine being thus 'emptied himself' he would cease to be divine. Practically speaking, abandonment of the doctrine of the incarnation as traditionally conceived would greatly facilitate dialogue between Christians and members of other faiths, and would make possible an approach to them as having something to give Christians we well as to receive from them. To retain the doctrine more or less inevitably entails a dismissive attitude to the non-Christian religions, as mere errors based on rejection or neglect of the one true revelation of God in Jesus Christ.[8]

Now it is true enough that formulae of the type used by the Greek councils, and taken for granted by later theologians, do not occur in the New Testament. And the authors' ways of expressing themselves, and more especially of speaking about Jesus, certainly differ a great deal. But the crucial question is of what, if any, unity of meaning there is to be found among the admitted variety of expression.[9] Shakespeare's *The Tempest* is a real rag-bag of heterogeneous elements; but it does not immediately follow that it lacks the fundamental unity of meaning and intention which would make it a great poetic drama. Traditional Christianity is based on the assumption that there is an analogous unity in the New Testament, and more especially in the account which it gives of Jesus Christ. All of the authors, in their different ways and from their different points of view, bring out that Jesus Christ was a man, and yet one whose relation to God was unique in kind rather than merely in degree.[10] If the stories in the Gospels are seen in the context of the Old Testament, the reader is given to understand that Jesus feeds the people 'in a desert place' as God fed their fathers in the wilderness; that he rules the sea as God rules the sea according to the Psalmist; that he gives a new law 'on the mount' as God gave the old law on Sinai.[11] He himself speaks with divine authority, without the 'Thus saith the Lord' of the prophets; he calls God his Father in a manner which implies that the relationship sets him apart from other men. The unique nature and status attributed to Jesus are brought out in a concrete and dramatic form in the Gospel narratives; the Epistles, especially those of Paul, and the Johannine discourses, tend to express the matter more abstractly and in a manner which anticipates later theology. Here also we find so-called 'Trinitarian formulae', with God (the Father) and Jesus (the Son) mentioned together with a mysterious third being, 'the Spirit'. This being, as one learns especially from *Acts*, is supposed somehow to have come to possess Jesus' followers and to direct their mission.

The process of explication and abstraction, which was already beginning in the New Testament period, went on well beyond it. For what the disciples had to say about Jesus, for all the width of its appeal to all sorts and conditions of men, even the simplest and most ignorant, is not without problems to tax the educated and reflective mind. Here is talk of a man who was born, grew up, walked about, talked, and was finally executed in an agonizing and humiliating way; and yet who is supposed to be the unique Son of God, 'the express image of his glory', equal with him, associated with him in creation, and himself actually divine. What was to be made of all this? Was Jesus just a man with a special divine commission? This by no means seems to do justice to all that is stated and implied about him the New Testament. Was he then only apparently a man, and really simply God? This is inconsistent with the emphasis by the New Testament writers, and especially the authors of the Gospels, on the reality of his human feeling, that he could be hungry and thirsty, and undergo grief and pain. Nor does it explain the fact that it is always implied that God (the Father) is in some sense at least *another than* Jesus Christ. Is he then the first and greatest of creatures, but less than truly divine? This conflicts with the direct assertion of the divinity of Christ in some texts, and the broad hints of it in many others. In considering solutions like this to the problem of who or what Jesus Christ was, and successively rejecting them as inadequate to him who was proclaimed in the New Testament and worshipped within the Church, the Fathers gradually worked out the classical formulations of Christian doctrine—for example, that the Son (Jesus) is 'consubstantial' with the Father (God), and that he is one 'Person' in two 'Natures'.[12]

Now it is worth noting that these two doctrines are just the opposite of 'mythological', in one sense at least of that ambiguous and much-abused term. The great myths speak dramatically and imaginatively about gods and heroes; they eschew technical precision. But the whole point of the doctrines worked out at Nicea and Chalcedon was to state an exact meaning which would rule out ambiguity; which would leave no loophole for Arianism in the former case, for Nestorianism in the latter. Once the Arian controversy was under way, it was clear that there was a logical issue which could no longer be evaded. Christ, the Son or Word of God, must either be divine strictly speaking, with all that *that* entailed, or a creature; he could not be both. The Arians chose the latter way out. Christ was strictly speaking a creature, and consequently had been brought into existence in time; 'there was when he was not'. The party which posterity has called orthodox grasped the other horn of the dilemma; Christ was strictly speaking divine, and therefore eternal. To set out with the greatest possible precision what they meant, the Fathers took over a term from the ordinary language of the time—apparently it had no currency in philosophy then—and employed it in a technical sense. To say that the Son is consubstantial with the Father, as is brought out with especial clarity by Athanasius, is to submit to a rule which governs discourse about the Son in relation to discourse about the Father. Whatever you attribute to the Father in respect of his divinity, you must also attribute to the Son; if the Father is eternal, the Son is too; if the Father is Creator, so is the Son; and so on.[13]

In more contemporary terms, one might express the matter as follows:

> B is consubstantial with A in respect of φ-ness iff in all respects that A is φ, B is φ.

Such a rule is worth making if *A* is the paradigm case of something that is φ, as God the Father is of deity, and there is a question whether *B* has φ-ness *really* and *literally* or not. In the case at issue, the point to be made is that Christ (the Son) is divine literally and unequivocally just as God (the Father), the one whom everyone agrees to be divine, is so. Similarly, for Jesus Christ to be consubstantial with us in respect of his humanity is for him to be really and truly a human being, in just the same sense as you and I are.

In the development of doctrine, to settle one problem is to give rise to others. Once it is determined that the Son is consubstantial with the Father, the question arises whether the Holy Spirit is so. And, notoriously, there is the problem of how it can be that while there are three distinct beings who are God, there is only one God.[15] On the matter of the Person of Christ, it is evident that humanity and divinity are not easily attributable to the same being; and one may be tempted to resolve the difficulty by suggesting that strictly speaking there are two beings, the human individual Jesus and the divine Word of God, who are united with one another in a specially intimate way. At the Councils of Ephesus and Chalcedon, the Church rejected this way out as well, on grounds, once again, which are fundamentally Biblical. As the New Testament writers present the matter, it is the divine Word himself which *becomes* a man among men, rather than simply *possessing* or *adopting* a man and using him as its instrument. Here too the Fathers used technical terms to clarify a point which had to be made with logical precision—this time the terms already had a certain philosophical currency, though the Fathers used them in their own way and for their own purposes. For Christ to be one 'Person' in two 'natures', in the sense intended by the Council of Chalcedon, is nothing more nor less than for him to be one individual with two sets of properties, by virtue of one of which he is really and truly God or divine, by virtue of the other of which he is really and truly human or a man.

Whenever anything radically new has to be said, available language has to be stretched to accommodate it; it is only in the course of time that a language emerges to state literally and unambiguously what at first has to be conveyed by parable and symbol. The message about Jesus, about his astounding words and mighty deeds, was conveyed to its first hearers in terms of a collocation of images and titles derived from the Old Testament; in course of time, as the news was brought to men of different backgrounds, and as the inevitable questions of interpretation were asked, it seemed to emerge that the news was of another divine being besides the Father who had lived as a man among men. The classical expressions of the doctrine of the Incarnation merely state this and draw out its full implications.

It is perfectly true that the Fathers were recasting a faith which was originally preached in terms of one cultural background into terms which would make sense in another. But they were also up to something of more permanent significance which may easily be overlooked. For the fact is that the concepts involved in the conciliar definitions have a bearing which is trans-cultural. To say that someone is the Messiah, or the Son of David, or the Son of Man as alluded to by Daniel, is to say something which makes sense against an Old Testament background, but hardly otherwise. But to say that someone is both human and divine is to say something which makes sense, whether one believes it or not, wherever there is a concept of man and a concept of God—in other words, virtually in all human cultural contexts whatever. The point of the technical terms employed by the Councils, as I have already tried to show, is merely to ensure that those who maintain this belief should accept the full consequences of doing so.[14]

The authors of *The Myth of God Incarnate* rightly point out the variety of expression among the New Testament writers and the early Church Fathers, and to the absence from their writings of the formulae which were imposed by the conciliar definitions. But the question between their position on the one hand, and that which I have just sketched on the other, is not whether there is uniformity of expression, but whether there is, underlying the admitted variety of expression, a basic unity of meaning and intention; and whether it was this meaning which was worked out, through a series of questions and answers, by the Fathers of the early Church, and unambiguously asserted by the ecumenical Councils.

I do not find, in Frances Young's very detailed account of the period of theological development to which I have been alluding, any reason to reject the view that I have set forward of what was going on. She has not *noted* it and *advanced arguments against* it; she has apparently not adverted to it at all. An impressive pile of data is assembled by her in support of the thesis that 'the traditional formulations of Christology, so far from enshrining revealed truth, are themselves the product of witness and confession in a particular historical environment'.[15] But what is the basis of the assumption that 'the product of witness and confession in a particular historical environment' must be incapable of 'enshrining revealed truth'? It would be easy to make a show of invalidating the claims of any scientist, or perhaps even of some New Testament historians, by just the same kind of argument. Does not the whole practice of science and of history presuppose that theories and judgments which are the products of their particular social environments may yet represent the truth about what is quite independent of those social environments? If it could be shown that the later formulations were not an explication of what was in some way implicit in the earlier confessions and acts of witness, the case would be a reasonable one. But this possibility does not seem really to have been attended to, let alone refuted, by the authors of this book.[16] And if it should by any chance be the truth, then what was going forward was by no means a matter of haphazard intellectual bungling, as one would think on their account, but rather takes on the complexion of a single 'research programme' directed towards understanding the mystery attested by the New Testament and entrusted to the Church.

That it is the case neither that there are 'two roots' of a fundamentally disparate nature to the New Testament Christology, nor that it is a 'tangled mass',[17] but that it is an organic whole related to a single, central meaning, may be argued, as Harald Riesenfeld has suggested, on historical grounds.[18] It seems on the face of it unlikely that the preaching of the early Church should have had the impact that it did if it were really such a collection of utterly diverse ideas as is supposed by some of these authors. Nor can one easily explain how ordinary Christians throughout the ages have apprehended the New Testament as an overall unity in the way that they have. That such apprehension of unity of fundamental meaning and intention *need* not be destroyed, but may actually be enhanced, by historical scholarship, is evident if one appeals to some writers.[19] Perhaps, as Riesenfeld says, some scholars miss a unity of meaning that there really is within the New Testament and its witness to Jesus Christ, owing to their preoccupation with minute detail.[20]

One particular point about historical scholarship is worthy of special emphasis in this context. The view current in many circles since the middle of the last century, that the Fourth Gospel is as a whole a theological reflection on aspects of the tradition about Jesus preserved in the other Gospels, with little or no value as a historical source about Jesus in its own right, has been subjected to a large number of well-argued attacks, on *historical* and not merely on *dogmatic* grounds, during the last half century.[21] Raymond Brown has gone so far as to complain of the 'uncritical traditionalism'[22] with which the older view is often unquestioningly propounded, in defiance of the large and increasing heap of evidence against it, by some authorities.[23] Now I am not suggesting that the authors of *The Myth of God Incarnate* should concede that what has been called 'the new look' in Johannine scholarship is *justified*; but I do think they are somewhat at fault for not even adverting to its *existence*. The view of Jesus which the authors wish to maintain seems to depend rather heavily on the material peculiar to John's Gospel, and the Johannine elements in the Synoptics, being secondary, having no direct basis in the preaching and activity of Jesus himself or of the early Christian community. The more historicity there is attributed to these, the stronger is the case for holding that Jesus, whether rightly or wrongly, had the kind of conception of his own person and significance that is presupposed by traditional Christendom. That he was not in the habit of stating the Aramaic equivalents of 'I am God' or 'I am incarnate God', is certainly true; but does not, as I have argued already, by itself determine the issue. Now whether we particularly *want* Jesus to have been such a man, and whether the supposition of his being so consorts well with the pluralism or religious tolerance which may be deemed fashionable or desirable, has no bearing on the immediate point, which is simply historical.

Still, if it were really true that it would be better for mutual understanding between the religions if Jesus was as the liberal theologians present him, rather than as the traditional faith entails, it would be right to say so. However, it seems to me more or less the reverse of the truth. The idea of a divine incarnation is very strikingly analogous to motifs in Hinduism, Mahayana Buddhism, and many primitive religions; to dispense with this element in Christianity would rather be to destroy than to create possibilities of mutual understanding here.[24] And if attribution of divinity to any human being is offensive to Muslims and Jews, is not belief in the personal and transcendent God common to Christians, Muslims and Jews a stumbling-block to Theravada Buddhists, Jains, Marxists and liberal humanists? Such phenomena as the sacral kingship of the ancient world, the idolatry of political leaders in the modern, and indeed the deification of great religious leaders like Muhammad and the Buddha in flat opposition to their own claims and wishes, are hints of a hunger for an incarnate God widespread among mankind.[25] It seems likely that the appeal of Christianity is due largely to the manner in which it meets this need. If this is so, then these authors are suggesting that we rid Christianity of precisely the element which has made it the special force in human civilization that it is.[26] To retain belief in the incarnation, in its traditional sense and with its traditional implications, is by no means to be constrained to have a dismissive or contemptuous attitude to other religions. If *traditional* Christianity is true, other religions are to be respected and even revered as proclaiming or at least hinting at aspects of the truth, in a manner from which Christians are likely to have a great deal to learn. The tenets of Theravada Buddhism may bring home to the Christian the vanity of self-assertion and self-indulgence, those of Chinese religion the existence of a right order in the world and in himself which he is called upon to maintain, those of Islam the presence of the transcendent God, those of Vaishnavite Hinduism the mystery of his incarnation, those of primitive religions the life and deeds of the representative hero whose fate is shared and symbolically re-enacted in a cult.[27]

It may be insisted that even if the historical and practical objections to the doctrine of the incarnation can be rebutted, the philosophical objections remain fatal to it. Granted that Jesus did act and speak in a way which can be best accounted for as an implicit claim to be divine, and granted that the notion has the immense religious value and attractiveness that history would seem to suggest; it may yet be the case that the claim that any individual is at once human and divine is a contradiction. Now *how* those things can be so which the Church proclaims to be so is the concern of systematic theology; and a good deal of attention to this matter has been given by theologians in the course of the history of the Church. That such solutions to the present problem as were put forward, for example in the Middle Ages,[28] have to be confronted with the results of modern scholarship, is true and important; but it does not immediately follow that they should be rejected. One suggestion made quite recently, which takes medieval systematics and modern scholarship into account, is to the effect that Christ was a single subject (Person) of a divine and human consciousness (natures), the human consciousness perpetually striving to express the mysteries of the divine in terms of concepts and meanings gained from experience and practice in the normal human way.[29] While such an explanation would hardly dissolve the mystery of Christ, it would at least go some way towards vindicating the claim that it is not merely a matter of conceptual confusion. It would also provide a viewpoint from which both biblical[30] and conciliar data could be appreciated as interlocking in a consistent and convincing manner, much as experimental and observational reports interlock with explanatory theories in the natural sciences.

Questions of a philosophical nature are raised by a number of the contributors to the book; some other similar questions may not unfairly be put *to* them. One wonders if some of them have adverted sufficiently to the difference between evincing an attitude on the one hand, and stating that something is the case, with the implication that it would have been the case even if one had never made the statement, on the other. Thus if I make the true statement that over half the surface of the earth is covered by water, or that the planet Neptune is further away from the sun than the planet Mercury, I state a fact which would have been the case if I had not stated it, and indeed might have been the case if I myself, or even mankind in general, had never existed at all. Frances Young contrasts 'the language of confession and testimony' with that appropriate in matters of philosophy or science;[31] but I wonder whether she does not overlook the fact that, if the term 'testimony' is understood in the usual sense, it is possible to give false testimony; and that testimony is commonly rather a matter of reporting what is so than of evincing an attitude. In many of his writings, John Hick has argued with cogency and brilliance against the attempt by some philosophers of religion effectively to make belief in God a mere feeling or practical attitude about the world and human life, as opposed to dependent for its truth on what is supposed to be so whatever the believer's attitude. And indeed it does seem pretty reasonable to suppose, on the face of it, that to claim that a Being exists who made the world out of nothing, and sustains it in existence at every moment, is not the same thing as for creatures occupying a tiny niche within space and time to evince a feeling or to display an attitude. But some of those who would be proud to think of themselves as Professor Hick's disciples with respect to belief in *God*, may wonder whether he has not too readily embraced with respect to belief in *Christ* assumptions which he has given excellent reasons for rejecting in the other context. If there are good reasons for rejecting an analysis of belief in God which is exclusively in terms of the feelings and attitudes of believers, do they not apply also to the analysis of belief in Christ? One should perhaps have some compelling reason for such drastically divergent accounts of these central and constitutive elements of Christian belief.

It is difficult not to feel some sympathy with those colleagues of one of the symposiasts who chide him for not giving up Christianity altogether.[32] If the foundation documents of Christianity are as full as is claimed of deliberate or unconscious misrepresentation, and traditional Christianity is based on so many misapprehensions, it seems curious to wish to identify oneself with the movement at all. The tactics which Dr. Goulder attributes to Paul, which he commends for their admirable flexibility, are such as I, and I think most people at most times, would call those of a most egregious liar.[33] Even if Paul were so dishonest as to corrupt the apostolic testimony about Jesus in order to outdo the claims made about Simon Magus,[34] it seems hard to believe that he could have imposed his misrepresentations on the early Christian community at large.[35] And if we are convinced that he did so, would it not be much more sensible to abandon Christianity, than to pin our faith on a hazardous reconstruction of the original message which has long been repudiated in substance by the Church at large?[36]

The point is made with great force in Dennis Nineham's Epilogue, that what many of these authors do believe about Jesus would be a natural inference from the traditional Christological dogmas, but without their underpinning seem quite arbitrary and unfounded.[37] Given that Jesus had the unique nature and status in relation to God the Father that one might infer from the New Testament as a whole, the perfection of his human nature follows more or less as a matter of course. But short of this, we simply do not have the necessary historical basis for maintaining that Jesus was the supreme instantiation of any moral ideal, unless—one thinks of phrases like 'the man for others'—this is couched in terms so vague as to be quite untestable. On the contrary, his persistent setting of himself above other men, his careless flouting of revered moral and religious customs on his own authority—which emerge from the documents if they can be relied on at all—are qualities which in most men would be the very opposite of morally admirable.

Professor Nineham maintains that modern historical studies have shown that the Christ of preaching is not identical with the historical

Jesus. The reason why Christians who are aware of this do not abandon their faith, he suggests in effect, is that they still find something of irreplaceable value for their lives in responding to the Christ of preaching. He admits that the status of this 'Christ' would be very difficult to make clear to an ordinary Christian; and that as a consequence it is no wonder that 'many preachers fall back on the assumption that the preached Christ and the historical Jesus are identical'.[38] However, I believe that devotion to the preached Christ, except on the assumption that he is really one and the same as the historical Jesus, is psychologically impossible except for a very few highly intellectual people; who can fall in love, or remain in love, with an abstraction or an admitted fiction?[39] The Christ of preaching has meant what he has to Christians at large on the assumption that he is the historical Jesus, and cannot long survive the abandonment of the assumption; here I think the ordinary believer, to whom the acute theological distinction just alluded to cannot be made clear or convincing, may well have more insight than some theologians.

However, as well be apparent from what I have argued earlier, I do not think one can take it as established beyond reasonable doubt that the Christ of preaching *is* distinct from the historical Jesus, at least in the relevant sense. The water perceived by the child or the savage, and what is represented by the chemical formula H_2O, seem very different things; but they are in fact the same substance envisaged from different points of view. Historical criticism does raise in an acute form the question *whether* the historical Jesus is really one and the same as the Christ of preaching; it does not immediately settle *that* they are different. If the relation between what the historical Jesus said and did, what the Evangelists reported, and what the Councils of the Church determined, is as I have argued, then one can indeed say that it is *one and the same* Jesus Christ who is proclaimed in one manner in the New Testament, in another manner in the later doctrines of the Church. But this presupposes a degree of historical reliability in the New Testament documents, and particularly in the Gospels, which would be denied by these authors, though by no means by all reputable historians.[40] If the historical Jesus and the Christ of faith are *not* one and the same in the sense that I have suggested, the conclusion drawn by these authors does seem to follow; that the doctrine of the incarnation is at best a *façon de parler* expressing what the historical Jesus, or the Christ of faith, or the latter on the assumption that he is identical with the former, have *meant to* Christians; and at worst false on the grounds that it has historical truth conditions which are unfulfilled.

The book as a whole does have the very great virtue of bringing out something which is too often forgotten by theologians; that belief in the traditional Christian faith cannot consort in the long run with denial of the historical presuppositions of that faith. But the authors gravely underestimate, as I have tried to show, the case for saying both that these historical presuppositions are true all the same, and that traditional Christian belief is coherent given that they are so. As to the alternative versions of Christianity which they offer, I see no reason to suppose that they are more than unstable compromises available temporarily to a few exceptional people, bound in the long run to yield to a form either of non-Christian theism or atheistic humanism.[41] Given the falsity of the historical presuppositions of traditional Christianity, these are so tied up with the whole way of talking, and of ritual and moral practice, which are constitutive of Christianity as a whole, that it would be much better frankly to abandon the whole thing. I have tried to show why I do not think it will do to plead, in the manner of some of these authors, that historical Christianity never really amounted to any palpable or coherent set of beliefs. Orthodox believers will continue to hope for and expect corroboration of their faith by unprejudiced historical inquiry. Fortunately for them, this does not all tend in the direction followed, not always too critically or circumspectly, by the authors of this book.

Department of Religious Studies,
University of Calgary

NOTES

[1] P. F. Strawson, *Individuals: An Essay in Descriptive Metaphysics*, Methuen, London, 1959, pp. 9–12.

[2] *God and the Soul*, Routledge and Kegan Paul, London 1969; *Providence and Evil*, Cambridge University Press, Cambridge, 1977; *The Virtues*, Cambridge University Press, Cambridge, 1977.

[3] Cf. especially the chapter on hell in *Providence and Evil*.

[4] Cf. especially 'Theology and Falsification' in A. G. N. Flew and A. C. MacIntyre (eds.), *New Essays in Philosophical Theology* (Library of Philosophy and Theology), SCM Press, London, 1955, pp. 96–99 and 106–108; A. G. N. Flew, *God and Philosophy*, Hutchinson, London, 1966.

[5] John Hick (ed.), *The Myth of God Incarnate*, SCM Press, London, 1977; referred to as *MGI* in subsequent notes.

[6] Cf. the statement that 'orthodoxy is a myth' (*MGI*, p. x). And yet the whole point of the book is that Christianity would be better off without that aspect of 'orthodoxy' which is belief in the Incarnation taken literally. If some belief which was not 'orthodox', but rather merely marginal or aberrant, were at issue, the matter would not be the important one which these authors rightly assume that it is.

[7] Cf. Maurice Wiles, 'Christianity without Incarnation?' (*MGI*, pp. 8–9).

[8] For historical reasons, cf. especially Frances Young in chapters 2 and 5, Michael Goulder in chapters 3 and 4, Leslie Houlden in chapter 6, and Don Cupitt in chapter 7; for the philosophical, Maurice Wiles in chapter 1 (especially pp. 4–6), Don Cupitt in chapter 7, and John Hick in chapter 9; for the practical, cf. the Preface and John Hick in chapter 9.

[9] That analysis of the meaning for its author of each particular New Testament text is necessary and inevitable (cf. Leslie Houlden, *MGI* p. 125), does not entail that synthesis of the meaning of the whole is not possible, or indeed absolutely necessary for the business of Christian theology. It is in the synthetic task that it may reasonably be doubted whether authors of this book have learned as much as they might have done from the work of earlier Christian generations.

[10] According to the Synoptics, Jesus refers to God as 'my Father', and 'your Father' but never as '*Our* Father', except in a context where he is telling his *disciples* how they ought to pray. To God the Father he is 'my beloved Son', and other men are commanded to listen to him (Mk. 9.7; Mt. 17.5; Lk. 9.35; cf. Mk. 1.11; Mt. 3.17; Lk. 3.22). Paul uses what is virtually a technical term *huiothesia*, to distinguish the sonship of the faithful from that of Christ (Gal. 4.5; Rm. 8.15 and 23, 9.4; Eph. 1.5). In the Gospel of John, Jesus is 'the only-begotten Son', and the faithful are called not *huioi* but *tekna theou*. Cf. B. Lonergan, *De Deo Trino*, I, apud aedes Universitatis Gregorianae, Rome, 1964, pp. 124–125.

[11] For the Old Testament background as giving meaning to the acts of Jesus, cf. E. C. Hoskyns and N. Davey. *The Riddle of the New Testament*, Faber and Faber, London, 1931, pp. 173–175.

[12] Cf. B. Lonergan, 'The Dehellenization of Dogma' in *A Second Collection*, Darton, Longman and Todd, London, 1974, pp. 11–32; *The Way to Nicea*, Darton, Longman and Todd, London, 1976.

[13] Cf. *The Way to Nicea*, Section IX; 'The Dehellenization of Dogma', pp. 22–23.

[14] The foregoing account owes an immense amount to the writings of Lonergan; cf. especially *Insight: A Study of Human Understanding*, Longmans, London, 1957, and *Method in Theology*, Darton, London, 1972. On the notion of a 'universal viewpoint' from which relativity to one's culture can in a manner be transcended, cf. *Insight*, xxiv, pp. 564–568, 738–739.

[15] *MGI*, p. 14. If the development of the doctrines of orthodox Christianity is to be understood in the way that I have suggested, it would not seem that they were essentially tied up with any political system or attitude (cf. Don Cupitt, *MGI*, pp. 139–146). I think that the use of titles with overtones of political authority for Christ has a significance more or less the reverse of what is suggested by Dr. Cupitt; Jesus Christ, and only Jesus Christ, is the divine leader that other men have blasphemously been supposed to be, and thus no merely human leader can claim absolute and unconditional authority.

[16] Don Cupitt's argument (*MGI*, p. 133ff.) seems vitiated by a failure to distinguish between mere innovation, and such innovation as makes explicit what was already in some sense implicit.

[17] Cf. the discussion between Michael Goulder and Frances Young in chapters 4 and 5 of *MGI* ('The Two Roots of the Christian Myth'; 'Two Roots or a Tangled Mass?').

[18] H. Riesenfeld, 'Reflections on the Unity of the New Testament' (*Religion* **3**, 1973, 35–51); cf. especially 40–42.

[19] Riesenfeld's own book *The Gospel Tradition* (B. Blackwell, Oxford, 1970) is a good example of this; so is *The Riddle of the New Testament*, (note 11 above), which has not 'dated' in relevant respects.

[20] 'Reflections on the Unity of the New Testament', 39–41.

[21] The evidence is conveniently summarised by Sydney Temple, *The Core of the Fourth Gospel*, Mowbray's, London and Oxford, 1975, pp. 3–7.

[22] Raymond E. Brown, *New Testament Essays*, Chapman, London and Dublin, 1965, p. 143.

[23] For example, in a letter to *The Times* by R.P.C. Hanson (11.5.1974). Here acceptance of the view that John is more or less without independent historical value is taken as the test of an informed historical attitude to the Bible. Frances Young will have it that the approach to the discourses attributed to Jesus in John which regards them as 'built out of homilies based on synoptic-type traditions' is 'most fruitful' (*MGI*, p. 15); but one may wonder whether the fruitfulness does not consist rather in the support it gives to the liberal theological consensus, than the support it finds in stringent historical inquiry. For argument to a different effect cf. C.H. Dodd, *Historical Tradition in the Fourth Gospel*, Cambridge University Press, Cambridge, 1963; and Leon Morris, *Studies in the Fourth Gospel*, W.B. Eerdmans Publishing Co., Grand Rapids, 1969.

[24] Cf. e.g., R.C. Zaehner, *Concordant Discord: The Interdependence of Faiths*, Clarendon Press, Oxford, 1970, p. 443. Dr. Cupitt's charge that Christianity was paganised in the early centuries (*MGI*, p. 138) is particularly interesting in this connection; orthodox Christianity may well be termed 'pagan'—just because of its close analogy with other world religions, as their fulfilment—if the cast of one's piety is essentially Islamic or unitarian.

[25] In the discourse on the Bread of Life in the sixth chapter of John, 'the feeding of the five thousand is . . . seen to point to Jesus as the ultimate answer to man's hunger' (F.N. Davey, in E.C. Hoskins and F.N. Davey, *The Fourth Gospel*, Faber and Faber, London, 1940, cited by Leon Morris, *op. cit.*, 57–58). It may, incidentally, be wondered how a historical Jesus whose view of himself was as proposed by these authors could have instituted a cult focussed on himself. But there is no historical proposition about Jesus more certain than that he did so.

[26] That belief in the incarnation as literally true, and not as a mere *façon de parler*, is central to the piety and devotion of ordinary Christians, is strongly suggested by the manner in which so many Christmas carols harp on the paradox that one and the same is God and man. 'Enough for him whom cherubim / Worship night and day / A breast full of milk / And a manger full of hay.' 'Lo! within a manger lies / He who built the starry skies.' Professor Wiles, I think, is the most sensitive of the authors to the elements of the traditional doctrine which capture the imagination; he feels that a frank unitarianism is comparatively impoverished (*MGI*, p. 6). But I cannot see how his own suggestions, which amount in effect to a combination of an imaginative picture of the incarnation

with a unitarianism of actual belief, can be very satisfactory in the long run from the point of view either of ordinary piety or of systematic coherence.

[27] A cumulative argument of this kind needs a mass of detail to be properly developed, and can hardly make its effect in a brief sketch. Zaehner's book cited above (note 24) is an example of how the thing should be done. For a comparatively simplified approach, cf. Hugo Meynell, 'Religious Disagreement', *Religious Studies* **9**, 1973, 427–435, and 'The Mechanics of Atonement', *Theology* **77**, 1974, 21–27.

[28] Cf. Thomas Aquinas, *Summa Theologica*, III, ix, 4. Aquinas' treatment of the problem was revolutionary for his time, in that he granted that Christ must at least in some sense have developed in respect to the knowledge which he had as a man.

[29] Cf. Bernard Lonergan, *De Constitutione Christi*, apud aedes Universitatis Gregorianae, Rome, 1956; *De Verbo Incarnato*, Pontifica Universitas Gregoriana, Rome, 1964. In the last few decades 'attention has increasingly turned to the consciousness of Christ.... The doctrine of one person with two natures transposes quite neatly into a recognition of a single subject of both a divine and a human consciousness' ('The Dehellenization of Dogma', p. 25). This seems to suggest the manner in which one might rebut the dilemma so skilfully set out by Dr. Cupitt (*MGI*, pp. 134–139). If Christ is divine, he cannot in any way be limited in knowledge; to appeal to a Christology of *kenosis* will not do, since for God to have 'emptied himself' to the extent of being ignorant in becoming incarnate would be for him to have ceased to be divine. But by applying the doctrine of the two natures in Christ in the manner suggested, one may perhaps combine conciliar orthodoxy with a critical view of the bible, without being forced into that form of the doctrine of 'kenosis' which Dr. Cupitt rightly charges with inconsistency.

[30] Leaving the Fourth Gospel out of account, one might cite Mt. 11, 27 (Lk. 10, 22) on the matter; also the use of the solemn *ego eimi* in some significant contexts in Mark (cf. E. Stauffer, tr. D.M. Barton, *Jesus and His Story*, SCM Press, London, 1960, VIII, p. 6). And the Epistle to the Colossians may, for all one knows, mean something when it declares that all the treasures of knowledge and wisdom are hidden in Christ (Col. 2, 3).

[31] *MGI*, p. 13. As Leslie Houlden says, 'Once conclude that descriptive, factual statement about God is to be disallowed, and the credal mode is done for' (*MGI*, p. 130). But at *that* rate, it is difficult to see how the Christianity of the future could amount to anything more than the evincing of emotions and the fostering of traditional liturgical habits.

[32] *MGI*, p. 48.

[33] *MGI*, p. 75.

[34] *MGI*, pp. 72–79.

[35] For a less imaginative account of the relation of Paul to his Christian contemporaries, cf. B. Gerhardsson, tr. E.J. Sharpe, *Memory and Manuscript* (Seminarium Neotestamenticum, Acta 22), University of Uppsala, Uppsala, 1961, pp. 272–318.

[36] To be exact, as Ebionism or Unitarianism. It does seem a defect in these authors that they seem to have a blind spot for unity within heresy, as well as in orthodoxy, in spite of changes in intellectual milieu and verbal expression over the centuries. One can see that it may be less embarrassing to identify oneself clearly and unambiguously with a position traditionally recognised as incompatible with mainstream Christianity, than to argue as though the latter never really existed.

[37] *MGI*, pp. 188, 194. Dr. Goulder seems very nearly to have grasped this point: 'It is impossible to justify claims . . . to Jesus' sinlessness, or his *complete* devotion to God's will, or his *invariable* attitude of love' (*MGI*, p. 53); but in that case what grounds can he have for saying that Jesus 'was not just one of a group of men of universal destiny . . . he is *the* man of universal destiny' (*MGI*, p. 57)? It is precisely the claim for the uniqueness of Jesus, which Dr. Goulder does think is essential to Christianity, which has no warrant on his general position.

[38] *MGI*, p. 200. Dr. Cupitt will have it that 'the core of a religion does not lie in the biography or personality of the founder, but in the specifically religious values to which, according to tradition, he bore witness' (*MGI*, p. 205). But this is simply false; traditional Christianity, with its belief in the incarnation, is the obvious counter-example—unless indeed one denies that it counts as a religion at all.

[39] Dennis Nineham and Maurice Wiles do concede that some continuity between the historical Jesus and the preached Christ is needed (*MGI*, pp. 200–201). But it seems to me that they can fairly be faced with a dilemma: either the theological Christ is just the historical Jesus considered from the systematic and transcultural viewpoint made possible by later developments, or he is not. If he is not, the needs not only of the orthodox theologian, but also of the simple believer, are not met.

[40] Two radically different positions, often unfortunately confused with one another, are to be distinguished on this; (i) that we cannot, at least at present, attain reasonable certainty as to how far the Gospels are historically accurate concerning, for example, the words of Jesus; (ii) that we can now be reasonably certain that a large proportion of the sayings, for example, the majority of those reported in the Fourth Gospel, were not due to Jesus, and indeed give a radically misleading picture of his view of his own person and work. The former position is of course quite compatible with Christian orthodoxy, a typical stance of which is that faith *goes beyond* what can be demonstrated by secular reasoning, but is perfectly *consistent* with it as far as it goes.

[41] The tendency of Don Cupitt and John Hick is towards a frankly Unitarian belief; that of Maurice Wiles towards an overlay of Unitarian belief—which is what is taken literally—with Christian symbolism—which justifies up to a point traditional Christian ways of talking and of liturgical practice. For Leslie Houlden and Frances Young, Christianity does not appear essentially to involve belief about what is so at all.

PETER GEACH

REPLIES

REPLY TO QUINE

I knew Quine only by correspondence till we met in 1951, when he had a visiting professorship in Oxford. He was characteristically pleased that the house he lived in was on the corner of Logic Lane. We were very glad to come to know Van and Marjorie, and have remained friends ever since, and have met whenever we could.

I found Quine's *Mathematical Logic* for myself. My father's logical teaching of me was sound but limited; he had not kept up his knowledge of the literature during his time in the part of the Indian Empire that is now Pakistan. He and Quine had both been taught by Whitehead—in Cambridge, Cambs., and Cambridge, Mass., respectively. It is thus not surprising that I found Quine's ways of thinking congenial; I greatly admired, and still do, the combination of neatness and rigour and clear exposition in that work. From comments on it (I believe originally on tricky points of what I call Inverted Commatology) there began our long correspondence.

Old sins have long shadows, says the proverb; I keep on being embarrassed by people citing as 'Geach's theory of definite descriptions' views that I expressed in a number of articles in the late 1940s and early 1950s. I then briefly lapsed from the sound Russellian theory of descriptions into a theory of 'presuppositions'. Historically, I think this doctrine arose from people's being excited by the appearance in English of Frege's article 'Ueber Sinn und Bedeutung'; certainly that was what worked with me. There has been an odd failure to recognise that Frege regarded the lapse of a sentence into truth-valuelessness because some 'presupposition' failed, not as part of the regular working of language, but as one of the many defects of natural language that a logician must try to avoid in his own work. My first published references to Quine's work were regrettably vitiated by this talk of 'presuppositions'; this is particularly true of my contribution to the 'Joint Session' programme of 1951, which also contained bad misunderstandings of Quine's paper 'On What There Is' (the title of the

Harry A. Lewis (ed.), Peter Geach: Philosophical Encounters, 247–306.

symposium was taken from this). Quine justly protested, in his contribution; but he in turn misunderstood my idea of 'presupposition'—not that I think it was all that worth understanding. The reason for mentioning the matter here is to emphasize that I have now long since recanted this 'presupposition' doctrine. In a paper of 1956 (*Logic Matters*, 1972, 2.5, p. 82) I already wrote: '... this idea of presupposition ... I cannot see ... has much significance for the philosophy of logic'; this negative verdict has continued to express my view. The removal of the muddle has helped subsequent communication between Quine and me.

Quine's deep understanding of Frege has been a strong intellectual bond between us. Already in my reading of *Mathematical Logic* I noticed his frequent references to Frege and his stress on the huge contribution of *Begriffsschrift* and *Grundgesetze* to the devices of logic and set theory. Like Frege, Quine has often emphasized the impossibility of regarding a class as somehow composed of its members. In my article 'Class and Concept' (*Logic Matters*, 1972, 6.3) I try to show that the differences between Quine and me about set theory are largely a matter of style. Those who on the contrary think of classes as built up out of their members nevertheless distinguish between any plural class and its unit class; now these are built up out of the same bricks, if we think that way, but must clearly not be identified; this objection, already raised by Frege, is just ignored.

The first part of our correspondence that Quine reproduces deals with problems about what is recognized in the grammar of some languages as the indirect reflexive pronoun; a pronoun in *oratio obliqua* answering to the first-person pronoun of *oratio recta*. It helps here to bring in the notion of *Shakespearean* predicables or contexts, even though the term does not figure in the parts of my text Quine quotes. A context is *Shakespearean* if it allows of mutual replacement *salva veritate* for proper names that name the same thing (for those who relativize identity, as I do, it must be added that both names have the same criterion of identity built into their sense). This will differ from the notion of an extensional context significantly if we cannot assimilate proper names and definite descriptions, treating definite descriptions as complex names, *à la* Frege, or proper names as replaceable by definite descriptions, *à la* Russell. As regards this I would lay stress on the idea that proper names are scopeless (an idea that I owe to Quine) whereas definite descriptions have scopes.

When we have a construction like 'It was N.N.'s aim (achievement) to bring it about that—should be so-and so', it is clear that we must consider the scope of any definite description inserted in the blank; it will make a difference whether this scope covers the whole sentence or only the *that* clause. But when an actual, rather than an intended, effect is in question, there will be a Shakespearean context for any name appearing in the *that* clause which expresses this effect. It is for this reason that we can with good conscience use e.g. such transitive verbs as 'catch' or 'kill' to interpret dyadic predicate letters; though such a transitive verb admits of analysis in the style '—brings it about that—so-and-so', the context of a proper name inserted in either blank will be Shakespearean; and as I maintained in *Logic Matters* (1972) 4.3, it is the Shakespearean character of a predicable that determines the application of natural-deduction rules.

As regards sentences where there is an indirect reflexive pronoun, I still think that the context for a proper name that is antecedent of such a pronoun is a Shakespearean context. But that does not mean that we may symbolize 'x believes that he himself is F' in such a way as '$Bx(Fx)$'; this is not a way of 'identifying variables' that we have under logical control. We need to get such sentences under logical control in order to get rules for such plainly valid inferences as this:

> Jones believes he can square the circle
> Ergo: There is somebody who Jones believes can square the circle.

Further, we need to disambiguate such examples as this (which I owe to Hector-Neri Castañeda):

> The richest man in the United States believes that the richest man in the United States hopes he will win a Nobel prize.

according as the antecedent of 'he' is the first or the second occurrence of the definite description, that is, according as the belief of the richest man is expressible thus:

> The richest man in the United States has this hope concerning *me*:
> Would that he won a Nobel prize!

or rather thus:

> The richest man in the United States hopes that *he himself* will win a Nobel prize.

(Obviously, the richest man need neither know nor believe that he is the richest man.) The clumsiness of this vernacular explanation cries out for some neat well-regulated symbolism; but I cannot supply that. The dubious identification of variables that I have mentioned is a blemish in some current work; see my answer to C. J. F. Williams. Some arguments of Arthur Prior's use this dubious device; though the fact that Shakespearean contexts are involved may mean that the broken links could be mended if we had a workable symbolic *Ersatz* for indirect reflexive pronouns. I have to confess that I now see this business not much more clearly than I did when corresponding with Quine twenty years ago.

I turn again to set theory. First of all, a word about the relation between set theory and modal logic. Quine dislikes modal logic; I must agree that some developments of modal logic and of related philosophy are highly objectionable—though the same holds, I think, as regards set theory. But some parts of modal logic are well under control; and in some applications we could use either modal logic or set theory. Consider the ordinary report of a chess game: it will mention not only actual moves but also possible variant lines of play. This could be set forth with use of modal logic. (We shall notice that there may be different individuals in the different possible histories of this 'chess world'; different captures, and different promotions of pawns, may be made.) I am sure Quine would wish to describe all this set-theoretically, and I am sure he could. But the ground for this is a similarity between modal logic and set theory in formal structure. If the similarity proves to be (limited) inter-translatability, then the question is which side explains which. Quine would regard translation into set-theoretical terms as making an honest theory of modal logic. But might one not regard expressibility in modal terms as justifying some part of set theory? A confidence that in set theory 'we know where we are' is often based on accepting the view of classes as built up out of their members; and this view, as Quine and I agree, is clearly wrong. Modal terms, on the other hand, are part of everybody's mother tongue: though of course this does not mean we really know our way about in using them.

We can get at a class in two ways: by giving a concept (or, in linguistic vein, a predicable) whose extension the class is, or by listing its members. The latter way leads us to consider the logic of lists; I owe to McTaggart my first interest in this subject. In ordinary language we often get contexts in which there may stand a list of proper names of indefinite length: '—all have a family likeness', '—all fit together to make a cube', '—all entered the Bank together and some of them were gangsters'. In *Reference and Generality* I introduced plural variables (as I may now call them), replaceable by lists of indefinite length. I have had hopes that a theory of such plural variables will throw light on set theory. I am extremely pleased to learn that George Boolos is working on such a theory.

As regards zigzag predicables, I am glad of the continuing agreement between Quine and me that zigzagness, not excessive size, is the source of antinomies. I may mention here a negative result of my own as to the possibility of a 'natural' set theory: there will be cases where *any* class of *F*s has *indenumerably* many *F*s falling outside it. The argument is very simple in its main lines. Let us use 'collection' for 'class not a member of itself'. Any class of collections must be a collection. Further, for any class *C* of collections, the class *C minus D*, of all members of *C* except *D*, will not be a member of *C*, unless it fulfils the odd condition that *C minus D* is the same class as *D*. For short, I call classes of natural numbers 'n.n.c.s'. It is easy to give set-theoretical accounts of natural numbers so that each natural number both is an n.n.c. and is a collection. Then any n.n.c., whether itself a natural number or not, will be a collection. Now consider any class *C* of collections. Either (a) indenumerably many n.n.c.s will not be members of *C*, or (b) indenumerably many n.n.c.s will be members of *C*; for otherwise there would not be indenumerably many n.n.c.s in all, as it is easy to prove there are. In case (a) there will be indenumerably many collections, namely certain n.n.c.s, that are not members of *C*. In case (b), for each n.n.c. *D* that is a member of *C*, there will be a collection *C minus D* that is *not* a member of *C*; the condition that *C minus D* = *D* cannot be met, for *C* is supposed to have indenumerably many members, whereas *D* has only denumerably many, being an n.n.c. (*C minus D* and *C minus D'* will always be distinct if *D* and *D'* are distinct members of *C*.)

This result came out in the *Journal of Philosophy* under the title 'Two Paradoxes of Russell's' (Geach, 1970; in fact it was suggested by a passage in *The Principles of Mathematics*). Thus it appears that not only is there not just one 'exceptional case' regarding the predicable 'collection', as Frege conjectured: any class of collections will fall short by indenumerably many exceptions from being the class of *all* collections. There are indeed certain theses of set theory used for the proof, but they are extremely weak. Only natural numbers are assumed to be members of classes; and as regards n.n.c.s it is assumed only that any predicable defined over natural numbers determines a n.n.c., not that a n.n.c. which is not a natural number is a *member* of a class. (Perhaps a predicable defined over natural numbers might be disqualified by reason of zigzagness from determining a n.n.c.; but this again would stop us from having a 'natural' set theory.) Long after Frege, Gödel expressed a hope that the paradoxes of set theory might turn out in the end to be like paradoxes about division by zero, resoluble by some simple restriction; I can see little ground for such hope.

I shall leave the exchange between me and Quine about his book *The Roots of Reference* to speak for itself, except for a brief further word about induction and natural selection. If human inductive habits had been genetically or culturally inherited, and had run very much against the grain of Nature, then we should not be here. But the degree of success in induction needed for human survival is very low. We do draw inductive conclusions about the very long ago and the very far away and the very minute with astounding confidence and success; our very success about the small-scale structure of bodies is positively a threat to survival; certainly men have managed to survive for millennia without such inductions. It is merely untrue that men cannot survive, or at least cannot build a high civilisation, without our sort of inductive canons; the Aztecs expected the unexpected every 52 years, when the end of a cycle might bring total novelties. The start of one new cycle brought the Spaniards with their horses and iron weapons and guns; the effect on Aztec morale was devastating. Of course once a new cycle had safely got under way the Aztecs' expectations as to the course of Nature would be much the same as the Spaniards'—for the next 52 years. The difference between the Aztecs and the Spaniards was, after all, theological; consciously or not, the Spaniards were heirs to a Judaeo-Christian tradition that God's covenant guarantees the order of Nature.

My intellectual debt to Quine is immeasurable. He has formed my mind negatively and positively: negatively, by quietly discouraging unsound lines of thought; positively, by setting me standards. It is a constant temptation and a common fault of thinkers to defend some thesis because they maintained it previously—perhaps by some twisty interpretation of the thesis; Quine is an impressive example of the contrary virtue. In the philosophy of logic we often agree; disagreements, even when not a matter of style, rest upon an underlying agreement which alone can make fruitful discussion possible.

REPLY TO DESMOND HENRY

The part-whole relation plays a vital role in the metaphysics of McTaggart; it is interesting to compare the way he worked out a general theory of this relation with the Polish mereology to which Henry resorts and the similar theory in Nelson Goodman's *The Structure of Appearance* (1951). When I followed McTaggart, I believed that there is *one* part-whole relation in many instances; 'part' has exactly the same sense when we speak of a scene as part of a play, a thought as part of a mind, a liver as part of a body, a man or a committee as part of a club... I have tried to present McTaggart's system faithfully and with imaginative sympathy in my book on him; but in my own person I have to say that this view now appears to me preposterous. I am afraid this means that I cannot take a lively interest in Abelard's alleged anticipations of mereological theory; in any event I lack such familiarity with Abelard's works as would be needed to appraise Henry's interpretations.

Let us say that two individuals *overlap* when the same individual is a (proper or improper) part of each of them. There is a general mereological thesis that for any individuals *a*, *b*, *c*,... there is a whole which overlaps just those individuals that either overlap *a* or overlap *b* or overlap *c* (or...); the individuals *a*, *b*, *c*,..., may then be said to *make up* this whole. Obviously if *a*, *b*, *c*,... are areas on a surface this comes out true, though the whole formed may be a scattered archipelago of islands. McTaggart and Nelson Goodman (I believe also the Polish mereologists) would hold that this holds quite generally; e.g. if *a* is an occurrent thought, *b* an attack of measles, and *c* a lump of cheese, there would be a whole made up precisely of *a*, *b* and *c*, though it would have no interest or importance except as a philosophical example. (McTaggart's own examples are sometimes equally whimsical.) One might try to make sense of this by talking about 'four-dimensional worms': *a*, an occurrent thought, would be that time-slice of a man which answers to the time he is thinking thus and so, and *b*, an attack of measles, would be the measly time-slice of a man, and then *a* and *b* would be of the same category as *c*, the four-dimensional worm vulgarly called a lump of cheese. But as Alice said, what dreadful nonsense we are talking!

Again, it might count as a theorem of general mereology that if *a* is a proper part of *b* there is another proper part of *b* that together with *a* makes up *b*. But in cases of intensive magnitude there seem to be exceptions to this. If a lesser light persists as part of a greater light, there is no identifiable individual which *is* the increment of light. McTaggart himself noticed this peculiarity of such cases (*The Nature of Existence*, 1927, vol. ii, section 568); and the point is of great importance for any worthwhile understanding of his *C* series. The various *C* series which underlie the appearance of temporal order were each to be an 'inclusion series' of perceptions ordered as more and less inclusive, precisely because one perception would differ from another in a theoretically postulated intensive magnitude, as a greater light may include a lesser. But McTaggart did not notice that *this* application of the terms 'part' and 'inclusion' was only *analogous* to other applications he made in his system.

In general, the prospects for a comprehensive theory of part and whole, covering all our varied applications of the words, seem to me very dim indeed. In my own mental development, Wittgenstein's treatment of complexity and simplicity in the *Philosophical Investigations* only reinforced a scepticism about such a project that came with my ceasing to be a McTaggartian.

REPLY TO KENNY

In the works Kenny has discussed I compared Aquinas's doctrine of the *modus significandi* of abstract nouns with dependent genitives to Frege's doctrine of the incompleteness of functional signs. I remain convinced that here I laid hold of an important clue; but with the passage of years I have come to think I did not get so far into the labyrinth as I once thought I had. My chief perplexities about the topic are on the logical or semantical level; as regards what distinctions must be recognised in the domain of what actually exists, my views have changed much less.

I find various difficulties in the general doctrine of functional signs to which I formerly made confident appeal. The definite descriptions of ordinary language are, I think, not syntactical units at all. Grammar-based intuitions may represent such descriptions as unitary 'referring expressions'; but I have shown in several papers that very often in such cases we demonstrably have no logical unit before us; a (broadly speaking) Russellian treatment of such phrases is here mandatory and always feasible, and this would always break up their ostensible syntactical coherence.

Mathematical expressions with functors in them, as in the equation '$2 + 5 \ln 3 = \ln(e^2.3^5)$', are a different case; it would be technically awkward to treat them as abbreviated definite descriptions and then dissolve the definite descriptions *à la* Russell. In ordinary mathematical practice simple and complex numerical expressions are freely substituted for one another; if we treated the complex expressions as short for Russellian definite descriptions, such substitution would need a complex mathematical justification. Frege's powerful theory of functors (*Functionsnamen*) and the functions they mean supplies a much simpler rationale of such substitutions. My scruples about it arise from attachment to the view that a name must be syntactically simple: a complex expression, which signifies only via its constituents, cannot stand to an object in the same relation as a name; or again, since no complexity a name may happen to have is needed for the naming relation, a name *qua* sign must be simple. But does a numerical sign, even if syntactically simple, relate to a number as name to thing named? To pursue these doubts would take us too far from Aquinas.

As regards such phrases as 'the wisdom of Socrates', we could easily dispense with the category of abstract nouns by using such a form as 'that by which Socrates is wise' (an echo of Aquinas's '*quo*') or 'that by which Socrates is made to be wise'. But these forms presuppose such constructions as '*x* is wise by *y*' or '*y* makes *x* to be wise'. The relevant senses of 'by' and 'makes', for formal causality, are not, I think, hopelessly obscure; but the logical relation between these complex forms and the simpler form '*x* is wise' is not clear to me. Here also there is a problem I can only indicate.

On the level of what actually exists, I have little to add to my essay 'What Actually Exists', in *God and the Soul* (1969). I cannot happily follow Aquinas in every case where he is prepared to recognise, besides the object *a* which *F*s, or is *F*, or can *F*, an actual individual *by* which *a F*s, or is *F*, or can *F*. My doubt arises from the unclarity, in some cases, of criteria for the identity or distinctness of such entities. But I am sure it is not always possible to explain away all ostensible talk about such entities as our being misled by reifications of sentences '*a* is *F*' or '*a Fs*' or '*a* can *F*'. For example, I see no way of eliminating *events* from our ontology; I have distinguished between mere Cambridge changes and actual changes, and in the latter case I think we must recognise a transitory actuality that *is* the change.

Despite Kenny's objections, I also remain convinced of the real distinction between that which exists and its *esse*. The case in which the distinction is easiest to establish is that of intensive magnitude. Kenny argues that in assigning to the side of *esse* the degree to which a quality is found in a subject, I am in fact not spelling out Aquinas's doctrine but diverging from it; but I cannot see that he has made this out. Just because the *esse* of an individual accident is for Aquinas its *inesse* or inherence, an increase of inherence *must* be an increase of *esse*. Of course when charity increases in intensity 'charity as such remains unchanged', in the sense that it does not change to something other than charity or another variety of charity. One class of cases in which there must be recognised a distinction, and a real distinction, between an individualized form and the *esse* relating to that form is enough to show that the distinction is not nonsensical or vacuous, and to open the question whether there is a real distinction in other cases; it is irrelevant to object that for other cases this argument about intensity cannot be used to establish a real distinction, for it may be establishable some other way.

Kenny has missed the form of my argument about *esse intentionale*. Obviously there is no question of the same *individualized* form's having *esse naturale* and *esse intentionale* simultaneously; but I was speaking of there being, for one and the same *form* (not *individualized* form), two *modes* of occurrence; for each occurrence there is a distinct *individualized* form. In a square body there is an individualized form of squareness that has *esse naturale*; in someone who apprehends squareness, whether in sense-perception or in thought, there is an individualized form where the same content occurs with *esse intentionale*. It is because the difference between the two individuations of squareness lies on the side of *esse*, that philosophers lacking a conception of *esse* have been embarrased to give an account of this difference; thus Hume writes (*Treatise*, ed. Selby-Bigge, 1888, p. 241):

> ... 'tis impossible our idea of a perception, and that of an object or external existence

can ever represent what are specifically different from each other. Whatever difference we may suppose between them, 'tis still incomprehensible to us...

and similarly Moore (*Philosophical Studies*, 1922, p. 25):

...the moment we try to fix our attention upon consciousness and to say *what*, distinctly, it is, it seems to vanish... When we try to introspect the sensation of blue, all we can see is the blue; the other element is as it were diaphanous. Yet it *can* be distinguished if we look attentively enough, and if we know that there is something to look for.

Both Hume and Moore look for difference where it is not to be found: on the side of content or essence, not on the side of *how* that content occurs.

In this way we can establish a general difference between an individualized form and its *esse*. If the distinction were not real, there could be no question of the same content's occurring not just multiply, but with different *modes* of occurrence; an individual square shape would differ no otherwise from someone's apprehension of squareness than from another individual square shape. Since there is no form that cannot be apprehended, there is no individualized form for which we may not relevantly distinguish between it and its *esse* in one or another mode. Aquinas held indeed that when the nature e.g. of a horse is apprehended by God there is no real distinction between an individual apprehension and God who apprehends nor between either and the Divine *esse*. It would take me too far to show how I think this doctrine could be defended; I have sketched a defence in the works Kenny cites. Even in this case, the individual occurrences of equine nature in a live horse and in God's mind *will* have distinct *esse*. For Aquinas, a horse in God's mind is a much finer thing than a horse in the paddock—being there really identical with God—but it is not so truly a horse.

I am sure my remaining argument about essence and *esse*, the one Kenny discusses first, would at best require much supplementation to make it convincing. I stand by the thesis that a kidney and a liver, say, in a live animal differ in their animate structure, but share one *esse* which is that of the animal (they are indeed not animals, but they are animate). But then, for Aquinas, the same *anima*, the same individualized life-form, is the life of kidney and liver, that by which each is animate; and it would take further argument to show that the shared *esse* is really distinct from the individualized life by which each gland lives. Kenny is right to find my argument on this topic unsatisfactory, and it may well be that what I said needs some correction as well as supplementation.

I had not any such thing as kidney transplants in mind when I wrote in *Three Philosophers* about kidneys; but I am tempted to say a word. I accept Aquinas's view that in an unmutilated animal the kidney has no individualized life of its own, but shares the life of the whole; and Kenny is clearly right in saying that a surgically removed kidney, even if still alive, no longer lives by *that* life, and that Aquinas would say it is only equivocally the same kidney. Subject to correction, I incline to say that a successfully transplanted kidney (except from an identical twin) does not live by the life of the body it lives in; the need to take continual precautions against rejection seems rather to show that there is an alien life against which the body reacts.

REPLY TO WOLNIEWICZ

Wolniewicz has made significant contributions to the study of Wittgenstein and of Frege as well by his excellent Polish translations from both authors and by the work on the *Tractatus, Rzeczy i fakty (Things and Facts*: 1968), that he wrote for his *habilitacja*. Before this study of the *Tractatus* had really begun, he sent me a small book he had written on Lenin's theory of perception; he asked me whether I thought he had the right sort of mind for Wittgensteinian studies. Wittgenstein himself had said that people were unlikely to understand the *Tractatus* unless they had already come on their own account to some of the thoughts it contained; this early work of Wolniewicz seemed to me to satisfy the precondition very clearly, and I strongly encouraged him to work steadily on the *Tractatus*. He did so with ever-growing enthusiasm, and the event has confirmed what I expected of him; his exposition of the theory of sentences and facts is further developed in his monograph *Ontologia sytuacji* (*The Ontology of Situations*: 1985).

I rather regret that when discussing the *Tractatus* in his essay, Wolniewicz follows Frege in using one and the same word for the relation of name to bearer, and the relation of a linguistic proposition to *something*; going by what Wolniewicz has said here, one might gather that the only difference of opinion on this matter between him and Frege is as to the terminus of the semantic relation for propositions. One fundamental lesson of the *Tractatus*, it seems to me, is that propositions do not stand to *anything* nonlinguistic as a name stands to its bearer: there is not even a close analogy. The semantics of propositions has a duality about it to which nothing in the semantics of names corresponds: to any proposition there is a contradictory proposition, names have no contradictories. Names are like dots on a map representing towns and villages; propositions are like roads marked on the map, and for any road there are two opposite ways of going along it. Obviously this is a metaphor, but a straightforward explanation of the difference is not to be asked for. Any intelligent human being knows in practice the difference between a name and a sentence; to bring this difference into clearer awareness, one can only look for some form of words that will evoke *anamnesis*. Wolniewicz might well agree to all this; even so, I regret that this important point does not at all emerge in his paper.

What Wolniewicz says about the *Investigations*, on the other hand, strikes me as an extraordinary failure of understanding.

The *Investigations* begins with a quotation from Augustine to illustrate a great mind held captive by the idea that a word's having meaning consists in its *naming* something. Some have protested that this is not fair to Augustine: a reading of his *De Magistro* convinces me that they are mistaken, for the view in question is developed there in explicit detail. The spell is strong even in our century; Russell's theory of meaning in *The Principles of Mathematics* shows this, and so does much of Fitch's logical work. In order to break the spell, Wittgenstein takes a very simple communication as an example: an order to a greengrocer, 'Five red apples'. (This story may be made a bit more plausible if we imagine the greengrocer to have an assistant, who, being a recent immigrant, has little command of English.) Even this communication is not a signal in Wolniewicz's sense: it has a syntax, and the three words all belong to different categories.

The full force of Wittgenstein's discussion will be missed if we do not think of Frege. Wittgenstein admired Frege to the end of his days, and he remarked (in a note now printed in *Zettel*: 712) that the influence of Frege is often important in remarks that do not clearly show it. So it is here. Consider what Frege would say about

There are five red apples on the counter.

Frege would say that what is here affirmed is that a second-level concept *there being five such* applies to a first-level concept *red apples on the counter*; 'five' does not modify 'red apples' in the same way as 'red' modifies 'apples'. Sixty years after Frege's death people will still tell us that 'five' and 'red' both apply simply to physical things like apples, but 'five' does so collectively like an attribution of weight or pattern to a group of apples. Reflections on what the shop-assistant is supposed to do may save us from this barbarism. He can count out apples only when he has the concept *red apple* (or some other such) to give him a *unit* to count by. Even from a grammarian's point of view numerals do not belong to the same category as adjectives like 'red'; any even half-decent grammar-book will give numeral words a special treatment. (This is conspicuously necessary for Slavonic languages.)

Moreover, we should note that the shop-assistant looks in the *apple* drawer and then checks the contents for *redness*: he has not a drawer full of *red articles* from which he pulls out objects answering to the concept *apple*. This part of the language-game, of the performance with words and activities plaited together, is also essential, and points to yet another category-difference. We can count apples in their own right; as Frege put it, the concept *apple* affords a unit for counting; but the concept *red apple* does so only because the concept *apple* does; the concept *red* affords no unit for counting—is a cube painted red *one* red thing, or rather *six* because of the six red surfaces? Frege was uneasy about this difference between concepts, and gave no clear account of it. In *Reference and Generality* (1962) I borrowed from Aquinas the terms *adjectival* and *substantival*, suggested by the rules of Latin grammar that adjectives agree in (grammatical) number with nouns; I have here gone into the matter further in my reply to Dummett.

I have commented at length on a short passage of the *Investigations* to illustrate how much the book repays reflection; how mistaken it is to read it as even programmatically, reducing language to 'signals'.

I turn to the one-word utterances that I have sometimes called 'simple acts of naming', by which I meant, not acts of *conferring* a name, but acts of *calling* by name. I remarked on the relation between such acts and such written signs as labels; the vocal speech acts are of course primary. When a man sees a plaque saying 'Woodhouse Lane' on a wall in Leeds, or a label 'Pepper' on a jar, he may easily realise that it would be correct (if not always socially appropriate) for him to utter the simple act of naming 'Woodhouse Lane!' or 'Pepper!' I distinguish such uses of names from merely elliptical uses, e.g. as one-word answers to questions stand in for complete sentences. In particular I rejected the view that such a use of a name is short for a sentence like 'This is Woodhouse Lane' or 'That is pepper'; I hold that logically these are not subject-predicate sentences at all, and assuredly, as Wittgenstein points out in the *Investigations*, demonstratives are not names. On the contrary, these demonstrative sentences are merely inflated forms of simple acts of naming.

My insistence that shared names and proper names alike may figure in simple acts of naming was directed against two classes of thinkers: those who follow Frege and Russell in rejecting shared names, and those who think that proper names are somehow devoid of sense and not really parts of the language. The latter error is a hangover from J. S. Mill's *System of Logic*, in many ways a disaster for English philosophy. Mill uses as an example of a 'meaningless mark', such as he takes proper names to be, the marking of a house-door to single out the house from other houses. But he oddly forgets that what the mark is meant to pick out is precisely a *house*; it might be meant to pick out a *door*, or a *plank* in a door, that some carpenter is to repair. The use of the mark depends on our attending to one or other criterion of identity; and so does the use of a proper name. As an elucidation of his denial of 'connotation' to proper names, Mill's story is remarkably inept.

Obviously proper names belong to a language, in that they enter into syntactical relations in sentences; even in English they take the possessive inflection, and to a speaker of Polish it is perfectly obvious that 'Warszawa' and 'Czarniecki' are Polish words, declinable by the rules of Polish accidence. Proper names can figure either in sentences or in simple acts of naming; one may link up the exclamatory utterance 'Woodhouse Lane!' with a sentence like 'A murder was done in Woodhouse Lane', e.g. by going on to say 'Now I've walked down the street where the murder was done'.

I have written more about these matters in Chapter 2 of *Reference and Generality* (better, I hope, in the third edition: 1980). The debt to Wittgenstein's later work ought to be obvious; I was however amused to find a reviewer who thought I had only been influenced by the *Tractatus*, because only that work is actually cited in *Reference and Generality*. While writing this chapter I never had a feeling of the supposed chasm between 'the two Wittgensteins'.

REPLY TO G.H. VON WRIGHT

Von Wright has given us a worthy piece of work in the area of tense logic; all his work in this area deserves careful study. I have worked myself in this area, though I have not published; my contributions (bricks towards a building) are mentioned in some of the now copious literature. It was Arthur Prior who first systematically developed the discussion of pastness and futurity in terms of propositional operators upon propositions, using the methods of modal logic. We became friends when he was giving his John Locke lectures *Time and Modality* (published in Prior, 1957); our friendship and co-operation lasted as long as he lived, and after his death I had the melancholy but rewarding task of editing his *Nachlass* (Prior, 1976a, b) along with Anthony Kenny.

Von Wright's work emphasizes notions that I am sure must enter importantly into a good overview of the subject: notions such as alternativeness and prevention. What is prevented is what *was* going to be; after the preventive action, the same course of events *is not* going to be. How should 'going to be' be explicated? It is not very difficult to draw a distinction, with modal-logical apparatus, between that future which is now-unpreventable, and thus in some sense *necessary*, and a merely *possible* future, allowing of alternative developments. But we need, I believe, a third sense of futurity: what is now 'going to be', in the sense that things are *now* all set for it to happen, but is yet *not* unpreventable. As yet I see no way to formalize this notion.

Another deficiency of tense logic as hitherto conducted is the failure to cover the distinction between what grammarians call perfective and imperfective verbs (or verbal 'aspects'). We may get 'Mandy will dance' and 'Mandy was dancing' as results of applying future-forming and past-forming operators to 'Mandy is dancing'; but if Big Bertha is a bomb, we have only 'Big Bertha (has) exploded' and 'Big Bertha will explode'—there is no proper present (if Big Bertha 'is exploding' she has already exploded.) The second can indeed be regarded as a 'futurition' of the first—'It will be that Big Bertha has exploded'; but the absence of a genuine present belongs to the *logic* of perfective verbs, and is not representable in what may be called standard tense logic. Antony Galton, till recently of the University of Leeds, has devoted a valuable book to these problems (Galton, 1984).

On a related point, I am sure that the difference found in some languages between an 'aorist', or 'preterite', and a 'perfect' form of a verb is *merely* what Arthur Prior called an idiotism of idiom. Languages that make no such distinction, like Latin and Slavonic languages, do not thereby lack expressive power; and languages that do make the distinction never make it the same way, as an Englishman learning French or German painfully learns. The dialogue 'Have you ever been in Vienna?' 'Yes, I lived there for some months' becomes in German 'Waren Sie einmal in Wien?' 'Ja, ich habe dort einige Monate gewohnt'; innumerable such illustrations could be given. The idiom is a little different even in U.S. and British dialects of English. Trying to extract something *logical* from this is like milking a bull.

REPLY TO ALTHAM

I have to thank Altham not only for his paper, but for much collaboration in recent years in the field of pleonotetic logic. Since Sir William Hamilton first pointed out that 'most' is a quantifier with the same right to logical recognition as 'all' and 'some', the subject has been curiously neglected. By virtue of known results about decidable theories, monadic pleonotetic logic is decidable in principle if the Universe is assumed finite; Quine in correspondence showed me an easy proof that the same holds for a denumerably infinite Universe. For larger Universes the very sense of numerical comparisons depends on matters of set theory. For example, suppose we use 'every-other' for the quantifier dual to 'most' or 'almost every' (the hyphen is to distinguish this quantifier from 'every other' in e.g. 'The Bey spared Basia but killed every other captive'.). Then whether ⌜Either every-other (P *aut* Q) is PQ' or every-other (P *aut* Q) is QP'⌝ is a theorem for such large Universes depends on the law of trichotomy, and thus on the axiom of choice. But no appeal to such set-theoretical questions is needed for a denumerable Universe; in this case logical rules that give a decision method for monadic pleonotetic logic are in principle formulable, though not yet discovered.

Some of the work Altham does with his decidedly unintuitive logical transcriptions of algebraic inequalities between integers can be done with much more intuitive means. A lot can be done with the principle:

I *From* ⌜Almost every P is RS'⌝, ⌜Almost every Q is RS⌝, *infer* ⌜Almost every (P *aut* Q) is R⌝

This is one of a family of principles with other quantifiers: e.g. the inference remains valid if we replace 'almost every' each time by 'every', whether or not we read 'every' as implying 'some'. From this we get the derived rule:

II *From* ⌜Almost every (*P aut Q*) is *Q*⌝, ⌜Almost every (*Q aut R*) is *R*⌝, *infer* ⌜Almost every (*P aut R*) is *R*⌝

This last rule might seem to turn on the transitiveness of *greater than*, in a way that made it impossible to derive it from I. But the derivation is in fact easy. ⌜Almost every (*P aut Q*) is *Q*⌝ is equivalent to ⌜Almost every (*P aut Q*) is *P'Q*⌝; similarly, ⌜Almost every (*Q aut R*) is *R*⌝, to ⌜Almost every (*Q aut R*) is *Q'R*⌝. These equivalences may well count as 'purely logical' in that they do not depend on *which* quantifier is used. Now the predicates ⌜*P'Q*⌝ and ⌜*Q'R*⌝ are replaceable by ⌜(*P'Q* v *Q'R*)*Q*⌝ and ⌜(*P'Q* v *Q'R*)*Q'*⌝ respectively: if we make this replacement, we get the premise-pair:

⌜Almost every (*P aut Q*) is (*P'Q* v *Q'R*)*Q*⌝
⌜Almost every (*Q aut R*) is (*P'Q* v *Q'R*)*Q'*⌝

which by I yield the conclusion:

⌜Almost every ((*P aut Q*) *aut* (*Q aut R*)) is (*P'Q* v *Q'R*)⌝

Here ⌜((*P aut Q*) *aut* (*Q aut R*))⌝ boils down to ⌜(*P aut R*)⌝: so we get

⌜Almost every (*P aut R*) is (*P aut R*) (*P'Q* v *Q'R*)⌝

A little Boolean algebra shows that ⌜(*P aut R*) (*P'Q* v *Q'R*)⌝ boils down to ⌜*P'R*⌝: hence:

⌜Almost every (*P aut R*) is *P'R*⌝

which is equivalent to: ⌜Almost every (*P aut R*) is *R*⌝. *Q.E.D.*

None of this working depends on a sidelong glance at numerical inequalities for its intuitive justification; the principles employed are both quite intuitive in their own right, and 'purely logical' in the sense that they are parallel to principles holding for other, more familiar, quantifiers. *E.g.* the equivalence of ⌜*(*P aut Q*) is *Q*⌝ to ⌜*(*P aut Q*) is *P'Q*⌝ and of ⌜**S* is *T*⌝ to ⌜**S* is *ST*⌝ and the intersubstitution of complex terms that come out equivalent by Boolean algebra, hold good irrespective of the quantifier chosen to replace the asterisk. It should be a goal to break down all pleonotetic inferences into small intuitive 'purely logical' steps; my hunch is that goal can be reached, but I have no corpus of rules that can be proved to cover all the logical truths of pleonotetic logic. Such a breakdown *is* possible e.g. for the proofs of inconsistency in the 'Rev. J. Prodnose' problem, but too long to spell out here.

I have little to say in furtherance of Altham's ingenious work on quantifiers like 'very many'; bringing out that inexact propositions may have an exact logic he has worthily advanced our knowledge. To him and others I offer the following Problem of the Paranoid Sarmatians.

> Out of very many Sarmatians, each one believes that most other Sarmatians are plotting against him.
>
> If very many Sarmatians are plotting against just one Sarmatian, then they are crazy.
>
> Any Sarmatian who believes that most other Sarmatians are plotting against him is himself crazy if in fact most other Sarmatians are not plotting against him.
>
> *Ergo*: Very many Sarmatians are crazy.

The difficulty here does *not* lie with the predicable '—believes that most other Sarmatians are plotting against him': this is, to my mind, both unequivocal and Shakespearean (see *Logic Matters*, 1972, 4.1 and 4.3), and may here be represented by a simple predicate-letter without concern for internal structure. The trouble rather arises for '—are plotting against—', where the first argument-place may congruously be filled with a list of proper names or with a quantified expression like 'very many Sarmatians': this is *not* an ordinary two-place predicable. The vagueness of 'very many' is *not* a headache: take any Sarmatian, and it is clear that if *there are* very many Sarmatians, then whatever is true of *most* Sarmatians *other than that one* is true of *very many* Sarmatians.

I have only a couple of remarks to add on the syntax of two-place quantifiers. First, we get a two-place quantifier in [Just one A *is* P and that A is Q]; the dangling pronoun 'that' which looks back to [just one A] rules out the reading of such propositions as merely conjunctive, and has rather the role of a bound variable. Arthur Prior in fact proposed for such cases notation like '(ιx) $(fx;gx)$, which brings out how we have here a two-place quantifier of the same category as Altham's examples; it was meant to replace Russell's 'g $(\iota x)(fx)$'.

Secondly, it is worth notice that the paraphrase of ⌜Just one *A* is *P* and that *A* is *Q*⌝ by ⌜Some *A* both is *P* and *Q*, and no more than one *A* is *P*⌝ still goes over perfectly well even if *Q* is not a predicable proper but a predicative fragment, with a dangling pronoun whose antecedent lies buried in *P*. For example, in a chosen Universe these two would be equivalent:

> Just one boy is reading a book and that boy is mutilating it.
> Some boy both is reading a book and is mutilating it, and no more than one boy is reading a book.

I do not know whether it would not be possible to construe Altham's puzzling sentence 'More people read a book than buy it' by some similar manoeuvre: by finding some formula that is equivalent to ⌜More people are *P* than are *Q*⌝ when ⌜are *P*⌝ and ⌜are *Q*⌝ are predicables properly so called, and then taking the equivalence *still* to hold when the second predicative expression is a fragment, with a dangling pronoun in it whose antecedent is buried in the first predicative expression.

This method of construal anyhow does work in other dangling-pronoun cases: e.g. the equivalence between ⌜The first *A* that was *P* was *Q*⌝ and ⌜Some *A* both was *P* and was *Q*; and no *A* was *P* before *that A* was *P*⌝ will work for the following case, if we go ahead mechanically, reading ⌜was *P*⌝ and ⌜was *Q*⌝ as past-tense predicative forms, and swallowing scruples about dangling pronouns:

> The first man who had a younger brother murdered him.
> Some man both had a younger brother and murdered him, and no man had a younger brother before *that* man had a younger brother.

I could not say how far this method of paraphrase is applicable or why it works. Gareth Evans, who was much concerned with such problems, might have told us much; like Arthur Prior and other logicians I have known, he is gone from us expectedly; but his difficult papers may yet give somebody the clue that eludes me.

REPLY TO ELIZABETH ANSCOMBE

The pattern of argument laid out at the beginning of Anscombe's paper depends, as in effect she points out, upon the supposition that we have here a single sense of 'if' throughout, and one for which *modus ponens* and *modus tollens* are both valid forms. I have heard her read drafts of this paper at meetings; rather to my surprise, sometimes people suggested that it is the material conditional that is here employed. This does not appear to be requisite for the validity of the argumentative steps: *modus ponens* and *modus tollens* are valid for other interpretations of 'if'.

Anscombe gives us a schema with propositional letters in it; how well do the concrete examples fit the pattern? At the end of the paper, Anscombe mentions the objection that the right way to represent the second premise as it comes in the examples would not be 'If *r*, then not (if *p*, then *q*)' but 'If *r*, then if *p*, then not-*q*'; she retorts, to my mind rightly, that for the cases considered the latter schema would represent too strong an assumption. But there is another schema whose fit to the concrete examples of the second premise seems to me to need consideration, namely 'Not: if both *r* and *p* then *q*'. In sentences of the vernacular involving 'if', 'and', and 'not', the scope of the operators may be unclear. It takes a little logical sophistication to distinguish between 'If both *r* and *s*, then *t*' and 'If *r*, then if *s* then *t*'; what is actually uttered may be of the form 'If *r*, and if *s*, then *t*', which cannot have clearly assigned to it either of these logical parsings. When there is also a negative in the sentence, it may be unclear just what is being negated.

Anscombe of course wanted to have a reading of the premises in her concrete examples such that they could reasonably all three be asserted together. Now is there not on the face of it an inconsistency between the first two premises if they have this form?

> If *p*, then *q*
> Not (If *r*, then if *p*, then *q*)

Not necessarily so, to my mind: they are compatible on what may be called a *ceteris paribus* reading of 'if'. From:

> If *p*, then other things being equal, *q*

we cannot infer:

> If both *r* and *p*, then, other things being equal, *q*.

For, one might say, the 'other things' that are supposed 'equal' on the supposition that *p* will not be the same as the 'other things' that have to be supposed 'equal' on the supposition that (both *r* and *p*). Of course we are just helping ourselves out with a comforting phrase when we say 'other things being equal'; in any concrete case there is no question of spelling out what these 'other things' would be. I am inclined to think that when Anscombe speaks of 'essential defeasibility', she is half-aware of this logical property of 'if' in the 'other things being equal' sense.

I think it is the *ceteris paribus* 'if' that occurs in a class of conditionals often called 'counterfactual'. The term is a bad one, for the contrary-to-fact character of the antecedent is not essential. 'If Tom were a man of sense he'd ask Mary to marry him' is not refuted, rather one's judgment is confirmed, if it turns out that Tom had plenty of sense and *has* asked Mary to marry him. 'Subjunctive conditionals', coming as it does from memories of how conditions are expressed in Latin, is a far better term; it may not help those who have no Latin, but at least it does not mislead.

Professor Duncan-Jones pointed out to me that in such a pair as this:

> If Parliament is dissolved in June, the Government will lose the election.
> If Parliament had been dissolved in June, the Government would have lost the election,

(assuming the same June to be meant) may express exactly the same judgment of a certain political situation. The former sentence obviously uses a *ceteris paribus* 'if'; the counterfactuality of the second is not logically essential.

The logic of *ceteris paribus* conditionals is very obscure. Several authorities (if one may use such a word when appeal to authority is so manifestly an infirm locus of argument) agree that this form of hypothetical syllogism is valid:

> If *p*, then *q*
> If both *p* and *q*, then *r*
> *Ergo*: If *p*, then *r*

But it is not difficult to construct a plausible counterexample. Those who favour the argument-form have counterfactuals in mind, but we shall see that this is not essential. I first set forth my counterexample in counterfactual style:

A. Tom wouldn't be able to raise money from his father. He's seeing his father tomorrow, and they're bound to have a flaming row; so if Tom asked his father for money, he wouldn't get it.

B. On the contrary, if Tom asked his father for money he would get it.

A. Why do you say that?

B. I know you study the logic of counterfactuals; I can prove that conclusion by a form of argument you accept from premises about the case that you accept.

A. How?

B. If Tom asked his father for money, and hadn't just had a flaming row with him, Tom would get the money. You grant that?

A. Yes.

B. But if Tom asked his father for money, Tom wouldn't just have had a flaming row with him. You know how proud Tom is.

A. Yes.

B. But from these premises, by a form of hypothetical syllogism that you accept, there follows the conclusion:

> If Tom asked his father for money, Tom would get the money.

The counterfactual guise of propositions in this argument is clearly inessential; if we transpose to another key *à la* Duncan-Jones, and change 'asked' to 'asks', 'would get' to 'will get', etc. this makes no difference to validity. But whichever way it is put, B's argument is unconvincing; if possible-worlds semantics of one or other variety would make it out valid, so much the worse for such semantics.

It is worth remarking that the quirks of English grammar seriously impede some Anglophone philosophy where they discuss conditionals. It is a mere accidental loss of inflexions (for all verbs except 'to be') that gives us an equiform antecedent in the pair:

> If I had decided wrongly, I paid a heavy penalty.
> If I had decided wrongly, the result would have been disastrous.

In Latin or German the antecedents would not be equiform; but I have seen serious philosophical argument based on the surface appearances of the antecedents in English (indeed, one elaborate monograph).

There is just logic, not English or German or Latin logic; such arguments are shown to be worthless by the fact that they will not translate, but they ought not to seem worth looking at anyhow.

Reverting to Anscombe's paper: I am inclined to think that the schema which actually fits her premises in the concrete examples is not the one she gives but rather:

If *p*, then *q*
Not: if *r* and *p*, then *q*
If not-*p*, then *r*

But from this there is no way to get the conclusion '*p* and *q*', if we read the 'if' as a *ceteris paribus* 'if' throughout. However, both the interpretation of vernacular conditionals and the logic of the *ceteris paribus* 'if' are too obscure to me to leave me happy that I have resolved the problem Anscombe sets us.

REPLY TO HINTIKKA

Hintikka commends (with some reserve) my style of approach to problems about ordinary-language quantifiers, but finds it cramped by hangovers of Fregean orthodoxy. As Pope's friend Bishop Warburton said, orthodoxy is my doxy, heterodoxy is the other man's doxy: I have been more faithful to my doxy than Hintikka allows.

My term 'predicable', which I introduced long ago, expresses no novel theory: I use it for linguistic entities of a certain sort, namely those which can be used as templates or stencils to form sentences out of names. Quine and Benson Mates and others have used 'predicate' in this sense; my motive for choosing 'predicable' was that I needed a word, and used the word 'predicate', to label a certain role that a predicable plays in some but not all sentences it is used to generate. I retain 'predicate' where it should strictly be 'predicable' in such compounds as 'predicate logic', 'predicate letter', out of respect for tradition and euphony; and in some papers where no ambiguity would arise I have kept the older term. I am puzzled at Hintikka's puzzlement over this terminological matter.

Hintikka seems to want some systematic way of construing ordinary-language sentences that I (and others) would regard as raising problems of scope; but we may have to wait a long time for this. It is easy to show by examples that no simple rule of thumb will do. The rule that the grammatical subject has the larger scope is refuted by the medieval example 'An eye is needed for Socrates to see', and by 'A Paragon shoe will fit any foot' (in its natural acceptation). Again, the rule that 'any' is a sign for the longest possible scope is refuted by this example: the right paraphrase of:

> I have a poison that will kill any rat.

is not:

> Take any rat, and I have a poison that will kill him.

but:

> I have a poison of which this holds: Take any rat, and it will kill him.

That this is the right reading is to be shown by appeal to a native speaker's intuition; similar appeal shows that my medieval sentence has no acceptable paraphrase of the form 'It holds good of an eye that...' (of which eye?), and my 'Paragon shoe' sentence has none of the form 'It holds good of a Paragon shoe that...'. Hintikka speaks of this ability to accept right paraphrases and reject wrong ones as *miraculous*—I fear, in the spirit of Matthew Arnold's 'Miracles do not happen'. But in fact native speakers do seem to have this ability: let us gratefully accept this gift of Providence and not look a gift horse in the mouth.

Hintikka rightly says that there is no sense, in the strict Fregean framework, in saying that a proposition is about some man, all animals, any boy, or most women. If he means, however, that Frege's theory would exclude reading the predicate letter '*f*' the same way in '*f*(Jack)' and '*f*(any boy)', then he is mistaken. Frege's analysis does not here differ from standard predicate logic in any relevant way; and standard predicate logic would so transform '*f*(any boy)' as to bring in a subclause '*f(x)*' in which the argument-place is filled by a name-variable rather than a name—but in which '*f*( )' is replaceable by an actual predicable exactly as in '*f*(Jack)'. As for myself I have repeatedly attacked the idea (what Frege calls *rohe, quantifizierende Auffassung*) that as 'Jack' refers to Jack so 'all men', 'some men', 'most men', 'no men', respectively refer to all men, some men, most men, and no men (or, to classes consisting of all men, some men, ...). Despite the term 'referring phrases', I do not *in propria persona* regard these phrases as having any referring role; in *Reference and Generality* (1962, p. 47), I explained that I was using this term only to expound other people's theories that *would* assign a referring role to such phrases—and I argued at length in the following chapter that all these theories are untenable.

Although in the past the idea of two operators operating simultaneously on a single operand has sometimes attracted me, it now seems to me doubtfully intelligible. The expression '$2 + 3 \times 5$' would ordinarily be disambiguated by brackets to show whether the function $\xi \times 5$ takes as argument a value of the function $2 + \xi$ or the other way round; if I am told 'No, no brackets: I mean the number that is obtained if both functions simultaneously take 3 as argument', then I have so far no idea what is wanted. An ingenious mathematician might *invent* some procedure that I could quite naturally take as the fulfilment of this instruction; but that does not mean the instruction already *had* a sense.

The relation between Hintikka's (3), (4), and (5) as readings of 'Most electors voted for most candidates' may be re-expressed as follows:

(3)′ For some group *A*, *A* includes a majority of electors, and, for any member *x* of *A*, it holds that for some group *B*, *B* includes a majority of candidates, and, for any member *y* of *B*, *x* voted for *y*.

(4)′ For some group *B*, *B* includes a majority of candidates, and, for any member *y* of *B*, it holds that for some group *A*, *A* includes a majority of electors, and, for any member *x* of *A*, *x* voted for *y*.

(5)′ For some group *A*, for some group *B*, *A* includes a majority of electors, and *B* includes a majority of candidates, and for any member *x* of *A* and any member *y* of *B*, *x* voted for *y*.

So far from opening up the perspective of a new logic that dispenses with the idea of scope, the best way to show the exact relations of these three readings of 'Most electors voted for most candidates' seems to be one *exploiting* that idea, as above. (5)′ is of course symmetrical, equivalent by ordinary quantification theory to:

(5)″ For some group *B*, for some group *A*, *B* includes a majority of candidates and *A* includes a majority of electors, and, for any member *y* of *B* and any member *x* of *A*, *x* voted for *y*.

And this does justice to the feeling of the two 'most' phrases operating at the same time. Hintikka may object to quantifying over groups; but *verbally* he himself does this in trying to make (5) intelligible; and if he does not mean what he seems to say, I do not know what he means.

REPLY TO MCCAWLEY

Since my first meeting with McCawley in the United States I have profited a great deal from his work. I have few cavils about the typically excellent piece he has contributed.

In trying to find logical patterns in ordinary vernacular reasoning that can usefully be described by the apparatus of 'natural deduction' systems, McCawley is mainly concerned with examples from propositional logic (though a brief example essentially involving quantifiers appears on his first page). The complexities involved in examples where we must consider not just propositions as units, but intra-propositional operators, are to my mind formidable; ordinary languages were not designed by or for logicians, and are cluttered with idioms of no logical interest. Articles in English, for example, may *often* be omitted without making the message in the least obscure or ambiguous; nor is there any logical significance in the syntactical change that may be needed in a Polish sentence because we replace the word for 'four' by the word for 'five', or use 'kilka' rather than 'niektóre' to mean 'several'. Here logic can give little help to grammar, or conversely.

Discourse analysis of non-inferential texts, and sometimes also of inferential passages, raises difficulties at which McCawley only hints here. A logician would like to break up a text into separate sentences, each expressing (as Frege would say) a new *Gedanke*; but if you make a serious attempt to do this with a straightforward passage of connected narrative, you may come to doubt if it is feasible—particularly if you have not muddled yourself with hazy notions about 'referring expressions'. In general, the liaisons bonding sentences together can neither be broken to get a *staccato* series of propositions out of the *legato* text, nor dealt with by any existing logical apparatus. This is not a pretext for despair, but a challenge to workers coming into the field; we need a major theoretical advance.

I am not disputing the success of McCawley's description of ordinary-language inferences by a formal apparatus. It is the character of that apparatus itself that he leaves in an obscurity that I think could in part be removed. The obscurity inheres in the description of the apparatus by formal logicians themselves; Fitch, on whom McCawley relies, is here entirely typical. McCawley's work does much to make perspicuous the syntactical structure of a complex natural-deduction argument; the obscurity I discern is a semantical one.

The rules of natural deduction are commonly listed as introduction rules and elimination (or exploitation) rules. Attending to the distinction, logicians have commonly failed to bring out a more important distinction, between truth-preserving or *schematic* and validity-preserving or *thematic* rules. (The short terms I use to mark the distinction are respectably antique: Aristotle wrote about schemata, Stoic logicians about themata.) The distinction is just as important as the distinction between truth and validity. A schematic rule directly gives you a pattern or schema such that if your premises and conclusion conform to it, the conclusion can be false only if some premise is. A thematic rule applies to derivations: it tells us how to construct a derivation from given premises which will be *valid* given that certain other related *derivations* are *valid*. The derivations to which a thematic rule is applied to yield a further derivation, valid if they are valid, are precisely what Fitch opaquely calls 'subordinate proofs'; proofs, in any respectable sense, they are not, not in general that is, for their premises may be false. The validity of a dilemmatic argument with a disjunctive premise is established by the validity of arguments in which each several disjunct appears as a premise; it is no matter that from our background information we may know that one or other of the disjuncts must be false; for the validity of the main argument depends only on the *validity* of the subordinate arguments, not upon their having *true* premises. Of course every schoolboy knows that an argument with a *false* premise may be *valid*. So, as Polonius said,

> Your bait of falsehood takes his carp of truth,
> And thus do we of wisdom and of reach . . .
> By indirections find directions out.

If this seems to some like a conjuring trick, we may suspect a hangover of the causal theory of inference: the theory that the truth of the conclusion in a valid argument is caused by the truth of the premises, so that a true conclusion results from false premises only *per accidens*. Obviously any lingering trace of this theory will make the working of thematic rules barely intelligible; particularly, the way *reductio ad absurdum* works. How can you claim to have proved something (so some will ask) if you use as a premise something that you intend later to discard as false? Of course the falsehood is used as a premise, not of the main argument, but of a subordinate argument, when *validity* yields the *validity* of the main argument. Addicts of the causal theory of inference will seriously ask whether it is the major or the minor premise whose truth causes the truth of a syllogistic conclusion (thus Maritain, and his hero John of St. Thomas); this absurd spectacle may make people laugh away these cobwebs. Without clarity here, you cannot understand even Aristotelian logic; Aristotle lays it down right at the beginning of the *Analytics* that the rules for valid syllogizing do not depend on the truth of the premises, or on whether we seriously assert them, or on how we got them, or on whether we think we know them.

I feel very doubtful whether anything sound can be salvaged from consideration of what Karttunen is reported as saying. What N. N. believes cannot be described, let alone philosophically elucidated, by the idea of a belief-world in which everything is just as N. N. *believes* things are ('are' in which world?). For in general a man's total system of beliefs is not consistent. A reasonably modest man will say, even of that fragment of his beliefs that comes in a book, that he is sure there are some mistakes; if so, there is certainly inconsistency in his corpus of beliefs. A man may indeed be conceited enough to say he is assured of having made no mistakes; but this assurance is no good warranty against inconsistency, any more than against error. Now if N. N.'s beliefs are inconsistent there can be no possible world in which everything *is* just as N. N. believes: not unless we are prepared to buy impossible possible worlds *à la* Hintikka. As for belief worlds accessible via other belief worlds, I can only echo Hume: we have got into fairyland before we are well started.

I do not see how *logic* can decide what are or are not implications of something said, from considering what would normally be said in given circumstances. What we say is guided by many considerations—prudence, charity, etiquette, to mention a few—that have little bearing on the truth of what might be said. The actual example under discussion, of the alleged 'pragmatic presuppositions' of sentences with the verb 'regrets' in them, is, I think, seriously open to exception. If the interest of the company is in John's attitude, N. N. may say 'John regrets that he never married', even though N. N. himself is aware of circumstances—say, peculiarities of the marriage law of Scotland—in which John could have got legally married though he doesn't know it; N. N. is not 'pragmatically' ruling out such possibilities; he simply does not see fit to mention them, for obvious reasons.

IDENTITY THEORY: REPLY TO MÜLLER, WILLIAMS, AND DUMMETT

When I wrote *Reference and Generality* I used an example in which 'the same' is used, to express diachronic identity, now in connexion with the count noun 'river' and now in connexion with the mass term 'water'. I fear I had not then sufficiently observed the peculiarity of mass terms. Aristotle used mass terms to interpret term letters in syllogisms without regard to this peculiarity; he gives 'swan' and 'snow' alike as terms to which 'white' can be truly attached, and similarly 'crow' and 'pitch' as terms to which 'black' can be truly attached. At the time I had not lost this Aristotelian innocence. I was following a long line from Aristotle to Neville Keynes and Lewis Carroll: the syllogism,

> Any gold is heavy; only gold will silence him; *ergo* only something heavy will silence him

is to be found (near enough) is Lewis Carroll. It has the surface form of a valid syllogism using exclusive propositions:

> Any *M* is *S*; only *M* is *P*; *ergo* only *S* is *P*

(Such syllogisms were studied by Neville Keynes and my Balliol predecessor Walter Burleigh.) But the reading of '*M*' as 'gold' raises problems I did not discern.

Critics of my book fastened upon this passage. Some reviews argued that the choice of example was essential to my reasoning; this was not so, for it was quite easy to reconstruct the reasoning with two count nouns (in the third edition, when resetting became possible). I was not helped when some alleged that a distinction I here overlooked, corresponding to different accentuations of the word 'some', had to do with this matter of mass terms and count nouns; it can be quickly shown that this is a false clue. One accentation comes alike in 'John drank some white wine' and 'John dined with some men'; another comes alike in 'Some white wine is undrinkable' and

'Some men are very silly'; *either* accentuation occurs *both* with mass terms *and* with count nouns, and thus no light is thrown on the category difference. I believe this accentual difference is one of those fiddly bits of English grammar to which importance is perversely attached; no corresponding difference is uses of 'some' shows up in a statement of the truth-conditions.

I had to attend more seriously to mass terms when I had the privilege of being on Helen Cartwright's committee for her brilliant, alas unpublished, doctoral dissertation at Ann Arbor in 1963. She carefully traces the complex semantical and syntactical similarities and differences between (a) count nouns in the singular (b) count nouns in the plural, (c) mass terms. There is now a growing literature on the topic; I hope progress may be made upon it. But I cannot claim to have contributed anything myself, or to have much to say.

For this reason I have no adequate comments to offer on Müller's paper. I do not think it would be possible to evaluate his arguments unless I had an adequate, or at least a fairly defensible, theory of mass terms; and I have none to offer. In Müller's paper we find quantification over individuals whose criterion for diachronic identity is expressed by such words as 'being the same water'. Such quantification raises problems to which, I fear, neither Müller nor I can offer a solution. Let us consider a portion of water set apart in a hermetically sealed vessel, let us say a cube. Geometrically, a cube divides into eight smaller cubes; is each of these cubes an individual of the sort Müller means? We know that the water in each will after a very brief time have partly diffused into the other cubes; so across time we cannot identify these water-individuals geometrically. We could do better if Dalton had been right; if in each cube, at the initial moment, there had been a definite set of water molecules, each one composed of eternal oxygen and hydrogen atoms, then the identity of a water individual *à la* Müller would consist in the identity of the molecules, and a molecule's identity would consist in certain definite oxygen and hydrogen atoms' being bonded together. But we know Dalton was wrong; and in modern physics the diachronic identity concept becomes more and more difficult to apply as we consider a smaller and smaller scale of phenomena. I do not think this is a reason for rejecting diachronic identity of macroscopic physical things; such a rejection would follow from the results of modern physics only if we naively accepted a premise that a body is built out of molecules as a clock is out of wheels and springs, etc.; and it is this premiss that we ought to question when the identities at a microscopic level look so dubious.

This is not the only problem that arises over diachronic material identity. Müller's text suggests another difficulty. To modify one of his examples a bit, suppose in a revolution the ancient holy crown of a country is melted down and made into a collection of medallions with the dictator's face on them; but the secretly loyal goldsmith who did the job keeps those medallions apart, and at the eventual restoration a crown is refashioned from them. We might then say: The same gold was first of all a crown, then many medallions, then a crown again. But ordinary quantificational logic is very ill adapted to express this sort of statement. We could hardly make sense e.g. of a formula in which the proper name of a crown, and a list of proper names of the medallions, stood on opposite sides of an identity predicable—whether this expressed absolute or relative identity!

In the third edition of *Reference and Generality* I distinguished two ways in which identity can figure in a proposition: we may speak of subject-identity, which comes out e.g. in the repeated (unequivocal) use of a name, and predicative identity, expressed by an identity predicable. Does this distinction help us? Could we conceive of giving a proper name to some continuing Müllerian gold-individual, and predicating of this individual that *it* was first all in one piece and in the shape of a crown and then in many pieces and in the shape of medallions, etc.? This might tie up, I think, with medieval talk about discrete quantity: the phrases 'in one piece' and 'in many pieces' would be predicated in the category of discrete quantity, and the continuing gold-individual would have ascribed to it changes in shape and in discrete quantity. Such a view might seem to conflict with Frege's insight about attributions of number; but I think the conflict is resoluble; the predicable 'being in *n* pieces', which applies (for a time) to the gold of the holy crown, is quite different from the second-level predicable 'there are *n*...', which applies to the concept *medallion made from the holy crown*. But we lack the logical apparatus to deal with discrete-quantity predicables; though I am rather strongly inclined to look for further insights in this direction.

What Müller has to say about such matters as surmen and different applications of the expression 'the same word', is rather slight; I shall deal with these topics in reply to Dummett. As I said, I hope that the present wave of interest in mass terms and in associated problems of diachronic identity will yield new light; but I do not like all the work of this sort that is done, and I want to put in some caveats (not addressed to Müller). First, I am sure there is no future for a concept of diachronic material identity that prescinds from down-to-earth applications such as we have in chemical experiments. Aquinas's conception of matter was closely tied up *e.g.* with his views on metabolism; he rightly compared the diachronic identity of a living body to that of a flame. Neoscholastics of our time have tried to formulate a 'metaphysical' conception of matter that is cut loose from any such coarse considerations; they regard themselves as getting something free from past scientific errors and immune to scientific criticism, but I fear what they come out with is pure nonsense. Secondly, since the logical distinction of mass terms and count nouns is suggested by grammar, logical treatment has been impeded by people's fossicking around with irrelevancies from the grammar of natural languages. One striking example will show the sort of thing I mean: more than one author has assimilated nouns like 'gold' and nouns like 'wisdom', because of such grammatical features as their going with 'much' and 'little' and 'more'!

Williams is mainly concerned with personal identity, though he seems to show some inclination to reduce this to a more general notion of identity: in the terms I have already used, to a sort of subject identity. Subject identity is not a relation, is not expressible by a relative term, and is tied to the repeated use of proper names and to (what is called in logic) the identification of (name) variables. Since our use of ordinary proper names for people is tied up with subject identity, I must clearly agree with Williams that no notion of a relation will serve as an *Ersatz*, particularly not one of a relation that is not even transitive. I have not read Parfit's works, and what Williams reports of them gives me little appetite to do so. Brain transplants and other such 'medical advances' are mere fantasies; and the demand that we should revise our conceptual apparatus, to fit the world as it would be if these fantasies were realized, is premature (to use no harsher word). In any case it must be nonsense to say 'identity doesn't matter for survival'. That both before and after some crisis *C* there is an *A* doesn't mean that some *A* survived *C*. Both before and after the revolution of 1688 there was a sovereign of England, in fact after it there were *two*, William and Mary; but that does not imply that there was some sovereign of England who survived that event (even though James II did so). For some *A* to survive the crisis *C*, the same *A* must have been there before and after *C*; and when one says 'the same *A*', that implies *identity* by some criterion; a relation, possibly not even transitive, holding between the *A* that is there before *C* and one or more *A*s that are there after *C*, will clearly not suffice.

Williams says little in his paper about predicative identity; given his main interest, perhaps he did not need to. Since I shall have much to say on that topic in reply to Dummett, I confine myself for now to pointing out that when such identity-predicables link ordinary proper names, the resulting propositions often are far from being uninformative or trivial—nor are they, disguisedly, *about* the use of the names that occur in them. Was Running Rein the same horse as Maccabeus? From acquaintance with both, Lord George Bentinck decided that indeed they were the same horse; to settle the matter, inspection of horses, not of racing men's linguistic habits, was called for. Similarly in many cases of human identity that give rise to litigation. (The names that would come in the atomic propositions of Wittgenstein's hypothetical language need not here concern us.) Certainly such questions of identity, of predicative identity, call for evidence and criteria.

Williams main concern is with propositions about self-regarding attitudes, like these:

> David judged that he himself deserved death.
> John Locke (at least seemingly) remembered himself seeing Noah's flood.

(Let nobody be distracted by the facts that the second is a bit different from Williams's similar example, and that it is presumably false.) With regard to such propositions, Williams recognises that '(he) himself' is what grammarians call an indirect reflexive pronoun, corresponding to an 'I' of directly reported speech—'I deserve death', 'I saw Noah's flood'; he also holds about them the two following theses:

(A) that from such a proposition we can extract a relationship expression,

(B) that the proposition affirms the relation of somebody to himself.

The relationship expressions for our two examples would be:

> —judged that—deserved death
> —(at least seemingly) remembered—seeing Noah's flood;

and to construe the examples as affirming someone's relation to himself is *one* way to read what a man is about if he is prepared to fill up the two blanks in each with equiform variables, say 'x'.

The justice of my ascribing theses (A) and (B) to Williams becomes clear from various passages, in particular this one:

> If David had been aware that *he* judged that *he* deserved to die, not only would he have been aware that someone judged that someone deserved to die, but he would also have been aware that, for some x, x judged that x deserved to die. Just so, the fact that it seems to John that *he* remembers that *he* witnessed Noah's flood entails, not only that it seems to John that someone remembers someone witnessing this, but that it seems to John that, for some x, x remembers x witnessing this. [Williams's italics]

Earlier, Williams says that the *someone* who does the apparent remembering is *designated* by the reflexive pronoun 'he'; this is not language I like, but perhaps it should not be pressed. But both theses, (A) and (B), appear clearly implied in a still earlier passage where Williams treats 'I remember that I witnessed Noah's flood' as entailing an identity proposition in the 'more basic' form '$\Sigma x \varphi xx$'.

I doubt both theses. William's examples (except perhaps the last cited) would not even if true imply that the *that* clauses expressed facts; but even where this would hold, which ought to make it *easier* to maintain thesis (A), (A) pretty clearly does not hold. From the proposition

> Lord George Bentinck knew that Maccabeus ran in the Derby

we *cannot* extract a complex relational expression

> —knew that—ran in the Derby;

for then we cannot say whether this relation does or does not hold between another racing man, say Sykes, and a certain horse Maccabeus *alias* Running Rein, if Sykes *did* know that Running Rein ran in the Derby, but *did not* know that Maccabeus ran in the Derby. The context is what I have called non-Shakespearean. (Notice that no such trouble arises for

> —fixed it that—ran in the Derby;

this *could* express a relation between a certain crook and a certain horse, no matter how they are styled, even though the entire clause abbreviated to 'p' would occur nonextensionally in 'a fixed it that p'.)

If there are no relational expressions to be extracted from such examples, then there is no question of reflexivizing the relations involved; so thesis (B) must collapse along with (A).

I am afraid Williams has *not* sufficiently grasped the puzzle about self-regarding attitudes: even though he refers to what McTaggart, Castañeda, and I myself have written on the topic. If we consider the three propositions:

(1) David judged that David was worthy of death,
(2) David judged that he himself was worthy of death,
(3) David judged concerning David that he was worthy of death,

we may see that here (3) follows from (1) and from (2), but not conversely, and that (1) and (2) are logically independent of each other. For (3) would be true if David's thought 'Worthy of death!' latched onto David under *any* guise; and (1) and (2) are logically independent because to pass between them we need the extra premise (again self-regarding in the problematic way) 'David knew that he himself was David'—no doubt true, but not as a matter of logic. (*Cf.* the senile Linnaeus reading his own books, and judging that Linnaeus was very clever, without realizing that he was Linnaeus.) As for the formula:

(4) For some (man) x, x judged that x was worthy of death,

with which we may think to help ourselves out, I am afraid in the present state of the art it only serves to cover up our confusion and ignorance. If (4) is regarded as related to (1), we are quantificationally handling a non-Shakespearean context in a way beyond our present powers. If (4) is regarded as related to (2), we are identifying variables in a way for which presently we have no rules; even the simpler inferences from (2), to (3) or again to:

(5) As regards somebody, David judged that he was worthy of death

are moves we can presently make only by guess and by God, as sailors say. And if we are not clear whether (4) relates to (1) or to (2), then we are confused. I am afraid I find this confusion in several pieces of work by Arthur Prior, and by Williams too in his contribution. I hope I have not myself added to confusion in this area; I have tried to show where the difficulties lie.

Williams several times refers to a man's power of identifying himself *without* a criterion of identity. Here we have to notice (as Williams does not explicitly do) an ambiguity in the term 'criterion of identity'. I may use a man's face as a criterion of identity, or again his voice; that is, I use such marks to tell that I am dealing with *this* man rather than another. But such marks form no part of what I mean when I say I am dealing today with the same man as I met twenty years ago; if meanwhile he has suffered an accident and surgery has altered his face and voice, that does not make him only a marginal or doubtful case of *being the same man*, it only removes one way of finding out that he is that man. Obviously a man does not (normally) recognise himself by *any* marks of identity; that is what makes it so comical for the old woman in the song to have tried to resolve her fear 'this be none of I' by finding out whether her dog would bark at her as at a stranger. But this does not mean that as regards the statement 'I am the same as I was forty years ago' the question 'same *how*? same *what*?' can be suppressed or dismissed; and when I have been writing in my own person about criteria of identity, I have meant: standards for making judgments as to identity, by supplying which we remove the ambiguity over how 'the same' is to be taken. And *here* I want to say: I am the same *man*, or *human being*; it is rather dangerous nowadays to use the good old word 'person'. The feminist use, as in 'personagement' and 'personhole', is merely comic; those on the other hand who talk of a man's becoming, or ceasing to be, a person (much as an iron bar may become, or cease to be, a magnet) are up to no good—but that is another story.

Dummett himself has used to expression 'a tangle' about the history of my own writings on identity and his responses to them; I am afraid this is all too apt. For fear of making the tale even more tangled, I have refrained in the past from direct comment on his writings, though I have naturally taken account of what he said when I was restating my views. I am conscious that I published on the matter when my mind was still in movement, and have had to revise my actual views, not just my way of putting them over. Apart from minor corrigenda, which I think do not affect the problem of identity, I stand by what I said in the *third* edition (1980) of *Reference and Generality* and in my Salzburg paper (1979); I shall try here to formulate the doctrine of identity stated in those works. How far I have recanted, and how far, retained, the doctrine of *earlier* writings, I must ask readers to judge for themselves. Several criticisms that Dummett has already published of my views are of later date than the piece he contributes to this volume; and there he engages himself with work of mine (Geach, 1962, 1979) that he does not mention here; it would give a wrong impression if readers did not bear this other published work of his in mind (especially Dummett, 1973, Chapter 16, 'Identity', and Dummett, 1981, Chapter 11, 'Geach on Identity'). I think Dummett has often quite mistaken my mind, but I shall not try to document this; controversy of the form 'What he says I said is . . . but what I did say is . . .' is rarely profitable. In fact, as in past writing on identity, I shall abstain from picking up Dummett's comments, though I have often benefited from them; here too, having given my advice on relevant material, I must ask readers to judge for themselves.

The notion of identity and of its twin sister difference pervades our discourse about all matters and is multiply expressed in language. Philosophical treatments often concentrate on one particular kind of expression: the identity proposition, in which 'is identical with' or 'is the same as' or 'is the same so-and-so as' occurs flanked by what people call referring expressions. The diet of examples is often even more restricted than this; people consider only *statements* of identity, i.e. *asserted* identity propositions, as if the same propositions could not occur non-assertorically; sometimes they even seem to forget that such assertions may be false!

It is worth while, then, to sketch the various other guises that identity and difference appear in; whereas use of the expressions just mentioned to form complete clauses is fairly rare, and identity statements are even rarer, other ways of appearance for these notions are ubiquitous in our language. Consider the pair of sentences 'Jim wounded a lion and Bill shot it dead', 'Jim wounded a lion and Bill shot another (lion) dead'. In the first, the *renvoi* of 'it' to the antecedent 'a lion' expresses identity; in the second, diversity is expressed, also by a kind of *renvoi*. Medieval logicians used the term '*relativa*' for pronouns with *renvoi*; in our first example there is a *relativum identitatis*, in the second a *relativum diversitatis*, with *renvoi* to 'a lion'. In neither case would any identity *predicable* be needed in a symbolic language; in the first case we could simply have the occurrence of equiform variables bound to the same quantifier; in the second case, as Wittgenstein pointed out, we could express difference simply by having two non-equiform variables: something like this:

> For some lion x, for some lion y, Jim wounded x and Bill shot y dead.

(It illustrates the shallowness of much that is written about identity, that Wittgenstein's proposal about variables is often represented as a proposal about proper names!)

Another way that difference gets expressed in language is by the particle 'else' going with quantifying pronouns like 'some(one)' and 'any(thing)'. Yet another way is by the use of certain prepositions, 'but', 'besides', 'except': 'every donkey but (except) Brownie' of course means 'every donkey that is a different donkey from Brownie'. Identity on the other hand gets expressed by the repeated use of a proper name; though here we have the complication that proper names get used and re-used for different things. For logic, this is merely a nuisance; we can easily imagine a tribe where the dead were honoured, not by naming people and places after them, but by a rigid taboo on such reuse of the names. This apart, repetition of a proper name shows forth identity; it involves, I shall argue, a criterion of identity. And yet another way that identity gets expressed in natural language is by the use of reflexive pronouns.

Faced with all this variety, one may well be perplexed where to begin; which thread of the tangled skein to begin pulling so as to unravel it. I shall begin by considering the connexion between identity and difference, on the one hand, and assignments of number (*Zahlangaben*, as Frege says) on the other. That there is a close connexion is obvious. As regards identity, the phrase 'one and the same', '*ein und dasselbe*', points the way; in fact, 'there is just one *A*' says the same thing as 'as regards something that is *A*, anything that is *A* is the same *A* as this *A*': *unum A est omne A*, to use a medieval formula. And if there are many *A*s, then obviously the number of *A*s is the number of *different A*s, and in assigning a number to each thing that is *A* we must not assign *the same* number to *different A*s.

All of this is, or rather ought to be, obvious; and of course it was from Frege that I gained what clarity I have in the matter. Several critics have on the contrary agreed that this doctrine is radically anti-Fregean: that it commits what for Frege is a capital sin, a confusion of levels. '"One and the same" forsooth! Doesn't Geach realise that for Frege assignment of the number *one* is second-level, whereas "the same" expresses a first-level relation?' I do not know how often I have seen this sort of thing; I have never thought it worth a reply. There is indeed a difference between my view and Frege's, but not one involving on my part what Frege would call a confusion of levels; my critics misunderstand both Frege and me.

What should be realized here is that *any* equivalence relation can serve as the basis of a count, of a *Zahlangabe*, for things in the field of the relation. For example, the question how many animals there are in a zoo, or how many words in a book, may be variously answered, according as we do or do not count the lions (say) as being all the same animal; or again, whether in the book what interests us is the total length, or the richness of the vocabulary. To take the general case: for *any* equivalence relation *R* we can take something from its field, and assign the number 1 to that and to anything else in the field that bears *R* to that; then we can take something else from the field, and assign the number 2 to *that* and to anything else in the field that bears *R* to *that*; and so on, till the field is exhausted and we have a count of *n*. Clearly this will work for any (finite) field and for any equivalence relation *R*. The matter can be controverted only through some confusion or prejudice, which I have tried to forestall by not talking (not *yet*) about relative identity, but just about equivalence relations. For me, a *Zahlangabe* latches onto an equivalence relation, expressed by a two-place predicable; for Frege, it latches onto a concept, expressed by a one-place predicable; this is certainly a difference, but *not* as regards our finding in a *Zahlangabe* something second-level ascribed to something first-level.

A relation may be an equivalence relation, may be reflexive, symmetrical and transitive, with reference to a given field but not all fields; however, its being an equivalence relation for a given field suffices to determine a count of that field in regard to that relation. Readers may find this thought puzzling because of a common confusion about the properties of relations. 'Surely, if a relation is transitive, that is a *logical* property of it; how can the field affect this?' All too often, people fail to distinguish two senses of 'logical property of the relation *R*': namely, that a property's holding of *R* can be spelled out in purely logical terms, and, that its holding of *R* is a logical truth. For example, *R*'s being transitive simply means that for no *a, b, c* do we get *aRb* and *bRc* but not *aRc*; *R*'s being transitive in a given field means that where our choice of *a, b, c,* is restricted to that field this condition holds. If transitiveness had to hold of *R* as a matter of logical truth, one might be puzzled at the restriction to a field; particularly if the field is specified by a *list*, since proper names cannot occur essentially (as Quine puts it) in a logically true statement. But transitiveness need not hold as a matter of logical truth. Quine has the elegant example of the relation *living within ten miles of*, which is transitive as between inhabitants of the western Azores; the reason is not logical but geographical, that the western Azores are eleven miles apart and the longer one only ten miles long.

Let us consider cases in which the field of a relation is specified by a finite list of proper names. Each proper name, I have maintained, has built into it such a criterion of identity; the bare intention to go on referring to *the same thing* by the proper name is, as lawyers say, void for uncertainty. (It is not only this expression, I think, that makes an intention void for uncertainty; 'the same material object' and 'the same event' are I think equally frustrating.) From this some have argued: Surely, counting things that are listed by name according to the criterion of identity that is built into each name will give us a count that is *distinguished*, as compared with counts that go by other equivalence relations that also apply in this field? Certainly: so long as all the names have *the same* criterion of identity built into their sense. But of course that need not be so. A list may include names with different criteria of identity built in, and is not then ill-formed. A list of names of material objects, let us say, may have very different criteria built into the sense of its several items. I shall return presently to this case of mixed lists.

An equivalence relation is often expressed in the form 'is the same so-and-so as'. This form of words has an unfortunate syntactical ambiguity in English, which I mention here, because it has got into the literature about identity, only in order to get it out of the road. The sentences 'Crimson is the same colour as scarlet' and 'My car is the same colour as your car' are syntactically different, and this involves there being different inferential moves from them that are allowable; from the first we may infer 'Crimson is a colour', from the second we may not infer 'My car is a colour'. I do not know how much ink has been split on this silly quirk of English: it is, so to say, just a syntactical pun, which is quite unreproducible, as you might expect, in most other languages. As regards my own work, there are critics who allege the 'car' type of example as an important counterexample to some principle I am supposed to have maintained. I do certainly wish to dismiss such examples, but not out of any puristic protest against solecism ('shouldn't it be: *of* the same colour'); I have, in fact, *never* tried to fault a philosophical example on the score of its not being quite correct English; I regard such fault-finding as contemptible. Neither on this ground, nor on any other, do I wish to ban anybody's saying 'My car is the same colour as your car', or the like; I simply do not want to discuss these examples, but only examples where from 'x is the same A as y' there does follow 'x is an A' (and of course also 'y is an A'). To *insist* on switching from these to the 'car' sort of example is, to put it politely, a distraction; and I must decline to be distracted.

When we have 'is an A' and 'is the same A as' standing in the relation that I am interested in, the question arises whether either is logically prior to the other, that is to say, whether it can serve for defining the other but not conversely. Here we are liable to be misled by the structures of natural languages. It may look as though 'longer than' were a more complex expression than 'long' and had to be defined in terms of it. But a moment's thought shows that 'This table is longer than that table' is more readily understood and established than 'This table is long'; it is at least arguable that 'long' calls for some explanation in terms of 'longer than'. As regards a term like 'father', the position is clear; 'is a father' as a monadic predicate has to be explained as 'is father of somebody'; it would be senseless to try to explain 'is father of John' in some such way as 'is a father and belongs to John'. The step I have taken in my writings on identity has been strictly analogous: I treat 'is an A' as derivative from 'is the same A as' namely as meaning 'is the same A as something'. In these last two cases we start, logically speaking, with a relative term, and proceed by *derelativizing* it (Quine's term). No logical procedure applied to a derelativized term regenerates the original relative term. If in a crowd of parents, teachers and children you know who is father of whom (let us for simplicity assume the teachers are all chaste celibates), then you know who are the fathers there; if you know who in that company are fathers, no logical procedure will tell you who is father of whom. (A correspondent suggested introducing by fiat an undefined operator which would make such a procedure possible; but logic is not magic.)

A protest I have encountered against the definitional pattern where 'x is an A' is explained as 'x is the same A as something' is that even if this works for 'x is a dog', say, it will surely not then work for 'x is a terrier'. That will be explained rather as 'x is a dog and has such-and-such points' (in the language of dog-shows); would it not be perverse to bring in the relative term 'is the same terrier as'? The objection overlooks the fact that restriction by added *Merkmale* (to use Frege's term) is possible for relations, not only for concepts. Suppose we define '*x* is the same terrier as *y*' to mean '*x* is the same dog as *y* and *x* and *y* both have such-and-such points'. If this shows how the relation *same terrier as* stands to the relation *same dog as*, then it is easy to show that if the concepts *is a terrier* and *is a dog* are treated as got by derelativizations, '*x* is a terrier' will indeed turn out to have the same truth-conditions as '*x* is a dog and has such-and-such points'; so its doing so nowise goes against my account. It is of course quite as vain to argue 'that isn't how we learn the concept *is a terrier*', as it would be to argue against the indisputably correct derelativization-account of 'is a father' by talk about how children first learn the word 'father'.

I have seen the suggestion that I should drop my proposal for terms like 'terrier' or even like 'dog', and confine it to terms of such generality as 'animal'. This would be a very unsound move on my part. I do not know if there even *is* a criterion for *being the same animal*, if 'animal' is used in so wide a sense as to encompass alike jelly-fish, amoebas, insects, and dogs; certainly framing and applying such a criterion would be, as Frege might say, *umständlich*. In a discussion I had with Dr. Bernard Harrison, he came up with a much better suggestions which I hope I am not misrepresenting by putting it this way: This term 'animal' answers, not to a single criterion of identity, but to *a test of* such criteria; a criterion that passes this test will be a criterion of identity for some kind of animal. Harrison's suggestion seems to me valuable, and throws new light on the old apparatus of genus, species, and differentia; but I fear I cannot discuss it further. ('Animals' of course has a sense, or more than one, to be found in popular talk, which is far more restricted; but so understood, the term is of no *special* logical interest, any more than 'dog' or 'terrier'. This is how e.g. I was talking about 'animals' in a zoo.)

One class of cases where '*x* is the same *A* as *y*' does entail '*x* is an *A*' has a special importance because there is no plausibility in taking 'is the same *A* as' to split up into 'is an *A* and is the same as': I mean, the cases where we could gloss 'the same *A*' as meaning 'the same kind of *A*'. Here, again, talk of solecism versus correct English may rear its ugly head; is not '*x* is the same kind of *A* as *y*' in such a case short for '*x* is OF the same kind of *A* as *y*', so that what is primarily in question is not an equivalence relation between *x* and *y*, but outright identity between a kind to which *x* belongs and a kind to which *y* belongs—and thus we are quantifying, not only over *x* and *y*, but over more abstract entities called kinds? This would display a misconstrual of how phrases with 'kind' or 'sort' have come to be used in English; very often, kinds or sorts are *not* our subject-matter, any more than spite is when we say 'in spite of'. Constructions with 'kind' and 'sort' make up for the absence in English of the series of pronominal adjectives that come in some other languages: Latin 'talis', 'qualis', 'aliqualis', correspondingly Polish 'takì, 'jaki', 'jakiś'.

Let us consider the sentence: 'The king wore a uniform at his coronation and he wore the same uniform at his son's wedding'. This calls for disambiguation; do we mean the king got *the very same* uniform out of mothballs for the wedding, or just that he wore *the same sort* of uniform, say that of a certain regiment? Like some other ways of speaking that serve for disambiguation, these expressions are rather clumsy. Latin and Polish would avoid the ambiguity; Latin could use the correlative pronouns 'is . . . qui . . .' (in the appropriate gender and case) for the 'very same' sense and 'talis . . . qualis . . .' for the 'same sort of' sense; Polish similarly distinguishes between 'ten sam' for 'the very same', and 'taki sam' for 'the same sort of' ('taki', as I said, is Latin 'talis'). In neither of these languages is there a verbal suggestion that in the 'same sort of' case we are quantifying over sorts or kinds; but a speaker of English need not be *very* strong-minded to resist this suggestion. If the king was wearing a sort of uniform, he was wearing a uniform, not some abstract entity, a *sort* of uniform. (In the literature I have encountered the protest that such a phrase as 'wearing a sort of uniform' is anyhow ungrammatical; we need not stop to consider this further.)

The distinction between 'the very same' and 'the same sort of' is made in clumsy homespun language, but is genuine enough; talk of numerical identity on the other hand is spuriously precise philosophical jargon. I suppose it is meant to suggest an identity we can count by; but clearly, as we have seen, it does not distinguish one equivalence relation from another that we can use it to count by. Let us revert to my example of counting animals in a zoo. We might attach one number to all the lions, and another to all the kangaroos, and so on, or on the other hand attach different numbers to different individual animals; these two ways of counting are cases where a Pole would use 'taki sam' or 'ten sam' respectively. It is not that one time we have a loose and popular use of 'the same', the other a strict and philosophical use. It is a folly to think you make your talk of identity strict by talking of strict identity. For example, it is useless to expound what is called the identity theory of body and mind by saying that an event in Jones's mind and an event in his brain are strictly identical; for 'the same event' expresses *no* criterion of identity, *no* equivalence relation, at all.

At this point someone may protest: How can this be? If 'the same event' expresses no criterion of identity, events will be uncountable; but e.g. chess moves are countable, and chess moves are events. This protest would involve what Frege would call a confusion of levels. It is a property of a concept, not of things falling under a concept, that is in question when we say that chess moves are countable or that events are not. And uncountability attaches to the concept *event* precisely because there is no corresponding relation *being the same event as*, nor yet *being the same kind of event as*.

Frege noticed that some concepts afforded no unit to count by (*Grundlagen* § 54), but could give no good explanation. If '*x* is the same *A* as *y*' were always analysable as '*x* is the same as *y* and *x* is an *A* (and *y* is an *A*)', then there could be no ground for any concept's being uncountable. But, as we have seen, some concepts that we can and do count by admit of no such analysis; if 'the same' here has the use of Polish 'taki sam' rather then 'ten sam', '*x* is the same *A* as *y*' definitely does not mean anything like '*x* is (of) the same kind as *y* and *x* is an *A* and *y* is an *A*'; there is in such cases no possibility of splitting apart the ascription of identity (in kind) and the predication of '*A*'. This may dispose us to look askance at the parallel analysis for 'ten sam' examples.

Acceptance of this point might still leave us holding the view that 'is the same — as' is a logical framework into which only those concept-words will fit whose corresponding concepts have (so to say) a certain definite logical shape, so as to fill the place of an argument neatly. People have come to call such concepts or concept-words *sortals*. But then the question arises how we recognise sortals; certainly ordinary grammar is no sure guide.

I leave the defence of such a view to its friends. Even today, more than two millennia since the *Euthydemus*, one sometimes meets with a corresponding view of kinship terms: that the monadic predicables '(is a) father, brother, uncle, etc.' are primary, and are of such a logical shape as to slot into the framework '— is . . . of —' and form a dyadic predicable. But the view that on the contrary the monadic predication of kinship terms is got by derelativizing the dyadic predicable is here completely convincing. And I maintain that the monadic predicables called sortals are likewise derelativizations. Here at last we get an explanation of why those and only those concepts are countable: they correspond to equivalence relations, and it is the equivalence relation that determines a number as result of a count. Moreover, two different relative terms standing for equivalence relations may have coincident derelativizations; *x* is the same individual animal as something or other iff *x* is the same kind of animal as something or other, so, as regards falling under the concept, it makes no difference which derelativization we take the concept *animal* to be. But as regards counting the animals in a zoo, it makes a great difference which equivalence relation we go by. This supports the view that in a *Zahlangabe* the number is being ascribed not to the concept, as Frege thought, but to an equivalence relation in a certain field.

In the past great advances of thought have been made because people exchanged the question what an *A* IS for the question what it is to be the same *A*. The Euclidean treatise on proportions contains one definition telling us what a ratio IS; this is perfectly futile and is never used; and then there is another definition giving the criteria of identity for ratios, which is a masterpiece, and shows how to deal with irrational ratios. Frege knew his Euclid well, and I feel this must have inspired him to cut short centuries of futile nattering about what a number IS by considering rather the criterion of identity for numbers. This is the way forward. I once read a complaint that some writing about Fregean *Gedanken* failed to supply 'a philosophical account of what thoughts ARE'; but if only we had a good criterion of identity for *Gedanken*, we could probably do without such an account.

For any proper name, I have argued, there is a criterion of identity built into its sense. If I am right, this criterion need not be a criterion of absolute identity; for a given use of a proper name, we must intend to keep on referring not just to the same *thing* but to the same *A*. But can any equivalence relation serve to found the use of a proper name? Suppose *x* and *y* are the same *A* but are two different *B*s. There is no logical impossibility about this; it just means that *x* and *y* come in the field of the equivalence relations *same A as* and *same B as*, so that each both is an *A* (the same *A* as something) and is a *B*, and *x* is the same *A* as *y* but not the same *B* as *y*. Then if an equivalence relation can found the use of a proper name, we can introduce a proper name *c* which is a name (as I have latterly put it) *for* an *A*, i.e. *same A* is the relevant criterion of identity; and each of *x* and *y*, which are two different *B*s, will be the same *A* as *c*.

The problem then arises: If we have *F*(*x*) but not *F*(*y*), do we have *F*(*a*) or not? If we tried to answer this question regardless of which predicable '*F*( )' indicates, and in which language, we should quickly get into contradictions. But in a regimented and extensional language we can specify by fiat the truth-conditions for occurrences of '*c*' as argument of primitive predicables, and then in Quine's phrase let the notes fall where they may for other cases via recursively specified truth-conditions. Such a method is logically not so neat as I could wish, but it is enough to show that this threat of contradiction can be nullified.

There is however another threat of contradiction which I think cannot be met so easily; it can be met only by my retreating from the view that I expressed in my Wolfson Lecture 'Names and Identity' (1975), and elsewhere, that *any* equivalence relation would serve to found the sense of a proper name. Let us take the example of horse, which I gave in that paper. If *x* is a horse and so is *y*, then *x* is the same sort of animal as *y*; could this relation, which is an equivalence relation, be embedded in the sense of a proper name? If it could, we might take 'Horse' to be this proper name. Each and every ordinary horse would then be the same kind of animal as each and every ordinary horse, but would also be the same kind of animal as Horse. I have already explained that this would not necessarily involve quantifying over abstract entities called kinds or sorts. We might indeed ask what 'Horse' was a name *for*; the preposition would here be used as it is where we say that 'Pegasus' is a name *for* a horse and so is 'Bucephalus' (but the latter only is a name *of* a horse, the former being empty). To this question we should have to answer: 'Horse' is a name *for* a kind of animal. But here again we are not quantifying over entities called kinds; it just means that *same kind of animal* is the criterion of identity built into 'Horse' as *same horse* is into 'Bucephalus'. Names *for* a horse do not include 'Horse' itself; Horse is the same kind of animal as any horse you like, but not the very same animal as any horse. However, since all horses stand to one another in the relevant equivalence relation, 'Horse' would be a *shared* name of each individual animal that is a horse.

This doctrine is perhaps reminiscent of some things Aristotle says about the species as a substance. To my mind there is nothing wrong with the doctrine of shared names; I have long defended it against the doctrine of Russell and the mature Frege that there are no shared names; those who ascribe the latter doctrine to *me*, as has several times been done, I can only describe as boldly mendacious. (The *young* Frege of *Begriffsschrift* allowed mass terms as substitutions for name-variables, and says nothing to indicate that here he is departing from the traditional view that each such term names all the several samples of the stuff concerned.) One sort of context in which I hold that a word like 'horse' is often parsable as a shared name is in propositions of the form '*F*(some horse)' or '*F*(every horse)'. In such a proposition, I further argued, what 'every' or 'some' logically modifies is not the term 'horse' but the predicable '*F*( )'; a complex, so to say adverbially modified, predicable '*F*(some—)' or '*F*(every—)' is attached to the name 'horse' as its subject.

I have no space here to defend this view: its defence may be read in *Reference and Generality* (1962). One thing that went closely with it was an explanation of how '*F*(some *a*)' and '*F*(every *a*)' should be read when '*a*' is a *proper* name: namely, that both of these should be read as equivalent to the plain '*F*(*a*)'. This could not easily be abandoned or modified, for it involves so much that I have said about how propositions with shared names and with proper names in them are mutually related. It follows, if 'Horse' is a proper name, that '*F*(every Horse)' will coincide with '*F*(some Horse)'—both coinciding with '*F*(Horse)'.

At this point breakdown is imminent. We are supposing that 'Horse' would be a shared name of each individual horse, as well as being (relative to its own built-in criterion of identity) a proper name. But 'horse' itself, the familiar term, is for me a shared name of every horse; and every horse is the same sort of animal as every other horse; so how does 'horse' differ in sense from 'Horse'? It is hard to see; and in fact in my paper 'Names and Identity' I make no such distinction. But if no distinction is made, then not only will '*F*(every Horse)' coincide with '*F*(some Horse)', but also '*F*(every horse)' will coincide with '*F*(some horse)': which is absurd.

Checkmate could no doubt be delayed by trying out such moves as complicating the rules of 'every' and 'some', or finding some effective distinction of sense after all between 'Horse' and 'horse'. But I prefer to abandon the reckless assumption that *any* equivalence relation can be built into the sense of a proper name. If not all can, then *which* can? To this I have no answer as yet; but one can live with an unsolved problem.

Now I can dig a mass grave for those unhappy children of mine, the sarmen. The criterion of identity I proposed for surmen was a perfectly respectable equivalence relation: x is the same surman as y iff x is a man and so is y and x and y have the same single surname. (I cannot imagine why some transatlantic critics thought this definition called for conjectural emendation of my text.) The question is whether I could go on to construct propositions of the forms '*F*(some surman)' and '*F*(every surman)'. By my account of the quantifying words 'every' and 'some', this would be legitimate only if there could also be propositions of the form '*F*(*a*)', where '*a*' is a proper name for a surman, a name with its built-in criterion of identity given by 'is the same surman as'. But without the unrestricted assumption that any old non-empty equivalence relation founds a class of proper names, there is not the faintest reason to believe such proper names could be given. Dummett and others have hotly attacked the poor surmen; I must abandon them to their doom.

There are, however, many sorts of word that on the face of it, by surface grammar, are proper names; some kinds could be explained away, many could not. The built-in criteria of identity are also very various. Dummett has rightly remarked that bad theories as to the meaning of proper names are the result of an unbalanced diet of examples, almost confined to personal and geographical names. If we wish for further light on the question which equivalence relations can be built into the sense of proper names, a consideration of all this variety is called for.

I shall here once more briefly consider one particular sort of proper names. In England the members of the College of Heralds bear such romantic names as 'Garter', 'Rouge Dragon', 'Portcullis'. Each of these is a name *for* a herald, not *for* a man; its use survives the death or resignation or promotion of the office-holder. Dummett, who admits quite a variety of grammatically proper names as being also logically so, oddly suspects these; Quine, as he has told me, does not. Quine and I would indeed differ about how to analyse such a proposition as 'From t_1 to t_2 Hilary Handel is Portcullis'. Quine would say that a temporal slice of the man Hilary Handel is identical with a temporal slice of the herald Portcullis, from t_1 to t_2 in each case; but he would fully admit 'Portcullis' as a non-empty proper name, of an object whose history consists of certain temporal slices of certain men. I dislike the notion of temporal slices: I prefer to say that from t_1 to t_2 Portcullis is the same man and the same herald as Hilary Handel, and thus for that time Hilary Handel is a herald and Portcullis is a man. This does not, I hold, commit me to saying that from '*F*(Portcullis)' we may in general pass to '*F*(some man)'; this would be lawful, by the account of restricted quantifiers in *Reference and Generality*, only if 'Portcullis' were a name *for* a man, as it is not. And so one may say e.g. that Portcullis was never born, but not that some man was never born; over the years Portcullis has successively been the same herald and the same man as various men, each of whom *was* born.

Let us use '*R*' as short for 'is the same man *and* the same herald as'. Suppose that at a certain date Hilary Handel *R* Portcullis and Quentin Scrope *R* Rouge Dragon. Now consider the list: 'H. Handel, Q. Scrope, Portcullis, Rouge Dragon'. How many is that? It we are counting men, two; if we are counting heralds, again two; by neither criterion, four. This is easy to work out if we remember that the *Zahlangaben* are determined by the equivalence relations *same man as* and *same herald as*. But now suppose an economizing government sacks Q. Scrope and decrees that henceforth H. Handel *R* Portcullis and H. Handel *R* Rouge Dragon too; now consider the list; 'H. Handel, Portcullis, Rouge Dragon'. How many is that? If we are counting men, just one. If we are counting heralds? It is *impossible* to count heralds in this field; for by the case supposed *same herald as* is *not* an equivalence relation in this field; H. Handel is the same herald as each of Portcullis and Rouge Dragon, but Portcullis is a different herald from Rouge Dragon. If we just keep our heads, we see there is no contradiction here. (Here the point made on p. 286 about a relation's having a property in a given field should be borne in mind.)

I have said nothing thus far to attack the notion, or as I would say, pseudo-notion, of absolute identity. There is a general problem of method: If somebody thinks he has to do with a pseudo-notion, how can he argue against it without at least provisionally using it, and thus in effect conceding that it is a genuine notion after all? I cannot here discuss this problem, which is of course important for the teaching of philosophy and its history. The alternative to banning some notions as pseudo-notions would be an uncontrolled flood of nonsense (though some would hold it could be got under control by a logic of nonsense!). I shall look the *general* difficulty firmly in the eye and pass it by.

Sometimes we are told identity is definable in second-order logic: for any *F*, *F*(*x*) iff *F*(*y*). But it is gravely doubtful whether such quantification is admissible if quite unrestricted: can a quantification cover *all* properties or concepts, including such as would be expressed by this very style of quantification? Dummett has himself expressed such doubts when talking about quantification. When he gets on to identity, it is another story; he sees nothing wrong with the second-order explanation of identity; and he certainly holds that when a concrete instance of a predicable is substituted for the '*F*( )', this may contain mention of identity; and thus *by implication* he employs quantifiers ranging over concepts that are themselves expressible by using just such quantifiers. This will never do: he cannot have it both ways.

Let us turn to the question whether there is an acceptable test for whether a predicable in a language *L* expresses absolute identity. There is an embarrassing confusion in several writers here, which I must do my best to clear up. There are two relativities in question: talk of 'the same', I hold, is relative to a criterion of identity, differing from case to case; talk of 'identity predicables' needs to be supplemented by mention of the language in which the predicables occur (a pretty obvious point). To talk of 'identity in (*or* for) a language *L*' is on the face of it a crossing of the wires. This talk could be defended if what were in question were criteria of identity *for expressions* of the language *L*; but I fear those who have used such-like phrases have not meant this.

With this confusion out of the way, I can state once more the point I have often made. No criterion has been given, or, I think, could be given for a predicable's being used in a language *L* to express absolute identity. The familiar axiom schemata for identity could at most guarantee, if satisfied, that the relative term under investigation will be true in *L* only of pairs that are indiscernible by descriptions framed in terms of the other predicables of *L*. This cannot guarantee that there is no proper extension of *L*, with extra predicables, that makes possible the discrimination of things which were indiscernible by the resources of *L*. If there is such a proper extension of *L*, obviously the friend of absolute identity will not be able to say that a predicable of *L* which in *L* satisfied the axiom schemata for identity was an absolute-identity predicable; for if so, then a pair which satisfied the predicable in *L*, but were descriptively discernible in the proper extension of *L*, would be absolutely identical! It is useless to protest that this *need* not happen; that such a situation *can* arise is already enough to show how inadequate a test it is for a predicable's expressing absolute identity in a language *L*, that when it is used in *L* the familiar axiom schemata for identity are satisfied. Sometimes people suggest that to get such a proper extension of *L*, related to *L* as I have described, the interpretations of the predicables in *L* must all be modified. This is wrong; the argument goes through even if we explicitly specify, as we ought, that any sentence in *L* retains its truth-conditions when *L* is thus extended by adding predicables.

What I have just said I have often said before. It is painful to repeat myself; I can only hope that by dint of repetition the message may at last get over. I am afraid I myself muddied the water by presenting this argument in association with the false doctrine, already recanted on previous pages, about which equivalence relations could found the use of some class of proper names—illustrated by my example of 'surman'. But it is unworthy of philosophers of logic to condemn the argument as guilty by association with the false doctrine; for the argument made no mention of proper names.

Could we conceive of an absolute identity predicable as a predicable that could be added to *any* language, and which when so added would guarantee, for any pair of which it held good, that they were indiscernible with regard to the other predicables? I do not see how this suggestion could be worked out without using a general notion, not tied to a particular language, of a predicable's being true of an object. The work of Tarski and others makes it at least doubtful whether we can use such a general notion without inconsistency.

Sometimes people cut the knot by saying: I mean by 'identical' just what is ordinarily meant in English—just as when I use other words! Of course they could not say: 'identical' in English always means absolute identity; they would not say this e.g. for the sentence 'She was wearing at the party the identical dress that I was'. So once again we come to the question of a test whether *absolute* identity is actually meant on a given occasion. I need not go into this; such an argument is anyhow not to be heard about any notion philosophically introduced and philosophically disputed. Somebody who doubts the soundness of the philosophical notion of facts ought not to be satisfied by such a dictum as: I mean by 'fact' just what the word means in the *Strand Magazine* (an actual dictum of John Wisdom's: see *Problems of Mind and Matter*, 1934, p. 20). More than once I have read the further explanation: I mean 'identical' as it is meant in the meta-language! I cannot see that this is anything but an incantation, a magic word supposed to *call up* the right meaning.

REPLY TO NORMAN MALCOLM

If everyone at some party has been drinking rather freely, it may become obvious that everyone is a bit confused in his apprehensions of things, though some are so to a much greater degree than others. Those who are least confused will then be able to discern that some of their companions are behaving in a dreadfully foolish way, and may even be able to restrain this folly by persuasion; but their success at this can only be partial, hampered as they are by their own semi-intoxication; and they may well not be too far gone to recognize this fact and the reason for the fact. But men are not confused in mind only by alcohol.

What is very manifestly so in the case described is in fact a general human predicament, even if it is often not clearly realized. In all of us there prevail, more or less, those illusions which Bacon labelled the idols of the tribe, the cave, the market-place, and the theatre; only in so far as a man overcomes these illusions in himself (and he will do this at best imperfectly) can he cope with other people's confusions and set about removing them. Of course I do *not* think this situation is *logically* inescapable; indeed, I believe that there has actually walked upon the earth that Paragon of whom I wrote, whose candid mind (*candor lucis aeternae*) was never clouded with our illusions; just as his pure will was never stained with our sins, even though he was in all points tempted like us.

The difficulty about understanding sin, I submit, is that sin as I have said involves self-deception, which is one form of mental confusion. The literature in recent years about self-deception shows that the difficulty in describing its nature is no fancy of mine.

Philosophical confusion is difficult to describe because on the face of it the confused philosopher is making a judgement, maintaining a thesis; it is often our duty as teachers to try and get over *what* judgement or thesis is ascribable to some historic figure, whom we judge to have been, on this or that point, badly confused; and yet, if the philosopher was indeed confused, what he was saying was perhaps not even a falsehood, but nonsense, and how can what one *judges* be *nonsense* (*Tractatus* 5.5422)? I do the best I can, in the practice of teaching, to deal with this difficulty in particular cases; but if I succeed, as I hope I sometimes do, I can certainly give no rationale of my success. Malcolm seems to think I ought not to worry about the obscurity of the matter, because such confusion is an all too familiar occurrence; isn't recognizing its familiarity enough? Not for me, not in my book! I dare say there is some philosophical confusion in what I have just said about the nature of philosophical confusion; if so, then *I* at least do well to worry, until I can get rid of this.

I am surprised at Malcolm's writing as if the problem how there *can* be sin presented itself only to Christians. It is very surprising that he should here forget the repeated appeals of the Psalmist to God as to why evildoing is allowed. But the perplexity about the existence of sin cannot always be represented, as the Psalmist's words could, just as an 'outcry': it can also be a genuine intellectual perplexity. The Stoic philosophers, for example, believed that the world is ruled by a wise and good Providence; they thought such belief to be *reasonable*, and nothing that Malcolm has said goes to show that they were wrong. Anyone who thinks there is a Providence has to face the *intellectual* challenge how to reconcile this belief with 'what men do in the folly of their minds'; the challenge is faced in Cleanthes' hymn to Zeus, and Cleanthes replies that evil deeds are allowed because God can make the odd even and the crooked paths straight; men's folly and wickedness are exploitable for God's overruling ends. Christian writers have said much the same.

Malcolm would have us distinguish between 'absolute metaphysical truth' and a 'Christian stand-point' that cannot be shown 'more true, more correct, more reasonable' than opposing 'world-pictures'. If I did not judge Christian belief to be 'more true, more correct, more reasonable' than modern atheistic humanism, then I could not be a Christian. Obviously we who say this are 'using our language-game as a basis' from which to judge what other people maintain; but anybody is bound to do this if he makes a judgement that other people contest. Malcolm implies, citing Wittgenstein as an authority, that there is something radically objectionable about such judgements that others are wrong: despite that authority, I find Malcolm's objection no more impressive than Toddy Beamish's discussion-stopping move 'So *you* say' in H. G. Wells's story 'The man who could work miracles'.

I still, to suppose it true, for my part
  See reasons and reasons; this, to begin:
'T is the Faith that launched point-blank its dart
  At the head of a lie—taught Original Sin,
  The Corruption of Man's Heart.

'This, to begin': it does not take Christian faith to see Original Sin in man's world. The unfeasibility of plans that ignore Original Sin is as well established by repeated experience as the impossibility of a perpetual motion machine or of an unbreakable gambling system. Schopenhauer, who did not even believe in God, knew all about Original Sin. He indeed saw in the lives of some Christian Saints the manifest signs of a deliverance which he thought the corrupt minds of the unconverted could not remotely understand; and others who have seen this evil face of man have had even less hope of deliverance than he had. Still, 'this to begin': someone who begins here may come to see 'reasons and reasons' for Christian belief. With someone rootedly convinced that there is no Original Sin I find much more difficulty in starting a discussion—even if he *calls* himself a Christian.

REPLY TO MEYNELL

Hugo Meynell's professional duties have obliged him to read the sort of literature that he here discusses. I am fairly confident that if I were to read such a work as *The Myth of God Incarnate*, I could only come up with much the same estimate as his. In fact I have read *no* book of this genre; however, from academic conversations and articles in periodicals, I know the sort of thing such authors say. My sampling of their wares, whose quality does not seem to improve over the years, does not incline me to read a whole book of such materials. There are no doubt those who would speak of my 'evading a challenge'. But I follow Newman in saying: 'Do not . . . challenge me to answer what I really have not the patience to read. Life is not long enough for such trifles'. I know I have left undone much that I ought to have done; I shall not be wasting any of the time and energy remaining to me on such books as this.

The prospects for a rethought or revised Christianity seem to me dim. I remember somebody once asked me whether some complex form of words did not express what we must 'now' mean by affirming that Jesus Christ was born of a Virgin. Naturally I dissented: what it means to us now to say a young woman is a virgin is not all that different, despite advances in physiology, from what the corresponding statement meant in the first century; whatever that man meant, it was certainly not this. I can see no point in a man's wanting to use the old words when he does not take them in the old sense: except indeed that he is trying to deceive himself and others. The response to my words on that occasion was not unexpected: 'So you're saying we have to be Fundamentalists!' Fundamentalism is a bogey regularly raised in such discussions; perhaps I ought to have asked the speaker what he understood by it. I could however point out, and did, that in the paper I had read I had not urged people to accept any positive dogma: not the Divine mission of Jesus Christ, let alone his Deity; not even the existence of a God. I had only been pleading for truth and honesty: if people no longer believe old dogmas, they should plainly say so, and not go on uttering the old affirmations. I know some non-Christians who were present both understood and agreed; I have no skill to get across to revisionary Christians even what my concern is.

To take another case: the opinion is gaining ground (to mimic the curious assertion sign that revisionary Christians use) that at the Last Supper Christ did not utter anything like the words variously reported in the Gospels and by St. Paul; these words are not historical, but come from liturgy, and the Scriptural narratives are of later compilation. I shall not here argue against this opinion, though one might well ask what *could* be evidence for it. What I want to draw attention to is the odd behaviour-pattern of clerics who defend this opinion in scholarly circles, but in their churches rehearse the 'words of institution' as having been said by Christ on the night be was betrayed. It ought not to take orthodox Christian faith to see that this is a deceit practised on the congregation, and a blasphemy against God if there is a God. My father's teacher McTaggart described this sort of thing with words borrowed from Dickens's Silas Wegg: with the sole exception of the salary, they resign the entire situation.

Hobbes thought all Christians could accept as a firm foundation of faith the New Testament credal formula: Jesus is the Christ, the Son of God. The event has shown that people will claim the name of Christian even when they *reject* Hobbes's minimum creed; for such Christians the idea of a Christ, a Son of man who shall come one day to judge the world, is an outworn dream of Israelite nationalism, and they would have us give up thinking (though perhaps in church continuing to *talk*) of the Son of God. I cannot imaginatively fit myself into such a frame of mind.

The attitude of revisionary Christians to their Master is indeed hard to understand. On the one hand they view with considerable scepticism the reports of his teaching in the Gospels; on the other hand they are prepared, when occasion offers, to reject some of his views as mistaken, e.g. concerning the Old Testament or the reality of devils and hell. If the Gospels are historically as unreliable as they hold them to be, clearly they are on shaky ground in ascribing these 'mistaken' teachings to the alleged source. Whether the Gospels are unreliable or not, it is absurd of such thinkers to claim to follow a teacher whose mind was either so gravely in error, or so gravely misrepresented, as they suppose; although through the mists there might dimly shine a noble figure, the case would be like that of Socrates. Julian the Apostate indeed, if my memory serves, spoke of 'being saved by Socrates through philosophy'; but the salvation Christians have looked for is different; if someone thinks that is not to be had, his profession of Christianity is merely farcical.

There is one area in which Meynell's opinions and attitudes appear to differ greatly from mine: on the matter of what are called 'great religions' or 'world religions'. These concepts are for me a bit obscure; perhaps a 'great' religion is one generally known to be the religion of a very large number of human beings, and a 'world' religion one that has millions of followers in many countries; by these criteria, Judaism would be neither a 'great' religion nor a 'world' religion, and Hinduism would clearly not be a 'world' religion, since it has few followers outside the Indian subcontinent. Anyhow, Meynell seems to think that a religion's falling under one or the other of these vague concepts is a *prima facie* ground for a Christian to take it seriously rather than religions that do not; though Judaism is of course in a special position for Meynell, as it must be for any Christian. What I fail to understand is why the religion of a Zoroastrian (Parsee) or of a Sikh (say) should bulk less large in a Christian's mind than Hinduism or Buddhism; why should either numbers or dispersion be significant?

Meynell's own Christianity is traditional; but here he seems to depart from tradition in a startling way, and with no reason that is apparent in his paper. Both the Old and the New Testament continually denounce idolatry; the Old Testament does so with vituperation (e.g. Ezekiel regularly refers to the idols of apostate Israelites with a Hebrew word for turds), and in the New Testament the Apocalypse speaks similarly. The Epistles are not vituperative in the Old Testament style, but have a tone of grave warning—heathen ways are darkness and the end of them is death. The great mass of traditional Christians, both Catholic and Protestant, have adopted this attitude. For Meynell, it appears, this is a lesson we Christians must unlearn; we must rather regard Christianity as the 'fulfilment' of ancient pagan religious, as well as of Judaism; and apparently he holds that extant pagan cults, which do not even recognize one sole Creator as the unique object of adoration, are to be approached, not in the New Testament spirit of wishing to 'make disciples' and with God's help bring to an end 'the time of ignorance that God winked at', but in a recognition that *we* are ignorant and have to 'learn' from these religions insights that have thus far been lacking to us.

As I said, I find in Meynell's piece not a hint of *why* he thinks the tradition of his Christian forefathers, and their Jewish predecessors, is on this point hugely in error; if it could be, why does he not follow the modernists he criticizes, and doubt it elsewhere? Of course this appeal is no use except to a Christian reader; to non-Christians I am trying to make plain what one element in Christian tradition signifies, because Christians' humbug and dissimulation about this can only do harm.

The riches of the Christian past are great and varied. I do not find it a humble attitude, but quite the reverse, if a Christian decides that all of Christian spirituality can afford *him* only a one-sided diet; that *he* will suffer, so to say, from vitamin deficiencies if he does not supplement this diet with some spiritual reading or ascetic practices of other 'great religions'. I cannot imagine how Meynell fancies the mystery of the Incarnation could be 'brought home' to Christians by their reading pagan stories of 'incarnations': let us say, the story of Krishna's rogering milkmaids. (Compared with Krishna, Olympian Zeus was a model of married chastity; and let us remember that when educated Greeks would mention the amorous escapades of Zeus, they did so quite often with indignant rejection or in scornful satire, and certainly with no idea that such stories give us any knowledge of the Divine.)

It would be easy for someone to object that here I am reenacting the mistake made by many good Catholics in medieval times, including at least one Pope, when they viewed with suspicion or hostility the reformulation of Catholic thought in an Aristotelian conceptual framework. But this revolution in Christian thought is only superficially similar to the proposed new way of thinking about 'great' religions. Aristotle's thought is guided by a confidence in the power of the human mind to systematize its own intellectual operations, and to penetrate into the nature of the real world in which man lives and his own nature as a denizen of it. Of course Aristotle made many mistakes, but such a spirit is not alien to Christianity; by Christian belief, the world is made through the Logos, the Divine Reason, and is pervious to the human mind because man's reason and logic are images of the Logos. It is quite another thing when Christians are urged to cultivate friendly relations with cults that show a decided aversion to discourse of reason, and which indeed urge the pursuit of 'wisdom' by such mind-blowing techniques as drug-taking and peculiar breathing exercises and the thousandfold repetition of a word. Some cleric recently urged that Catholics should give hospitality in church halls to devotees of a cult for which smoking hashish is a sacred rite; he deprecated their being legally harassed over this rite, 'as sacred to them as Communion is to us'. Comment is needless. Mind-blowing practices in 'great' religions are perhaps not so grotesque; but not a bit more worthy of respect, if I may rely on accounts of them in English by people who wish to propagate them.

The University of Leeds

REFERENCES

Dummett, M.: 1973, *Frege: Philosophy of Language*, Duckworth, London (second edition, with index and textual references added, 1981).

Dummett, M.: 1981, *The Interpretation of Frege's Philosophy*, Harvard University Press, Cambridge, Massachusetts.

Galton, A.: 1984, *The Logic of Aspect*, Clarendon Press, Oxford.

Geach, P. T.: 1962, *Reference and Generality*, Cornell University Press, Ithaca (third edition, 1980).

Geach, P. T.: 1967, 'Identity', *Review of Metaphysics* **21**, 3–12. (Reprinted in Geach, 1972, pp. 238–247.)

Geach, P. T.: 1969, *God and the Soul*, Routledge and Kegan Paul, London.

Geach, P. T.: 1970, 'Two Paradoxes of Russell's', *Journal of Philosophy* **67**, 89–97.

Geach, P. T.: 1972, *Logic Matters*, Basil Blackwell, Oxford.

Geach, P. T.: 1975, 'Names and Identity', in S. Guttenplan (ed.), *Mind and Language: Wolfson College Lectures 1974*, Clarendon Press, pp. 139–158.

Geach, P. T.: 1979, 'Existential or Particular Quantifier?', in P. Weingartner and E. Morscher (eds.), *Ontology and Logic* (Proceedings of an International Colloquium, Salzburg, 21–24 September 1976), Duncker and Humblot, Berlin, pp. 137–151.

Geach, P. T.: 1980, *Reference and Generality* (third edition), Cornell University Press, Ithaca.

Goodman, N.: 1951, *The Structure of Appearance*, Harvard University Press, Cambridge Mass. (Third edition: 1977, Boston Studies in the Philosophy of Science **53**, Synthese Library **107**, D. Reidel, Dordrecht and Boston.)

Hume, D. (edited by L. A. Selby-Bigge): 1888, *A Treatise of Human Nature*, Clarendon Press, Oxford.

McTaggart, J. McT. E.: 1921 (Vol. 1) and 1927 (Vol. 2), *The Nature of Existence*, Cambridge University Press, Cambridge. (The second volume was edited by C. D. Broad.)

Moore, G. E.: 1922, *Philosophical Studies*, Routledge and Kegan Paul, London.

Prior, A. N.: 1957, *Time and Modality*, Clarendon Press, Oxford.

Prior, A. N. (edited by P. T. Geach and A. J. Kenny): 1976a, *The Doctrine of Propositions and Terms*, Duckworth, London.

Prior, A. N. (edited by P. T. Geach and A. J. Kenny): 1976b, *Papers in Logic and Ethics*, Duckworth, London.

Wisdom, John: 1934, *Problems of Mind and Matter*, Cambridge University Press, Cambridge.

Wolniewicz, B.: 1968, *Rzeczy i Fakty* [Things and Facts], Państwowc Wydawnictwo Nankowe, Warsaw.

Wolniewicz, B.: 1985, *Ontologia Sytuacji* [The Ontology of Situations], Państwowc Wydawnictwo Nankowe, Warsaw.

BIBLIOGRAPHY OF WORKS OF P.T. GEACH

1948

(a) 'Designation and Truth', *Analysis* **8**, 93–96. (Reprinted in 1972(a), pp. 189–193.)
(b) 'Mr Ill-Named', *Analysis* **9**, 14–16.
(c) 'Necessary Propositions and Entailment-statements', *Mind* **57**, 491–493. (Reprinted in 1972(a), pp. 201–203.)

1949

(a) 'If's and And's', *Analysis* **9**, 58–62. (Reprinted in 1972(a), pp. 194–198.)
(b) 'On Rigour in Semantics', *Mind* **58**, 518–522. (Reprinted in 1972(a), pp. 198–201.)
(c) Review of J. R. Weinberg, *Nicolaus of Autrecourt: A Study in 14th Century Thought, Mind* **58**, 238–245.

1950

(a) 'On Names of Expressions', *Mind* **59**, 388. (Reprinted in 1972(a), pp. 203–204.)
(b) 'Russell's Theory of Descriptions', *Analysis* **10**, 84–88. (Reprinted as 1956(c).)
(c) 'Subject and Predicate', *Mind* **59**, 461–482.
(d) Review of M. H. Carré, *Phases of Thought in England, Mind* **59**, 278–279.
(e) 'Designation and Truth—a Reply', *Analysis* **10**, 118–119. (Reprinted in 1972(a), p. 193.)

1951

(a) 'Frege's *Grundlagen*', *Philosophical Review* **60**, 535–544.
(b) 'On What There Is, I', *Proceedings of the Aristotelian Society* (Supplementary Volume) **25**, 125–136.
(c) Translation of G. Frege 'On Concept and Object', revised by M. Black, *Mind* **60**, 168–180. (Reprinted in 1952(a), pp. 42–55.)
(d) Review of W. V. O. Quine, *Methods of Logic, Mind* **60**, 424–426.

1952

(a) With M. Black. *Translations from the Philosophical Writings of Gottlob Frege*, Basil Blackwell, Oxford.

Harry A. Lewis (ed.), Peter Geach: Philosophical Encounters, 307–315.

1953

(a) 'Quine on Classes and Properties', *Philosophical Review* **62**, 409–412. (Reprinted in 1972(a), pp. 222–226.)
(b) 'A Note on the Reflexive Paradoxes', *Philosophical Review* **62**, 427–428.

1954

(a) With G. E. M. Anscombe. *Descartes: Philosophical Writings*, Nelson, London.
(b) Review of F. B. Fitch, *Symbolic Logic: an Introduction, Mind* **63**, 274–275.
(c) Review of R. I. Aaron, *The Theory of Universals, Philosophical Review* **63**, 267–269.

1955

(a) 'Class and Concept', *Philosophical Review* **64**, 561–570. (Reprinted in 1972(a), pp. 226–235.)
(b) 'Form and Existence', *Proceedings of the Aristotelian Society* **55**, 251–272. (Reprinted in 1969(a), pp. 42–64.)
(c) 'On "Insolubilia"', *Analysis* **15**, 71–72. (Reprinted in 1972(a), pp. 209–211.)
(d) Review of A. Moody, *Truth and Consequence in Medieval Logic, Mind* **64**, 558–561.
(e) Review of B. Mates, *Stoic Logic, Philosophical Review* **64**, 143–145.
(f) Review of R. W. Church, *An Analysis of Resemblance, Philosophical Review* **64**, 130–132.

1956

(a) 'Good and Evil', *Analysis* **17**, 33–42.
(b) 'On Frege's Way Out', *Mind* **65**, 408–409. (Reprinted in 1972(a), pp. 235–237.)
(c) 'Russell's Theory of Descriptions', in M. Macdonald (ed.), *Philosophy and Analysis*, Basil Blackwell, Oxford, 1956, pp. 32–36. (Reprint of 1950(b).)
(d) 'The Doctrine of Distribution', *Mind* **65**, 67–74.
(e) 'The Law of Excluded Middle, I', *Proceedings of the Aristotelian Society* (Supplementary Volume) **30**, 59–73. (Reprinted in 1972(a), pp. 74–87.)
(f) Review of D. J. B. Hawkins, *Being and Becoming: An Essay Towards a Critical Metaphysic, Mind* **65**, 417.
(g) 'The Third Man Again', *Philosophical Review* **65**, 72–82.

1957

(a) *Mental Acts, Their Content and Their Objects*, Routledge and Kegan Paul; Humanities Press, London; New York.
(b) 'On Beliefs About Oneself', *Analysis* **18**, 23–24. (Reprinted in 1972(a), pp. 128–129.)
(c) Review of a translation of Wittgenstein's *Tractatus Logico-Philosophicus* into Italian by G. C. M. Colombo, S. J., *Philosophical Review* **66**, 556–559.

1958

(a) 'Entailment, III', *Proceedings of the Aristotelian Society* (Supplementary Volume) **32**, 157–172.
(b) 'Imperative and Deontic Logic', *Analysis* **18**, 49–56. (Reprinted in 1972(a), pp. 270–278.)

1959

(a) 'Is It Right to Say Or Is a Conjunction?', *Analysis* **19**, 143–144. (Reprinted in 1972(a), pp. 204–205.)
(b) 'Russell on Meaning and Denoting', *Analysis* **19**, 69–72. (Reprinted in 1972(a), pp. 27–31.)

1960

(a) 'Ascriptivism', *Philosophical Review* **69**, 221–225. (Reprinted in 1972(a), pp. 250–254.)
(b) Review of E. Riverso, *Il Pensiero di Bertrand Russell: Esposizione Storicocritica, Philosophical Review* **69**, 277–278.
(c) 'Distribution: A Last Word', *Philosophical Review* **69**, 396–398. (Reprinted in 1972(a), pp. 62–64.)

1961

(a) With G. E. M. Anscombe. *Three Philosophers*, Cornell University Press, Ithaca, New York and London. (The essays on Aquinas and on Frege are by Peter Geach.)
(b) 'Ryle on Namely-riders', *Analysis* **21**, 64–67. (Reprinted in 1972(a), pp. 88–92.)
(c) Review of W. V. O. Quine, *Word and Object, Philosophical Books* **2**, 14–17.

1962

(a) *Reference and Generality: An Examination of Some Medieval and Modern Theories*, Cornell University Press, Ithaca, New York.
(b) 'Causality and Creation', *Sophia* **1**, 1–8.
(c) 'Causality and Creation, Part II', *Sophia* **1**, 27–30. (Both parts reprinted in 1969(a), pp. 75–85.)
(d) 'Namely-riders Again', *Analysis* **22**, 92–94. (Reprinted in 1972(a), pp. 92–95.)

1963

(a) 'Aristotle on Conjunctive Propositions', *Ratio* **5**, 33–45. (Reprinted in 1972(a), pp. 13–27.)
(b) 'Imperative Inference, II', *Analysis* **23** (Supplement), 37–42. (Reprinted in 1972(a), pp. 278–285.)

(c) 'Mr Strawson on Symbolic and Traditional Logic', *Mind* **72**, 125–128. (Reprinted in 1972(a), pp. 95–97.)
(d) 'What Are Referring Expressions?', *Analysis* **23**, 8–10. (Reprinted in 1972(a), pp. 95–97.)
(e) Review of B. Blanshard, *Reason and Analysis, Philosophical Review* **72**, 516–518.
(f) Review of the translation by D.F. Pears and B. McGuinness of L. Wittgenstein, *Tractatus Logico-Philosophicus, Philosophical Review* **71**, 264–265.

1964

(a) 'Nominalism', *Sophia* **3**, 3–14. (Reprinted in 1972(a), pp. 289–301.)
(b) 'Referring Expressions Again', *Analysis* **24**, 172–175. (Reprinted in 1972(a), pp. 97–102.)
(c) Review of Storrs McCall, *Aristotle's Modal Syllogistic, Ratio* **6**, 200–206.

1965

(a) 'Assertion', *Philosophical Review* **74**, 449–465. (Reprinted in 1972(a), pp. 254–269.)
(b) 'Complex Terms Again', *Journal of Philosophy* **62**, 716–717. (Reprinted in 1972(a), pp. 106–108.)
(c) 'Logical Procedures and the Identity of Expressions', *Ratio* **7**, 199–205. (Reprinted in 1972(a), pp. 108–115.)
(d) 'Man, Beast, and Philosophical Psychology', *British Journal of the Philosophy of Science* **16**, 246–248.
(e) 'On Complex Terms', *Journal of Philosophy* **62**, 5–7. (Reprinted in 1972(a), pp. 102–106.)

1966

(a) 'Plato's *Euthyphro*: An Analysis and Commentary', *Monist* **50**, 369–382. (Reprinted in 1972(a), pp. 31–44.)
(b) 'Comment on Mr Grimm', *Analysis* **26**, 146.
(c) 'Dr Kenny on Practical Inference', *Analysis* **26**, 76–79.
(d) 'Some Problems About Time', *Proceedings of the British Academy* **51**, 321–336. (Reprinted as 1968(c), and in 1972(a), pp. 302–318.)
(e) Review of A. Farrer, *The Freedom of the Will, Philosophical Review* **75**, 99–104.

1967

(a) 'Commentary on Aquinas', in D.R. Burrill (ed.), *The Cosmological Arguments*, Anchor Books, New York, 1967, pp. 57–82. (Reprinted from 1961(a), pp. 109–125.)
(b) 'Identity', *Review of Metaphysics* **21**, 3–12. (Reprinted in 1972(a), pp. 238–247.)
(c) 'Intentional Identity', *Journal of Philosophy* **64**, 627–632. (Reprinted in 1972(a), pp. 146–153.)

1968

(a) *A History of the Corruptions of Logic: An Inaugural Lecture*, Leeds University Press, Leeds. (Reprinted in 1972(a), pp. 44–61.)
(b) *Reference and Generality* (emended edition), Cornell University Press, Ithaca, New York.
(c) 'Some Problems About Time', in P. F. Strawson (ed.), *Studies in the Philosophy of Thought and Action*, Oxford University Press, Oxford, 1968, pp. 175–191 (Reprint of 1966(d).)
(d) 'What Actually Exists, I', *Proceedings of the Aristotelian Society* (Supplementary Volume) **42**, 7–16. (Reprinted in 1969(a), pp. 65–74.)
(e) Review of N. Rescher, *Galen and the Syllogism, Mind* **76**, 144–145.
(f) 'Mr Toms on Distribution', *Mind* **77**, 113–114. (Reprinted in 1972(a), pp. 65–66.)

1969

(a) *God and the Soul*, Routledge and Kegan Paul; Schoken Books, London; New York.
(b) 'A History of the Corruptions of Logic' (An Inaugural Lecture), *Leeds University Review* **12**, 42–58. (Reprint of 1968(a).)
(c) 'Contradictories and Contraries', *Analysis* **29**, 187–190. (Reprinted in 1972(a), pp. 70–74.)
(d) 'God's Relation to the World', *Sophia* **8**, 1–9. (Reprinted in 1972(a), pp. 318–327.)
(e) 'Konsekwentny Idealizm?', *Studia Filozoficzne* **59**, 93–102.
(f) 'Quine's Syntactical Insights', *Synthese* **19**, 118–129. (Reprinted as 1969(g), and in 1972(a), pp. 115–127.)
(g) 'Quine's Syntactical Insights', in D. Davidson and J. Hintikka (eds.), *Words and Objections: Essays on the Work of W. V. Quine*, Reidel, Dordrecht, 1969, pp. 146–157. (Reprint of 1969(f).)
(h) 'The Perils of Pauline', *Review of Metaphysics* **23**, 287–300. (Reprinted in 1972(a), pp. 153–165.)
(i) Review of D. F. Pears, *Bertrand Russell and the British Tradition in Philosophy, Ratio* **11**, 164–169.
(j) A Reply [to F. Feldman: 'Geach and Relative Identity', *Review of Metaphysics* **22** (1969), 347–355], *Review of Metaphysics* **22**, 556–559. (Reprinted, with some omissions, in 1972(a), pp. 247–249.)
(k) 'Should Traditional Grammar Be Ended or Mended? II', *Educational Review* **22**, 18–25.

1970

(a) With Christopher Coope, Timothy Potts and Roger White. *A Wittgenstein Workbook*, Basil Blackwell, Oxford.
(b) 'A Program for Syntax', *Synthese* **22**, 3–17. (Reprinted as 1972(c).)
(c) 'Arthur Prior: A Personal Impression', *Theoria* **36**, 186–188.
(d) 'Entailment', *Philosophical Review* **79**, 237–239. (Reprinted in 1972(a), pp. 186–188.)

(e) 'Two Paradoxes of Russell's', *Journal of Philosophy* **67**, 89–97.
(f) Review of A. Kenny, *The Five Ways, Philosophical Quarterly* **20**, 311–312.

1971

(a) 'Nazwy i Orzeczniki' ('Names and Predicables'), in J. Pelc (ed.), *Semiotyka Polska*, Pańtstwowez Wydawnictwo Nankowe, Warsaw, pp. 283–290.

1972

(a) *Logic Matters*, Blackwell, Oxford.
(b) '"Most-Most-Some" Arguments and the "Dictum de Omni"', *Philosophical Studies* **23**, 122–123.
(c) 'A Program for Syntax', in D. Davidson and G. Harman (eds.), *Semantics of Natural Language*, D. Reidel, Dordrecht and Boston, pp. 483–497. (Reprint of 1970(b).)
(d) 'Czemu Zdanie Nie Jest Nazwą' ('Why Sentences Are Not Names'), *Studia Semiotyczna* **3**, 13–21.
(e) 'Spinoza and the Divine Attributes', *Royal Institute of Philosophy Lectures* **5**, Macmillan Press Ltd, London and Basingstoke, 15–27.
(f) Review of R. Grossmann, *Reflections on Frege's Philosophy, Philosophical Review* **81**, 498–501.
(g) Untitled reply (to Bernard Gert and James A. Martin, 'Outcomes and Abilities', pp. 188–191 in same volume), *Analysis* **33**, 191–192.

1973

(a) 'Omnipotence', *Philosophy* **48**, 7–20. (Reprinted in 1977(a), pp. 1–28.)
(b) 'An Irrelevance of Omnipotence', *Philosophy* **48**, 327–333. (Reprinted in 1977(a), pp. 29–39.)
(c) 'O Rzekomo Określonej Przyszłości' ('The Alleged Determinateness of the Future'), *Ruch Filozoficzny* (*The Philosophical Movement*) **31**, 99–104.
(d) 'On Certain Modal Systems Connected with S4.3', *Bulletin of the Section of Logic* **2**, 8–11.
(e) 'Ontological Relativity and Relative Identity', in M. K. Munitz (ed.), *Logic and Ontology*, New York University Press, New York, 1973, pp. 287–302.
(f) Review of D. P. Henry, *Medieval Logic and Metaphysics, Medium Aevum* **42**, 269–271.
(g) (Untitled) Comment on B. Gert and J. A. Martin, "'What a man does he can do'" (*Analysis* **33** [1973], 168–173), *Analysis* **33**, 173.

1974

(a) 'Entailment Once Again', *Philosophical Review* **83**, 238–239.
(b) 'Fallacy and Proof and the Rationality of Religion', *Sophia* **13**, 1–4.

1975

(a) 'Back-Reference', *Philosophia (Israel)* **5**, 193–206.
(b) 'Names and Identity', in S. D. Guttenplan (ed.), *Mind and Language: Wolfson College Lectures 1975*, Clarendon Press, Oxford, pp. 139–158.
(c) 'Nazwy a Identyczność', *Studia Semiotyczne* **6**, 125–131.
(d) 'On Entailment', *Analysis* **35**, 186–187.
(e) 'Teleological Explanation', in S. Körner (ed.), *Explanation*, Basil Blackwell, Oxford, pp. 76–95.
(f) 'Reply to Comments', in S. Körner (ed.), *Explanation*, Basil Blackwell, Oxford, pp. 112–117.

1976

(a) *Reason and Argument*, Blackwell, Oxford.
(b) 'Back-Reference', in A. Kasher (ed.), *Language in Focus*, D. Reidel, Dordrecht, pp. 25–39. (Reprint of 1975(a).)
(c) 'Distribution and "Suppositio"', *Mind* **85**, 432–435.
(d) 'Morally Significant Theses', *Open Mind* **4**, 5–12.
(e) 'Murder and Sodomy', *Philosophy* **51**, 346–348.
(f) 'Saying and Showing in Frege and Wittgenstein', in K. J. J. Hintikka (ed.), *Essays on Wittgenstein, in Honour of G. H. von Wright* (Acta Philosophical Fennica, 28), North–Holland, Amsterdam, pp. 54–70.
(g) 'Two Kinds of Intentionality?', *Monist* **59**, 306–320.
(h) 'Why Logic Matters', in H. D. Lewis (ed.), *Contemporary British Philosophy* (Fourth Series), George Allen and Unwin, London, pp. 86–99.
(i) Critical notice of M. Dummett, *Frege: Philosophy of Language, Mind* **85**, 436–449.
(j) Review of A. R. Anderson and N. Belnap, *Entailment: the Logic of Relevance and Necessity*, Vol. 1, *Philosophy* **52**, 493–495.
(k) 'Comment' (to F. B. Fitch, 'The Relation Between Natural Languages and Formalized Languages'), in S. Körner (ed.), *Philosophy of Logic*, Basil Blackwell, Oxford, pp. 191–196.

1977

(a) *Providence and Evil: The Stanton Lectures, 1971–2*, Cambridge University Press, London.
(b) *The Virtues: The Stanton Lectures 1973–4*, Cambridge University Press, London.
(c) 'Again the Logic of "Ought"', *Philosophy* **52**, 473–476.
(d) Review of Michael Durrant, *The Logical Status of 'God', Second Order* **6**, 111–114.
(e) Review of Nicholas Griffin, *Relative Identity, Times Literary Supplement*, Issue No. 3946, November 11, 1329.
(f) Review of R. Campbell, *From Belief to Understanding, Philosophy* **52**, 234–236.
(g) 'Can God Fail to Keep Promises?', *Philosophy* **52**, 93–95.

1978

(a) 'Evans on Quantifiers', Canadian Journal of Philosophy **8**, 375–379.
(b) 'Intentionality of Thought versus Intentionality of Desire', in R. M. Chisholm (ed.), *Die Philosophie Franz Brentanos*, Rodopi, Amsterdam, pp. 131–138.
(c) 'Russell on Denoting', *Analysis* **38**, 204–205.

1979

(a) *Truth, Love and Immortality: An Introduction to McTaggart's Philosophy*, University of California Press, Berkeley.
(b) 'Existential or Particular Quantifier?', in P. Weingartner and E. Morscher (eds.), *Ontology and Logic: Proceedings of an International Colloquium* (Salzburg, 21–24 September 1976), Duncker und Humblot, Berlin, pp. 137–151.
(c) 'Kinds of Statement', in Cora Diamond (ed.), *Intention and Intentionality*, Cornell University Press, Ithaca, pp. 221–235.
(d) 'Names and Predicables', in J. Pelc (ed.), *Semiotics in Poland*, D. Reidel, Dordrecht, pp. 240–246. (Translated by the author from 1971(a).)
(e) 'On Teaching Logic', *Philosophy* **54**, 5–17.
(f) 'Russell and Frege Again', *Analysis* **39**, 159–160.
(g) 'Some Hobbesian Theses about Justice', *Dialectics and Humanism* **6**, 45–50.

1980

(a) With Max Black. *Translations from the Philosophical Writings of Gottlob Frege* (Third Edition), Basil Blackwell, Oxford.
(b) *Reference and Generality* (Third Edition), Cornell University Press, Ithaca, New York.
(c) 'Some Problems about the Sense and Reference of Proper Names', *Canadian Journal of Philosophy* (Supplementary Volume) **6**, 83–96.
(d) 'Strawson on Subject and Predicate', in Zak van Straaten (ed.), *Philosophical Subjects*, Clarendon Press, Oxford, pp. 174–188.
(e) With Gilbert Fulmer. 'An Exchange between Peter Geach and Gilbert Fulmer', *Southwestern Journal of Philosophy* **11**, 165–170.
(f) Review of Gerd Brand, *The Central Texts of Ludwig Wittgenstein, History and Philosophy of Logic* **1**, 237–238.

1981

(a) 'Wittgenstein's Operator, *N*', *Analysis* **41**, 168–170.
(b) 'La Cuantificacion de Segundo Orden de Frege', *Teorema* **11**, 167–178.
(c) 'On a Claim of Hare's', *Analysis* **41**, 63.
(d) 'Some Hobbesian Theses about Justice', *Etyka* **19**, 69–75. (Reprint of 1979(a).)
(e) 'The Religion of Thomas Hobbes', *Religious Studies* **17**, 549–558.

1982

(a) 'More on Wittgenstein's Operator *N*', *Analysis* **42**, 127–128.
(b) 'Moral Autonomy Still Refuted', *Philosophy* **57**, 127–129.
(c) 'Truth and God, I', *Proceedings of the Aristotelian Society* (Supplementary Volume) **56**, 83–97.
(d) 'Whatever Happened to Deontic Logic?', *Philosophia* (Israel) **11**, 1–12.
(e) 'Reply to Lowe's "Paradox of the 1,001 Cats"', *Analysis* **42**, 30.
(f) 'Reply to Lowe's Reply on "The Paradox of the 1,001 Cats"', *Analysis* **42**, 32.

1983

(a) 'Abstraction Reconsidered', in Carl Ginet (ed.), *Knowledge and Mind: Philosophical Essays*, Oxford University Press, New York, pp. 160–173.
(b) 'Comparatives', *Philosophia* (Israel) **13**, 235–246.
(c) 'On Teaching about Distribution', *Philosophy* **58**, 539–540.

1985

(a) Dummett on Frege: a Review Discussion, *The Thomist* **49**, 116–121.

1987

(a) 'Reference and Buridan's Law', *Philosophy* **62**, 7–15.
(b) 'Relative Identity', in *Proceedings of the 30th Conference on the History of Logic, Dedicated to Roman Suszko, 1984*, Jagiellonian University, Cracow.

1988

(a) (Editor) *Wittgenstein's Lectures on Philosophical Psychology, 1946–47*, University of Chicago Press, Chicago.

1989

(a) 'Amor y Eternidad: la Filosofia Idealista de McTaggart', *Annuario Filosofico* **22**, 27–34.
(b) Review of J. McT. E. McTaggart, *The Nature of Existence* Volumes i and ii, *Times Literary Supplement*, Issue No. 4969, June 2–8, 599–600.

INDEX

SYNTHESE LIBRARY

Studies in Epistemology, Logic, Methodology, and Philosophy of Science

1. J. M. Bochénski, *A Precis of Mathematical Logic*. 1959 ISBN 90-277-0073-7
2. P. Guiraud, *Problèmes et méthodes de la statistique linguistique*. 1959 ISBN 90-277-0025-7
3. H. Freudenthal (ed.), *The Concept and the Role of the Model in Mathematics and Natural and Social Sciences*. 1961 ISBN 90-277-0017-6
4. E. W. Beth, *Formal Methods*. An Introduction to Symbolic Logic and to the Study of Effective Operations in Arithmetic and Logic. 1962 ISBN 90-277-0069-9
5. B. H. Kazemier and D. Vuysje (eds.), *Logic and Language*. Studies dedicated to Professor Rudolf Carnap on the Occasion of His 70th Birthday. 1962 ISBN 90-277-0019-2
6. M. W. Wartofsky (ed.), *Proceedings of the Boston Colloquium for the Philosophy of Science, 1961–1962*. [Boston Studies in the Philosophy of Science, Vol. I] 1963 ISBN 90-277-0021-4
7. A. A. Zinov'ev, *Philosophical Problems of Many-valued Logic*. A revised edition, edited and translated (from Russian) by G. Küng and D.D. Comey. 1963 ISBN 90-277-0091-5
8. G. Gurvitch, *The Spectrum of Social Time*. Translated from French and edited by M. Korenbaum and P. Bosserman. 1964 ISBN 90-277-0006-0
9. P. Lorenzen, *Formal Logic*. Translated from German by F.J. Crosson. 1965 ISBN 90-277-0080-X
10. R. S. Cohen and M. W. Wartofsky (eds.), *Proceedings of the Boston Colloquium for the Philosophy of Science, 1962–1964*. In Honor of Philipp Frank. [Boston Studies in the Philosophy of Science, Vol. II] 1965 ISBN 90-277-9004-0
11. E. W. Beth, *Mathematical Thought*. An Introduction to the Philosophy of Mathematics. 1965 ISBN 90-277-0070-2
12. E. W. Beth and J. Piaget, *Mathematical Epistemology and Psychology*. 1966 ISBN 90-277-0071-0
13. G. Küng, *Ontology and the Logistic Analysis of Language*. An Enquiry into the Contemporary Views on Universals. 1967 ISBN 90-277-0028-1
14. R. S. Cohen and M. W. Wartofsky (eds.), *Proceedings of the Boston Colloquium for the Philosophy of Sciences, 1964–1966*. In Memory of Norwood Russell Hanson. [Boston Studies in the Philosophy of Science, Vol. III] 1967 ISBN 90-277-0013-3
15. C. D. Broad, *Induction, Probability, and Causation*. Selected Papers. 1968 ISBN 90-277-0012-5
16. G. Patzig, *Aristotle's Theory of the Syllogism*. A Logical-philosophical Study of *Book A* of the *Prior Analytics*. 1968 ISBN 90-277-0030-3

17. N. Rescher, *Topics in Philosophical Logic.* 1968 ISBN 90-277-0084-2
18. R. S. Cohen and M. W. Wartofsky (eds.), *Proceedings of the Boston Colloquium for the Philosophy of Science, 1966–1968, Part I.* [Boston Studies in the Philosophy of Science, Vol. IV] 1969 ISBN 90-277-0014-1
19. R. S. Cohen and M. W. Wartofsky (eds.), *Proceedings of the Boston Colloquium for the Philosophy of Science, 1966–1968, Part II.* [Boston Studies in the Philosophy of Science, Vol. V] 1969 ISBN 90-277-0015-X
20. J. W. Davis, D. J. Hockney and W. K. Wilson (eds.), *Philosophical Logic.* 1969 ISBN 90-277-0075-3
21. D. Davidson and J. Hintikka (eds.), *Words and Objections.* Essays on the Work of W. V. Quine. 1969 ISBN 90-277-0074-5; Pb 90-277-0602-6
22. P. Suppes, *Studies in the Methodology and Foundations of Science. Selected Papers from 1911 to 1969.* 1969 ISBN 90-277-0020-6
23. J. Hintikka, *Models for Modalities.* Selected Essays. 1969 ISBN 90-277-0078-8; Pb 90-277-0598-4
24. N. Rescher *et al.* (eds.), *Essays in Honor of Carl G. Hempel.* A Tribute on the Occasion of His 65th Birthday. 1969 ISBN 90-277-0085-0
25. P. V. Tavanec (ed.), *Problems of the Logic of Scientific Knowledge.* 1969 ISBN 90-277-0087-7
26. M. Swain (ed.), *Induction, Acceptance, and Rational Belief.* 1970 ISBN 90-277-0086-9
27. R. S. Cohen and R. J. Seeger (eds.), *Ernst Mach: Physicist and Philosopher.* [Boston Studies in the Philosophy of Science, Vol. VI]. 1970 ISBN 90-277-0016-8
28. J. Hintikka and P. Suppes, *Information and Inference.* 1970 ISBN 90-277-0155-5
29. K. Lambert, *Philosophical Problems in Logic.* Some Recent Developments. 1970 ISBN 90-277-0079-6
30. R. A. Eberle, *Nominalistic Systems.* 1970 ISBN 90-277-0161-X
31. P. Weingartner and G. Zecha (eds.), *Induction, Physics, and Ethics.* 1970 ISBN 90-277-0158-X
32. E. W. Beth, *Aspects of Modern Logic.* 1970 ISBN 90-277-0173-3
33. R. Hilpinen (ed.), *Deontic Logic.* Introductory and Systematic Readings. 1971 ISBN 90-277-0167-9; Pb 90-277-1302-2
34. J.-L. Krivine, *Introduction to Axiomatic Set Theory.* 1971 ISBN 90-277-0169-5; Pb 90-277-0411-2
35. J. D. Sneed, *The Logical Structure of Mathematical Physics.* 2nd rev. ed., 1979 ISBN 90-277-1056-2; Pb 90-277-1059-7
36. C. R. Kordig, *The Justification of Scientific Change.* 1971 ISBN 90-277-0181-4; Pb 90-277-0475-9
37. M. Čapek, *Bergson and Modern Physics.* A Reinterpretation and Re-evaluation. [Boston Studies in the Philosophy of Science, Vol. VII] 1971 ISBN 90-277-0186-5
38. N. R. Hanson, *What I Do Not Believe, and Other Essays.* Ed. by S. Toulmin and H. Woolf. 1971 ISBN 90-277-0191-1
39. R. C. Buck and R. S. Cohen (eds.), *PSA 1970.* Proceedings of the Second Biennial Meeting of the Philosophy of Science Association, Boston, Fall 1970. In Memory of Rudolf Carnap. [Boston Studies in the Philosophy of Science, Vol. VIII] 1971 ISBN 90-277-0187-3; Pb 90-277-0309-4
40. D. Davidson and G. Harman (eds.), *Semantics of Natural Language.* 1972 ISBN 90-277-0304-3; Pb 90-277-0310-8

41. Y. Bar-Hillel (ed.), *Pragmatics of Natural Languages*. 1971
ISBN 90-277-0194-6; Pb 90-277-0599-2
42. S. Stenlund, *Combinators, λ-Terms and Proof Theory*. 1972 ISBN 90-277-0305-1
43. M. Strauss, *Modern Physics and Its Philosophy*. Selected Paper in the Logic, History, and Philosophy of Science. 1972 ISBN 90-277-0230-6
44. M. Bunge, *Method, Model and Matter*. 1973 ISBN 90-277-0252-7
45. M. Bunge, *Philosophy of Physics*. 1973 ISBN 90-277-0253-5
46. A. A. Zinov'ev, *Foundations of the Logical Theory of Scientific Knowledge (Complex Logic)*. Revised and enlarged English edition with an appendix by G. A. Smirnov, E. A. Sidorenka, A. M. Fedina and L. A. Bobrova. [Boston Studies in the Philosophy of Science, Vol. IX] 1973 ISBN 90-277-0193-8; Pb 90-277-0324-8
47. L. Tondl, *Scientific Procedures*. [Boston Studies in the Philosophy of Science, Vol. X] 1973 ISBN 90-277-0147-4; Pb 90-277-0323-X
48. N. R. Hanson, *Constellations and Conjectures*. Ed. by W. C. Humphreys, Jr. 1973
ISBN 90-277-0192-X
49. K. J. J. Hintikka, J. M. E. Moravcsik and P. Suppes (eds.), *Approaches to Natural Language*. 1973 ISBN 90-277-0220-9; Pb 90-277-0233-0
50. M. Bunge (ed.), *Exact Philosophy*. Problems, Tools and Goals. 1973
ISBN 90-277-0251-9
51. R. J. Bogdan and I. Niiniluoto (eds.), *Logic, Language and Probability*. 1973
ISBN 90-277-0312-4
52. G. Pearce and P. Maynard (eds.), *Conceptual Change*. 1973
ISBN 90-277-0287-X; Pb 90-277-0339-6
53. I. Niiniluoto and R. Tuomela, *Theoretical Concepts and Hypothetico-inductive Inference*. 1973 ISBN 90-277-0343-4
54. R. Fraissé, *Course of Mathematical Logic* – Volume 1: *Relation and Logical Formula*. 1973 ISBN 90-277-0268-3; Pb 90-277-0403-1
(For *Volume 2* see under Nr. 69).
55. A. Grünbaum, *Philosophical Problems of Space and Time*. 2nd enlarged ed. [Boston Studies in the Philosophy of Science, Vol. XII] 1973
ISBN 90-277-0357-4; Pb 90-277-0358-2
56. P. Suppes (ed.), *Space, Time and Geometry*. 1973
ISBN 90-277-0386-8; Pb 90-277-0442-2
57. H. Kelsen, *Essays in Legal and Moral Philosophy*. Selected and introduced by O. Weinberger. 1973 ISBN 90-277-0388-4
58. R. J. Seeger and R. S. Cohen (eds.), *Philosophical Foundations of Science*. [Boston Studies in the Philosophy of Science, Vol. XI] 1974
ISBN 90-277-0390-6; Pb 90-277-0376-0
59. R. S. Cohen and M. W. Wartofsky (eds.), *Logical and Epistemological Studies in Contemporary Physics*. [Boston Studies in the Philosophy of Science, Vol. XIII] 1973 ISBN 90-277-0391-4; Pb 90-277-0377-9
60. R. S. Cohen and M. W. Wartofsky (eds.), *Methodological and Historical Essays in the Natural and Social Sciences. Proceedings of the Boston Colloquium for the Philosophy of Science, 1969–1972*. [Boston Studies in the Philosophy of Science, Vol. XIV] 1974 ISBN 90-277-0392-2; Pb 90-277-0378-7
61. R. S. Cohen, J. J. Stachel and M. W. Wartofsky (eds.), *For Dirk Struik. Scientific, Historical and Political Essays*. [Boston Studies in the Philosophy of Science, Vol. XV] 1974 ISBN 90-277-0393-0; Pb 90-277-0379-5

62. K. Ajdukiewicz, *Pragmatic Logic*. Transl. from Polish by O. Wojtasiewicz. 1974
ISBN 90-277-0326-4
63. S. Stenlund (ed.), *Logical Theory and Semantic Analysis*. Essays dedicated to Stig Kanger on His 50th Birthday. 1974 ISBN 90-277-0438-4
64. K. F. Schaffner and R. S. Cohen (eds.), *PSA 1972. Proceedings of the Third Biennial Meeting of the Philosophy of Science Association*. [Boston Studies in the Philosophy of Science, Vol. XX] 1974 ISBN 90-277-0408-2; Pb 90-277-0409-0
65. H. E. Kyburg, Jr., *The Logical Foundations of Statistical Inference*. 1974
ISBN 90-277-0330-2; Pb 90-277-0430-9
66. M. Grene, *The Understanding of Nature*. Essays in the Philosophy of Biology. [Boston Studies in the Philosophy of Science, Vol. XXIII] 1974
ISBN 90-277-0462-7; Pb 90-277-0463-5
67. J. M. Broekman, *Structuralism: Moscow, Prague, Paris*. 1974
ISBN 90-277-0478-3
68. N. Geschwind, *Selected Papers on Language and the Brain*. [Boston Studies in the Philosophy of Science, Vol. XVI] 1974 ISBN 90-277-0262-4; Pb 90-277-0263-2
69. R. Fraissé, *Course of Mathematical Logic* – Volume 2: *Model Theory*. 1974
ISBN 90-277-0269-1; Pb 90-277-0510-0
(For *Volume 1* see under Nr. 54)
70. A. Grzegorczyk, *An Outline of Mathematical Logic*. Fundamental Results and Notions Explained with All Details. 1974 ISBN 90-277-0359-0
71. F. von Kutschera, *Philosophy of Language*. 1975 ISBN 90-277-0591-7
72. J. Manninen and R. Tuomela (eds.), *Essays on Explanation and Understanding*. Studies in the Foundations of Humanities and Social Sciences. 1976
ISBN 90-277-0592-5
73. J. Hintikka (ed.), *Rudolf Carnap, Logical Empiricist*. Materials and Perspectives. 1975 ISBN 90-277-0583-6
74. M. Čapek (ed.), *The Concepts of Space and Time*. Their Structure and Their Development. [Boston Studies in the Philosophy of Science, Vol. XXII] 1976
ISBN 90-277-0355-8; Pb 90-277-0375-2
75. J. Hintikka and U. Remes, *The Method of Analysis*. Its Geometrical Origin and Its General Significance. [Boston Studies in the Philosophy of Science, Vol. XXV] 1974 ISBN 90-277-0532-1; Pb 90-277-0543-7
76. J. E. Murdoch and E. D. Sylla (eds.), *The Cultural Context of Medieval Learning*. [Boston Studies in the Philosophy of Science, Vol. XXVI] 1975
ISBN 90-277-0560-7; Pb 90-277-0587-9
77. S. Amsterdamski, *Between Experience and Metaphysics*. Philosophical Problems of the Evolution of Science. [Boston Studies in the Philosophy of Science, Vol. XXXV] 1975 ISBN 90-277-0568-2; Pb 90-277-0580-1
78. P. Suppes (ed.), *Logic and Probability in Quantum Mechanics*. 1976
ISBN 90-277-0570-4
79. H. von Helmholtz: *Epistemological Writings. The Paul Hertz/Moritz Schlick Centenary Edition of 1921 with Notes and Commentary by the Editors*. Newly translated by M. F. Lowe. Edited, with an Introduction and Bibliography, by R. S. Cohen and Y. Elkana. [Boston Studies in the Philosophy of Science, Vol. XXXVII] 1975 ISBN 90-277-0290-X; Pb 90-277-0582-8
80. J. Agassi, *Science in Flux*. [Boston Studies in the Philosophy of Science, Vol. XXVIII] 1975 ISBN 90-277-0584-4; Pb 90-277-0612-2

81. S. G. Harding (ed.), *Can Theories Be Refuted?* Essays on the Duhem-Quine Thesis. 1976 ISBN 90-277-0629-8; Pb 90-277-0630-1
82. S. Nowak, *Methodology of Sociological Research.* General Problems. 1977 ISBN 90-277-0486-4
83. J. Piaget, J.-B. Grize, A. Szemińska, and V. Bang, *Epistemology and Psychology of Functions.* 1977 ISBN 90-277-0804-5
84. M. Grene and E. Mendelsohn (eds.), *Topics in the Philosophy of Biology.* [Boston Studies in the Philosophy of Science, Vol. XXVII] 1976 ISBN 90-277-0595-X; Pb 90-277-0596-8
85. E. Fischbein, *The Intuitive Sources of Probabilistic Thinking in Children.* 1975 ISBN 90-277-0626-3
86. E. W. Adams, *The Logic of Conditionals.* An Application of Probability to Deductive Logic. 1975 ISBN 90-277-0631-X
87. M. Przełęcki and R. Wójcicki (eds.), *Twenty-Five Years of Logical Methodology in Poland.* 1976 ISBN 90-277-0601-8
88. J. Topolski, *The Methodology of History.* 1976 ISBN 90-277-0550-X
89. A. Kasher (ed.), *Language in Focus: Foundations, Methods and Systems.* Essays dedicated to Yehoshua Bar-Hillel. [Boston Studies in the Philosophy of Science, Vol. XLIII] 1976 ISBN 90-277-0644-1; Pb 90-277-0645-X
90. J. Hintikka, *The Intentions of Intentionality and Other New Models for Modalities.* 1975 ISBN 90-277-0633-6; Pb 90-277-0634-4
91. W. Stegmüller, *Collected Papers on Epistemology, Philosophy of Science and History of Philosophy.* 2 Volumes. 1977 Set ISBN 90-277-0767-7
92. D. M. Gabbay, *Investigations in Modal and Tense Logics with Applications to Problems in Philosophy and Linguistics.* 1976 ISBN 90-277-0656-5
93. R. J. Bogdan, *Local Induction.* 1976 ISBN 90-277-0649-2
94. S. Nowak, *Understanding and Prediction.* Essays in the Methodology of Social and Behavioral Theories. 1976 ISBN 90-277-0558-5
95. P. Mittelstaedt, *Philosophical Problems of Modern Physics.* [Boston Studies in the Philosophy of Science, Vol. XVIII] 1976 ISBN 90-277-0285-3; Pb 90-277-0506-2
96. G. Holton and W. A. Blanpied (eds.), *Science and Its Public: The Changing Relationship.* [Boston Studies in the Philosophy of Science, Vol. XXXIII] 1976 ISBN 90-277-0657-3; Pb 90-277-0658-1
97. M. Brand and D. Walton (eds.), *Action Theory.* 1976 ISBN 90-277-0671-9
98. P. Gochet, *Outline of a Nominalist Theory of Proposition.* An Essay in the Theory of Meaning. 1980 ISBN 90-277-1031-7
99. R. S. Cohen, P. K. Feyerabend, and M. W. Wartofsky (eds.), *Essays in Memory of Imre Lakatos.* [Boston Studies in the Philosophy of Science, Vol. XXXIX] 1976 ISBN 90-277-0654-9; Pb 90-277-0655-7
100. R. S. Cohen and J. J. Stachel (eds.), *Selected Papers of Léon Rosenfield.* [Boston Studies in the Philosophy of Science, Vol. XXI] 1979 ISBN 90-277-0651-4; Pb 90-277-0652-2
101. R. S. Cohen, C. A. Hooker, A. C. Michalos and J. W. van Evra (eds.), *PSA 1974. Proceedings of the 1974 Biennial Meeting of the Philosophy of Science Association.* [Boston Studies in the Philosophy of Science, Vol. XXXII] 1976 ISBN 90-277-0647-6; Pb 90-277-0648-4
102. Y. Fried and J. Agassi, *Paranoia.* A Study in Diagnosis. [Boston Studies in the Philosophy of Science, Vol. L] 1976 ISBN 90-277-0704-9; Pb 90-277-0705-7

103. M. Przełęcki, K. Szaniawski and R. Wójcicki (eds.), *Formal Methods in the Methodology of Empirical Sciences*. 1976 ISBN 90-277-0698-0
104. J. M. Vickers, *Belief and Probability*. 1976 ISBN 90-277-0744-8
105. K. H. Wolff, *Surrender and Catch*. Experience and Inquiry Today. [Boston Studies in the Philosophy of Science, Vol. LI] 1976 ISBN 90-277-0758-8; Pb 90-277-0765-0
106. K. Kosík, *Dialectics of the Concrete*. [Boston Studies in the Philosophy of Science, Vol. LII] 1976 ISBN 90-277-0761-8; Pb 90-277-0764-2
107. N. Goodman, *The Structure of Appearance*. 3rd ed., 1977 [Boston Studies in the Philosophy of Science, Vol. LIII] 1977 ISBN 90-277-0773-1; Pb 90-277-0774-X
108. Kazimierz Ajdukiewicz, *The Scientific World-Perspective and Other Essays, 1931-1963*. Edited and with an Introduction by J. Giedymin. 1978 ISBN 90-277-0527-5
109. R. L. Causey, *Unity of Science*. 1977 ISBN 90-277-0779-0
110. R. E. Grandy, *Advanced Logic for Applications*. 1977 ISBN 90-277-0781-2
111. R. P. McArthur, *Tense Logic*. 1976 ISBN 90-277-0697-2
112. L. Lindahl, *Position and Change*. A Study in Law and Logic. 1977 ISBN 90-277-0787-1
113. R. Tuomela, *Dispositions*. 1978 ISBN 90-277-0810-X
114. H. A. Simon, *Models of Discovery and Other Topics in the Methods of Science*. [Boston Studies in the Philosophy of Science, Vol. LIV] 1977 ISBN 90-277-0812-6; Pb 90-277-0858-4
115. R. D. Rosenkrantz, *Inference, Method and Decision*. Towards a Bayesian Philosophy of Science. 1977 ISBN 90-277-0817-7; Pb 90-277-0818-5
116. R. Tuomela, *Human Action and Its Explanation*. A Study on the Philosophical Foundations of Psychology. 1977 ISBN 90-277-0824-X
117. M. Lazerowitz, *The Language of Philosophy*. Freud and Wittgenstein. [Boston Studies in the Philosophy of Science, Vol. LV]. 1977 ISBN 90-277-0826-6; Pb 90-277-0862-2
118. Not published
119. J. Pelc, *Semiotics in Poland, 1894–1969*. 1979 ISBN 90-277-0811-8
120. I. Pörn, *Action Theory and Social Science*. Some Formal Models. 1977 ISBN 90-277-0846-0
121. J. Margolis, *Persons and Mind*. The Prospects of Nonreductive Materialism. [Boston Studies in the Philosophy of Science, Vol. LVII]. 1977 ISBN 90-277-0854-1; Pb 90-277-0863-0
122. J. Hintikka, I. Niiniluoto, and E. Saarinen (eds.), *Essays on Mathematical and Philosophical Logic*. 1979 ISBN 90-277-0879-7
123. T. A. F. Kuipers, *Studies in Inductive Probability and Rational Expectation*. 1978 ISBN 90-277-0882-7
124. E. Saarinen, R. Hilpinen, I. Niiniluoto and M. P. Hintikka (eds.), *Essays in Honour of Jaakko Hintikka on the Occasion of His 50th Birthday*. 1979 ISBN 90-277-0916-5
125. G. Radnitzky and G. Andersson (eds.), *Progress and Rationality in Science*. [Boston Studies in the Philosophy of Science, Vol. LVIII] 1978 ISBN 90-277-0921-1; Pb 90-277-0922-X
126. P. Mittelstaedt, *Quantum Logic*. 1978 ISBN 90-277-0925-4
127. K. A. Bowen, *Model Theory for Modal Logic*. Kripke Models for Modal Predicate Calculi. 1979 ISBN 90-277-0929-7

128. H. A. Bursen, *Dismantling the Memory Machine*. A Philosophical Investigation of Machine Theories of Memory. 1978 ISBN 90-277-0933-5
129. M. W. Wartofsky, *Models*, Representation and the Scientific Understanding. [Boston Studies in the Philosophy of Science, Vol. XLVIII.] 1979 ISBN 90-277-0736-7; Pb 90-277-0947-5
130. D. Ihde, *Technics and Praxis*. A Philosophy of Technology. [Boston Studies in the Philosophy of Science, Vol. XXIV] 1979 ISBN 90-277-0953-X; Pb 90-277-0954-8
131. J. J. Wiatr (ed.), *Polish Essays in the Methodology of the Social Sciences*. [Boston Studies in the Philosophy of Science, Vol. XXIX] 1979 ISBN 90-277-0723-5; Pb 90-277-0956-4
132. W. C. Salmon (ed.), *Hans Reichenbach: Logical Empiricist*. 1979 ISBN 90-277-0958-0
133. P. Bieri, R.-P. Horstmann and L. Krüger (eds.), *Transcendental Arguments in Science*. Essays in Epistemology. 1979 ISBN 90-277-0963-7; Pb 90-277-0964-5
134. M. Marković and G. Petrović (eds.), *Praxis*. Yugoslav Essays in the Philosophy and Methodology of the Social Sciences. [Boston Studies in the Philosophy of Science, Vol. XXXVI]. 1979 ISBN 90-277-0727-8; Pb 90-277-0968-8
135. R. Wójcicki, *Topics in the Formal Methodology of Empirical Sciences*. 1979 ISBN 90-277-1004-X
136. G. Radnitzky and G. Andersson (eds.), *The Structure and Development of Science*. [Boston Studies in the Philosophy of Science, Vol. LIX] 1979 ISBN 90-277-0994-7; Pb 90-277-0995-5
137. J. C. Webb, *Mechanism, Mentalism and Metamathematics*. An Essay on Finitism. 1980 ISBN 90-277-1046-5
138. D. F. Gustafson and B. L. Tapscott (eds.), *Body, Mind and Method*. Essays in Honor of Virgil C. Aldrich. 1979 ISBN 90-277-1013-9
139. L. Nowak, *The Structure of Idealization*. Towards a Systematic Interpretation of the Marxian Idea of Science. 1980 ISBN 90-277-1014-7
140. C. Perelman, *The New Rhetoric and the Humanities*. Essays on Rhetoric and Its Applications. 1979 ISBN 90-277-1018-X; Pb 90-277-1019-8
141. W. Rabinowicz, *Universalizability*. A Study in Morals and Metaphysics. 1979 ISBN 90-277-1020-2
142. C. Perelman, *Justice, Law and Argument*. Essays on Moral and Legal Reasoning. 1980 ISBN 90-277-1089-9; Pb 90-277-1090-2
143. S. Kanger and S. Öhman (eds.), *Philosophy and Grammar*. Papers on the Occasion of the Quincentennial of Uppsala University. 1981 ISBN 90-277-1091-0
144. T. Pawlowski, *Concept Formation in the Humanities and the Social Sciences*. 1980 ISBN 90-277-1096-1
145. J. Hintikka, D. Gruender and E. Agazzi (eds.), *Theory Change, Ancient Axiomatics and Galileo's Methodology*. Proceedings of the 1978 Pisa Conference on the History and Philosophy of Science, Volume I. 1981 ISBN 90-277-1126-7
146. J. Hintikka, D. Gruender and E. Agazzi (eds.), *Probabilistic Thinking, Thermodynamics, and the Interaction of the History and Philosophy of Science*. Proceedings of the 1978 Pisa Conference on the History and Philosophy of Science, Volume II. 1981 ISBN 90-277-1127-5
147. U. Mönnich (ed.), *Aspects of Philosophical Logic*. Some Logical Forays into Central Notions of Linguistics and Philosophy. 1981 ISBN 90-277-1201-8

148. D. M. Gabbay, *Semantical Investigations in Heyting's Intuitionistic Logic*. 1981 ISBN 90-277-1202-6
149. E. Agazzi (ed.), *Modern Logic – A Survey*. Historical, Philosophical, and Mathematical Aspects of Modern Logic and Its Applications. 1981 ISBN 90-277-1137-2
150. A. F. Parker-Rhodes, *The Theory of Indistinguishables*. A Search for Explanatory Principles below the Level of Physics. 1981 ISBN 90-277-1214-X
151. J. C. Pitt, *Pictures, Images, and Conceptual Change*. An Analysis of Wilfrid Sellars' Philosophy of Science. 1981 ISBN 90-277-1276-X; Pb 90-277-1277-8
152. R. Hilpinen (ed.), *New Studies in Deontic Logic*. Norms, Actions, and the Foundations of Ethics. 1981 ISBN 90-277-1278-6; Pb 90-277-1346-4
153. C. Dilworth, *Scientific Progress*. A Study Concerning the Nature of the Relation between Successive Scientific Theories. 2nd, rev. and augmented ed., 1986 ISBN 90-277-2215-3; Pb 90-277-2216-1
154. D. W. Smith and R. McIntyre, *Husserl and Intentionality*. A Study of Mind, Meaning, and Language. 1982 ISBN 90-277-1392-8; Pb 90-277-1730-3
155. R. J. Nelson, *The Logic of Mind*. 2nd. ed., 1989 ISBN 90-277-2819-4; Pb 90-277-2822-4
156. J. F. A. K. van Benthem, *The Logic of Time*. A Model-Theoretic Investigation into the Varieties of Temporal Ontology, and Temporal Discourse. 1983 ISBN 90-277-1421-5
157. R. Swinburne (ed.), *Space, Time and Causality*. 1983 ISBN 90-277-1437-1
158. E. T. Jaynes, *Papers on Probability, Statistics and Statistical Physics*. Ed. by R. D. Rozenkrantz. 1983 ISBN 90-277-1448-7; Pb (1989) 0-7923-0213-3
159. T. Chapman, *Time: A Philosophical Analysis*. 1982 ISBN 90-277-1465-7
160. E. N. Zalta, *Abstract Objects*. An Introduction to Axiomatic Metaphysics. 1983 ISBN 90-277-1474-6
161. S. Harding and M. B. Hintikka (eds.), *Discovering Reality*. Feminist Perspectives on Epistemology, Metaphysics, Methodology, and Philosophy of Science. 1983 ISBN 90-277-1496-7; Pb 90-277-1538-6
162. M. A. Stewart (ed.), *Law, Morality and Rights*. 1983 ISBN 90-277-1519-X
163. D. Mayr and G. Süssmann (eds.), *Space, Time, and Mechanics*. Basic Structures of a Physical Theory. 1983 ISBN 90-277-1525-4
164. D. Gabbay and F. Guenthner (eds.), *Handbook of Philosophical Logic*. Vol. I: Elements of Classical Logic. 1983 ISBN 90-277-1542-4
165. D. Gabbay and F. Guenthner (eds.), *Handbook of Philosophical Logic*. Vol. II: Extensions of Classical Logic. 1984 ISBN 90-277-1604-8
166. D. Gabbay and F. Guenthner (eds.), *Handbook of Philosophical Logic*. Vol. III: Alternative to Classical Logic. 1986 ISBN 90-277-1605-6
167. D. Gabbay and F. Guenthner (eds.), *Handbook of Philosophical Logic*. Vol. IV: Topics in the Philosophy of Language. 1989 ISBN 90-277-1606-4
168. A. J. I. Jones, *Communication and Meaning*. An Essay in Applied Modal Logic. 1983 ISBN 90-277-1543-2
169. M. Fitting, *Proof Methods for Modal and Intuitionistic Logics*. 1983 ISBN 90-277-1573-4
170. J. Margolis, *Culture and Cultural Entities*. Toward a New Unity of Science. 1984 ISBN 90-277-1574-2
171. R. Tuomela, *A Theory of Social Action*. 1984 ISBN 90-277-1703-6

172. J. J. E. Gracia, E. Rabossi, E. Villanueva and M. Dascal (eds.), *Philosophical Analysis in Latin America*. 1984 ISBN 90-277-1749-4
173. P. Ziff, *Epistemic Analysis*. A Coherence Theory of Knowledge. 1984 ISBN 90-277-1751-7
174. P. Ziff, *Antiaesthetics*. An Appreciation of the Cow with the Subtile Nose. 1984 ISBN 90-277-1773-7
175. W. Balzer, D. A. Pearce, and H.-J. Schmidt (eds.), *Reduction in Science*. Structure, Examples, Philosophical Problems. 1984 ISBN 90-277-1811-3
176. A. Peczenik, L. Lindahl and B. van Roermund (eds.), *Theory of Legal Science*. Proceedings of the Conference on Legal Theory and Philosophy of Science, Lund, Sweden, 11–14 December 1983. 1984 ISBN 90-277-1834-2
177. I. Niiniluoto, *Is Science Progressive?* 1984 ISBN 90-277-1835-0
178. B. K. Matilal and J. L. Shaw (eds.), *Analytical Philosophy in Comparative Perspective*. Exploratory Essays in Current Theories and Classical Indian Theories of Meaning and Reference. 1985 ISBN 90-277-1870-9
179. P. Kroes, *Time: Its Structure and Role in Physical Theories*. 1985 ISBN 90-277-1894-6
180. J. H. Fetzer, *Sociobiology and Epistemology*. 1985 ISBN 90-277-2005-3; Pb 90-277-2006-1
181. L. Haaparanta and J. Hintikka, *Frege Synthesized*. Essays on the Philosophical and Foundational Work of Gottlob Frege. 1986 ISBN 90-277-2126-2
182. M. Detlefsen, *Hilbert's Program*. An Essay on Mathematical Instrumentalism. 1986 ISBN 90-277-2151-3
183. J. L. Golden and J. J. Pilotta (eds.), *Practical Reasoning in Human Affairs*. Studies in Honor of Chaim Perelman. 1986 ISBN 90-277-2255-2
184. H. Zandvoort, *Models of Scientific Development and the Case of Nuclear Magnetic Resonance*. 1986 ISBN 90-277-2351-6
185. I. Niiniluoto, *Truthlikeness*. 1987 ISBN 90-277-2354-0
186. W. Balzer, C. U. Moulines and J. D. Sneed, *An Architectonic for Science*. The Structuralist Program. 1987 ISBN 90-277-2403-2
187. D. Pearce, *Roads to Commensurability*. 1987 ISBN 90-277-2414-8
188. L. M. Vaina, *Matters of Intelligence*. Conceptual Structures in Cognitive Neuroscience. 1987 ISBN 90-277-2460-1
189. H. Siegel, *Relativism Refuted*. A Critique of Contemporary Epistemological Relativism. 1987 ISBN 90-277-2469-5
190. W. Callebaut and R. Pinxten, *Evolutionary Epistemology*. A Multiparadigm Program, with a Complete Evolutionary Epistemology Bibliograph. 1987 ISBN 90-277-2582-9
191. J. Kmita, *Problems in Historical Epistemology*. 1988 ISBN 90-277-2199-8
192. J. H. Fetzer (ed.), *Probability and Causality*. Essays in Honor of Wesley C. Salmon. 1988 ISBN 90-277-2607-8; Pb 1-5560-8052-2
193. A. Donovan, L. Laudan and R. Laudan (eds.), *Scrutinizing Science*. Empirical Studies of Scientific Change. 1988 ISBN 90-277-2608-6
194. H.R. Otto and J.A. Tuedio (eds.), *Perspectives on Mind*. 1988 ISBN 90-277-2640-X
195. D. Batens and J.P. van Bendegem (eds.), *Theory and Experiment*. Recent Insights and New Perspectives on Their Relation. 1988 ISBN 90-277-2645-0

196. J. Österberg, *Self and Others.* A Study of Ethical Egoism. 1988 ISBN 90-277-2648-5
197. D.H. Helman (ed.), *Analogical Reasoning.* Perspectives of Artificial Intelligence, Cognitive Science, and Philosophy. 1988 ISBN 90-277-2711-2
198. J. Wolenski, *Logic and Philosophy in the Lvov-Warsaw School.* 1989 ISBN 90-277-2749-X
199. R. Wójjcicki, *Theory of Logical Calculi.* Basic Theory of Consequence Operations. 1988 ISBN 90-277-2785-6
200. J. Hintikka and M.B. Hintikka, *The Logic of Epistemology and the Epistemology of Logic.* Selected Essays. 1989 ISBN 0-7923-0040-8; Pb 0-7923-0041-6
201. E. Agazzi (ed.), *Probability in the Sciences.* 1988 ISBN 90-277-2808-9
202. M. Meyer (ed.), *From Metaphysics to Rhetoric.* 1989 ISBN 90-277-2814-3
203. R.L. Tieszen, *Mathematical Intuition.* Phenomenology and Mathematical Knowledge. 1989 ISBN 0-7923-0131-5
204. A. Melnick, *Space, Time, and Thought in Kant.* 1989 ISBN 0-7923-0135-8
205. D.W. Smith, *The Circle of Acquaintance.* Perception, Consciousness, and Empathy. 1989 ISBN 0-7923-0252-4
206. M.H. Salmon (ed.), *The Philosophy of Logical Mechanism.* Essays in Honor of Arthur W. Burks. With his Responses, and with a Bibliography of Burk's Work. 1990 ISBN 0-7923-0325-3
207. M. Kusch, *Language as Calculus vs. Language as Universal Medium.* A Study in Husserl, Heidegger, and Gadamer. 1989 ISBN 0-7923-0333-4
208. T.C. Meyering, *Historical Roots of Cognitive Science.* The Rise of a Cognitive Theory of Perception from Antiquity to the Nineteenth Century. 1989 ISBN 0-7923-0349-0
209. P. Kosso, *Observability and Observation in Physical Science.* 1989 ISBN 0-7923-0389-X
210. J. Kmita, *Essays on the Theory of Scientific Cognition.* 1990 ISBN 0-7923-0441-1
211. W. Sieg (ed.), *Acting and Reflecting.* The Interdisciplinary Turn in Philosophy. 1990 ISBN 0-7923-0512-4
212. J. Karpiński, *Causality in Sociological Research.* 1990 ISBN 0-7923-0546-9
213. H.A. Lewis (ed.), *Peter Geach: Philosophical Encounters.* 1991 ISBN 0-7923-0823-9
214. M. Ter Hark, *Beyond the Inner and the Outer.* Wittgenstein's Philosophy of Psychology. 1990 ISBN 0-7923-0850-6
215. M. Gosselin: *Nominalism and Contemporary Nominalism.* Ontological and Epistemological Implications of the Work of W.V.O. Quine and of N. Goodman. 1990 ISBN 0-7923-0904-9
216. J.H. Fetzer, D. Shatz and G. Schlesinger (eds.): *Definitions and Definability.* Philosophical Perspectives. 1991 ISBN 0-7923-1046-2

Previous volumes are still available.

KLUWER ACADEMIC PUBLISHERS – DORDRECHT / BOSTON / LONDON

GPSR Compliance

The European Union's (EU) General Product Safety Regulation (GPSR) is a set of rules that requires consumer products to be safe and our obligations to ensure this.

If you have any concerns about our products, you can contact us on ProductSafety@springernature.com

In case Publisher is established outside the EU, the EU authorized representative is:

Springer Nature Customer Service Center GmbH
Europaplatz 3
69115 Heidelberg, Germany

Batch number: 10809439

Printed by Printforce, the Netherlands